APPLICATIONS OF FUNCTIONAL FOODS IN DISEASE PREVENTION

AAP Advances in Nutraceuticals

APPLICATIONS OF FUNCTIONAL FOODS IN DISEASE PREVENTION

Edited by

Raj K. Keservani, MPharm
Eknath D. Ahire, MSPharm

AAP | APPLE ACADEMIC PRESS

First edition published 2024

Apple Academic Press Inc.
1265 Goldenrod Circle, NE,
Palm Bay, FL 32905 USA

760 Laurentian Drive, Unit 19,
Burlington, ON L7N 0A4, CANADA

CRC Press
2385 NW Executive Center Drive,
Suite 320, Boca Raton FL 33431

4 Park Square, Milton Park,
Abingdon, Oxon, OX14 4RN UK

Library and Archives Canada Cataloguing in Publication

Title: Applications of functional foods in disease prevention / edited by Raj K. Keservani, MPharm, Eknath D. Ahire, MSPharm.
Names: Keservani, Raj K., 1981- editor. | Ahire, Eknath D., editor.
Series: AAP advances in nutraceuticals.
Description: First edition. | Series statement: AAP advances in nutraceuticals | Includes bibliographical references and index.
Identifiers: Canadiana (print) 20230497985 | Canadiana (ebook) 20230498027 | ISBN 9781774913611 (hardcover) | ISBN 9781774913604 (softcover) | ISBN 9781003395737 (ebook)
Subjects: LCSH: Functional foods. | LCSH: Medicine, Preventive.
Classification: LCC QP144.F85 A67 2024 | DDC 613.2—dc23

Library of Congress Cataloging-in-Publication Data

CIP data on file with US Library of Congress

ISBN: 978-1-77491-361-1 (hbk)
ISBN: 978-1-77491-360-4 (pbk)
ISBN: 978-1-00339-573-7 (ebk)

ABOUT THE BOOK SERIES: AAP ADVANCES IN NUTRACEUTICALS

Series Editor:
Raj K. Keservani, MPharm
Faculty of B. Pharmacy, CSM Group of Institutions, Allahabad, India
Telephone: +91-7897803904 / Email: rajksops@gmail.com

Rajesh K. Kesharwani, PhD
Nehru Gram Bharati (Deemed to be University), Prayagraj, India

Anil K. Sharma, PhD, MPharm
Department of Pharmacy, School of Medical and Allied Sciences,
GD Goenka University, Gurugram, India

In the modern era, mankind has witnessed a paradigm shift with respect to fundamental eating behavior. The lack of physical workouts and busy schedules at offices and in households have promoted the consumption of junk foods, which eventually results in numerous diseases and disorders since the nutritional content from fast food is inadequate. Plumpness and obesity have become a global health threat. The leading causes of death from most developing nations are noncommunicable diseases such as cardiovascular diseases, cancer, arthritis, osteoporosis, and liver toxicity. Patients suffering from such lifestyle ailments are a bit apprehensive towards prolonged use of costly modern therapeutics, which encourages instead the use of alternative approaches for management of such diseases and disorders.

The emerging sector of the nutraceutical industry encompasses products derived from nature, dietary supplements, and functional foods. Nutraceuticals are used for treatment and prevention of a broad range of diseases, such as the common cold, arthritis, sleep-related disorders, cancers, cardiovascular complications, metabolic disorders, and others.

The research on nutraceuticals is increasing day by day, considering the beneficial effects of food or food supplements in the management of diverse diseases. The issue of paramount concern is standardization and establishment of clinical efficacy of nutraceuticals, which is in fact a challenge for researchers around the globe.

This book series aims at realizing the significance of the variety of nutraceuticals for human well-being. The books in the series will also emphasize the role of dieticians and nutritionists for the prescription of judicious eating habits. The key food components such as carbohydrates, proteins, and lipids as well as micronutrients (vitamins, minerals) are demonstrated to maintain good health, obviating the need for medicines. Thus, nutraceuticals that are indeed derived from food ingredients are believed to be potential alternative therapeutics.

Asymptomatic diseases/disorders necessitate proper diagnosis. Stressed circumstances have been reported to cause weight loss. Workplace stress is a key variable in the genesis and propagation of metabolic disorders. The nuances of the underlying mechanisms are still to be deciphered. Poor lifestyle trends have been attributed along with the consumption of junk foods in several such instances. Sugar-rich carbonated beverages, the use of contraband drugs, and the consumption of liquor in excess are assigned as risk factors. This book series will cover advances and applications in addition to providing the basics of nutraceuticals.

The books in this series may be valuable resources for industry professionals to design and develop quality products for end users. In addition, the book series will be fortified by ideas of innovation, concept building, manufacturing aspects, quality control, and regulatory status of nutraceuticals.

Ostensibly, business has seen a rise in the nutraceutical consumption in recent years. Pure, safe, efficacious products are the need of the hour globally. There are a number of books on nutraceuticals in the market, yet these are only a few books presenting certain aspects of nutraceuticals therein. This book series aims to offer a holistic view of this promising strategy, which is picking up pace with time. This book series will be a unique endeavor to bring manufacturing, research and development, and marketing strategies under a single umbrella.

CURRENT BOOKS IN THE SERIES

- **Micronutrients and Macronutrients as Nutraceuticals**
  Editors: Prakash Chandra Gupta, PhD, Sayan Bhattacharyya, MBBS, MD, Nisha Sharma, PhD, MPharm, Rajesh K. Kesharwani, PhD, and Raj K. Keservani, MPharm

- **Formulations, Regulations, and Challenges of Nutraceuticals**
  Editors: Tingirikari Jagan Mohan Rao, PhD, Rajesh K. Kesharwani, PhD, Raj. K. Keservani, MPharm, and Anil K. Sharma, PhD
- **Food Supplements and Dietary Fibers in Health and Disease**
  Editors: Bhushan R. Rane, PhD, MPharm, Raj K. Keservani, M Pharm, Durgesh Singh, DPhil, Nayan A. Gujarathi, PhD, and Ashish S. Jain, PhD
- **Nutraceuticals in Cancer Prevention, Management, and Treatment**
  Editors: Raj K. Keservani, MPharm, Bui Thanh Tung, PhD, Sippy Singh, DPhil, and Rajesh K. Kesharwani, PhD
- **Advances in Flavonoids for Human Health and Prevention of Diseases**
  Editors: Nisha Sharma, PhD, Deepika Saini, PhD, Rajesh K. Kesharwani, PhD, Prakash Chandra Gupta, and Raj. K. Keservani, MPharm
- **The Flavonoids: Extraction and Applications**
  Editors: Deepika Saini, PhD, Rajesh K. Kesharwani, PhD, and Raj K. Keservani, MPharm
- **Immune-Boosting Nutraceuticals for Better Human Health: Novel Applications**
  Editors: Urmila Jarouliya, PhD, Raj K. Keservani, MPharm, Rajesh K. Kesharwani, PhD, Virendra K. Patel, PhD, and Adi D. Bharti
- **Applications of Functional Foods in Disease Prevention**
  Editors: Raj K. Keservani, MPharm, and Eknath D. Ahire, MSPharm
- **Plant Metabolites and Vegetables as Nutraceuticals**
  Editors: Raj K. Keservani, MPharm, Bui Thanh Tung, PhD, Rajesh K. Kesharwani, PhD, MPharm, and Eknath D. Ahire
- **Preventive and Therapeutic Role of Vitamins as Nutraceuticals**
  Editors: Khemchand R. Surana, Eknath D. Ahire, Raj K. Keservani, MPharm, and Rajesh K. Kesharwani, PhD
- **Nutraceuticals and Bone Health**
  Editors: Deepak Sharma, PhD, Madan Mohan Gupta, PhD, Anil K. Sharma, PhD, MPharm, Raj K. Keservani, MPharm, and Rajesh K. Kesharwani, PhD
- **Flavonoids as Nutraceuticals**
  Editors: Rajesh K. Kesharwani, PhD, MTech, Deepika Saini, PhD, Raj K. Keservani, MPharm, and Anil Kumar Sharma, PhD, MPharm

- **Herbals as Nutraceuticals: Their Role in Healthcare**
  Editors: Raj K. Keservani, MPharm, Rajeshwar Kamal Kant Arya, PhD, MPharm, and Rajesh K. Kesharwani, PhD, MTech

- **Nutraceutical Fruits: Overview and Disease Prevention**
  Editors: Raj K. Keservani, MPharm

- **Nutrigenomics and Nutraceuticals**
  Editors: Raj K. Keservani, MPharm, Eknath D. Ahire, Shubham J. Khainar, PhD, Sanjay J. Kshirsagar, PhD, and Rajesh K. Kesharwani, PhD

- **Nutraceuticals in Cardiac Health Management**
  Editors: Raj K. Keservani, MPharm, Rajesh K. Kesharwani, PhD, Praveen Kumar Jain, PhD, Bhushan R. Rane, PhD, Nayan A. Gujarathi, PhD

- **Nutraceuticals for the Treatment and Prevention of Sexual Disorders**
  Editors: Raj K. Keservani, MPharm, Sharangouda J. Patil, PhD, and Ivan Aranha, PhD

- **Antioxidants as Nutraceuticals**
  Editors: Nayan A. Gujaratih, PhD, Raj K. Keservani, MPharm, Rajesh K. Kesharwani, PhD, Bhushan R. Rane, PhD, and Yogeeta Sameer Goyal, PhD

- **The Nature of Nutraceuticals: History, Properties, Sources, and Nanotechnology**
  Editors: Rajesh K. Kesharwani, PhD, Prashant Kumar, PhD, and Raj K. Keservani, MPharm

- **Nutraceuticals in Arthritis and Psoriasis**
  Editors: Meenakshi Jaiswal, PhD, Raj K. Keservani, MPharm, Rajesh K. Kesharwani, PhD, and Swati G. Talele, PhD

ABOUT THE EDITORS

Raj K. Keservani, PhD, MPharm
Associate Professor, Faculty of B. Pharmacy, CSM Group of Institutions, Prayagraj, India

Raj K. Keservani, PhD, MPharm, is working as Associate Professor in the Faculty of B. Pharmacy at the CSM Group of Institutions, Prayagraj, India. He has more than 12 years of academic (teaching) experience from various institutes of India in pharmaceutical education. He has supervised more than 25 undergraduate students for their research work. He has published over 30 peer-reviewed papers in the field of pharmaceutical sciences in national and international journals, six patents, over 70 book chapters, three co-authored books, and over 40 edited books. He is also active as a reviewer for several international scientific journals. His research interests include nutraceutical and functional foods, novel drug delivery systems (NDDS), transdermal drug delivery/drug delivery, health science, cancer biology, and neurobiology. Dr. Keservani graduated with a pharmacy degree from the Department of Pharmacy, Kumaun University, Nainital, Uttarakhand, India. He received his Master of Pharmacy (MPharm) (specialization in pharmaceutics) from the School of Pharmaceutical Sciences, Rajiv Gandhi Proudyogiki Vishwavidyalaya, Bhopal, India. He earned his PhD degree from IKG Punjab Technical University, Jalandhar, Punjab, India. Dr. Keservani is a life member of the Society of Pharmaceutical Education and Research (SPER).

Eknath D. Ahire, MSPharm
Assistant Professor, Department of Pharmaceutics, MET's, Institute of Pharmacy, Bhujbal Knowledge City, Adgaon, Nashik, Maharashtra, India

Eknath D. Ahire, MSPharm, is currently working as an Assistant Professor in the Department of Pharmaceutics at MET's, Institute of Pharmacy, Bhujbal Knowledge City, Adgaon, Nashik, and affiliated with Savitribai Phule Pune University, Pune. He earned his graduation degree from Savitribai Phule

Pune University, India. He received his master's degree (MSPharm) in Pharmaceutics from NIPER-Ahmadabad. He has a total of five years of experience, including four years of academic experience as an Assistant Professor and one year as an Industrial Formulation Scientist II with Torrent Research Center, Gandhinagar, Gujarat. He is working as reviewer for several reputed international journals. He was awarded an AICTE Sponsored P.G. Research Scholar by the Govt. of India. He has more than 25 publications (research, review, and book chapters) published in different national and international reputed indexed journals alongwith five textbooks. He has worked on nanotechnology-based drug delivery systems, nutraceuticals, and novel drug delivery systems.

CONTENTS

CONTRIBUTORS

Pankaj R. Aher
Department of Pharmaceutical Chemistry, Loknete Dr. J. D. Pawar College of Pharmacy, Nashik, Maharashtra, India

Eknath D. Ahire
Department of Pharmaceutics, MET's Institute of Pharmacy, Affiliated to Savitribai Phule Pune University, BKC, Nashik, Maharashtra, India

Manjusha R. Ahire
Department of Pharmacology, R. C. Patel Institute of Pharmaceutical Education and Research, Shirpur, Maharashtra, India

Kajal A. Bendale
Department of Pharmacology, R. C. Patel Institute of Pharmaceutical Education and Research, Shirpur, Maharashtra, India

Indranil Chanda
Assam Science and Technology University, Jalukbari, Guwahati, Assam, India

Smriti Rekha Chanda Das
Girijananda Chowdhury Institute of Pharmaceutical Science, Guwahati, Assam, India

Trishna Das
Girijananda Chowdhury Institute of Pharmaceutical Science, Guwahati, Assam, India

Shiwani A. Diwane
Department of Pharmacology, R. C. Patel Institute of Pharmaceutical Education and Research, Shirpur, Maharashtra, India

Sachin Gunjal
Department Pharmaceutics, Shreeshakti Shaikshanik Sanstha, Divine College of Pharmacy, Nashik, Maharashtra, India

Mohini R. Jagtap
JMCTE Institute of Pharmacy, Nashik, Maharashtra, India

Pankaj G. Jain
Department of Pharmacology, R. C. Patel Institute of Pharmaceutical Education and Research, Shirpur, Maharashtra, India

Shubham J. Khairnar
MET's Institute of Pharmacy, Bhujbal Knowledge City, Nashik, Maharashtra, India

Aditi M. Kokande
Department of Pharmaceutical Chemistry, Oriental Education Society College of Pharmacy, Navi Mumbai, Maharashtra, India

Sanjay J. Kshirsagar
Department of Pharmaceutics, MET's Institute of Pharmacy, BKC, Affiliated to Savitribai Phule Pune University, Nashik, Maharashtra, India

Tushar Lokhande
Department of Pharmaceutical Chemistry, Mahatma Gandhi Vidyamandirs Pharmacy College Panchavati, Nashik, Maharashtra, India

Sunil K. Mahajan
Department of Pharmaceutical Chemistry, Mahatma Gandhi Vidyamandirs Pharmacy College Panchavati, Nashik, Maharashtra, India

Shanti Bhushan Mishra
Department of Pharmacognosy, United Institute of Pharmacy, Prayagraj, Uttar Pradesh, India

Khushali Pagar
Department Pharmaceutics, Shreeshakti Shaikshanik Sanstha, Divine College of Pharmacy, Satana, Nashik, Maharashtra, India

Pratiksha M. Pandagale
Department of Pharmacology, Oriental College of Pharmacy, Navi Mumbai, Maharashtra, India

Vaishali N. Pardeshi
Department of Pharmaceutical Chemistry, Abhinav Education Society's College of Pharmacy, Pune, Maharashtra, India

Afsar S. Pathan
Department of Pharmacology, R. C. Patel Institute of Pharmaceutical Education and Research, Shirpur, Maharashtra, India

Chandrashekhar D. Patil
Department of Pharmacology, Shreeshakti Shaikshanik Sansthas, Divine College of Pharmacy, Satana, Nashik, Maharashtra, India

Dhananjay M. Patil
Department of Pharmaceutics, Shreeshakti Shaikshanik Sanstha, Divine College of Pharmacy, Satana, Nashik, Maharashtra, India

Shital J. Patil
Department of Pharmaceutical Chemistry, MGV's Pharmacy College, Panchvati, Nashik, Maharashtra, India

Mayuri P. Pol
Department of Pharmaceutics, Shreeshakti Shaikshanik Sanstha, Divine College of Pharmacy, Satana, Nashik, Maharashtra, India

Mithun Rudrapal
Rasiklal M. Dhariwal College of Pharmacy, Chinchwad, Pune, Maharashtra, India

Dipankar Saha
Girijananda Chowdhury Institute of Pharmaceutical Science, Guwahati, Assam, India

Tanmay Sarma
Girijananda Chowdhury Institute of Pharmaceutical Science, Guwahati, Assam, India

Divya Rani Sharma
Department of Pharmacology, United Institute of Pharmacy, Prayagraj, Uttar Pradesh, India

Ashwini Shelke
Department of Pharmaceutical Chemistry, Sandip Institute of Pharmaceutical Sciences, Nashik, Maharashtra, India

Amit Kumar Singh
Department of Pharmaceutics, United Institute of Pharmacy, Prayagraj, Uttar Pradesh, India

Anil Kumar Singh
Department of Pharmaceutics, United Institute of Pharmacy, Prayagraj, Uttar Pradesh, India

Shradhanjali Singh
Department of Pharmaceutical Chemistry, United Institute of Pharmacy, Prayagraj, Uttar Pradesh, India

Ganesh B. Sonawane
Department of Pharmaceutical Chemistry, Shreeshakti Shaikshanik Sanstha, Divine College of Pharmacy, Satana, Nashik, Maharashtra, India

Khemchand R. Surana
Department of Pharmaceutical Chemistry, Shreeshakti Shaikshanik Sanstha, Divine College of Pharmacy, Satana, Nashik, Maharashtra, India

Pradip P. Wagh
Department of Pharmacology, R. C. Patel Institute of Pharmaceutical Education and Research, Shirpur, Maharashtra, India

ABBREVIATIONS

AAFC	Agriculture and Agri-Food Canada
AHA	American Heart Association
AHZ	b-alanyl-L-histidina to zinc
AP1	activator protein 1
ARE	antioxidant response element
ASD	autism spectrum disorder
Ats	autoantibodies
BBJ	blueberry juice
BITC	benzyl ITC
BMD	bone mineral density
BMI	body mass index
BPD	brief psychotic disorder
CAP	cellulose acetate phthalate
CHD	coronary heart disease
CLA	conjugated linoleic acid
COVID-19	coronavirus disease 2019
CR	C-reactive protein
CS	chondroitin sulfate
CVD	cardiovascular disease
DALYs	disability-adjusted life year
DHA	docosahexaenoic acid
DM	diabetes mellitus
DRI	dietary reference intake
DU	duodenal ulcers
DXA	dual-emission x-ray absorptiometry
EC	epicatechin
ECG	epicatechin 3-gallate
EE	energy expenditure
EGC	epigallocatechin
EGCG	epigallocatechin gallate
EI	energy intake
EPA	eicosapentaenoic acid
EVOO	extra-virgin OO
FD	functional dyspepsia

FDA	Food and Drug Administration
FFC	functional food center
FFs	functional foods
FNB	Food and Nutrition Board
FOS	fructooligosaccharides
FOSHU	food for specified health use
GABA	gamma amino butyric acid
GADA	glutamic acid decarboxylase
GAO	general accounting office
GERD	gastroesophageal reflux disease
GI	gastrointestinal/glycemic index
GLP-1	glucagon-like peptide-1
GLUT2	glucose carrier
GRAS	generally recognized as safe
GTCs	green tea catechins
HDL-C	HDL cholesterol
HOB	head of the bed
HSV	herpes simplex virus
IBD	inflammatory bowel disease
IBS	irritable bowel syndrome
IFIC	International Food Information Council
ILSI	International Life Sciences Institute
INI	Integrity Nutraceuticals International
IU	international units
LAB	lactic acid bacteria
LDL	low-density lipoprotein
LES	lower esophageal sphincter
LGG	lactobacillus rhamnosus GG
MBJ	mulberry juice
MC	Momordica charantia
MCTs	medium-chain triacylglycerols
MI	myocardial infarction
MODY	maturity-onset diabetes of the young
NDM	neonatal diabetes mellitus
NF	nuclear factor
NHP	natural health products
NO	nitric oxide
OO	olive oil
OVX	ovarian ovariectomy

PAC	proanthocyanidins
PKC	protein kinase C
PLA2	phospholipase A2
PLP	pyridoxal phosphate
PPAR	peroxisome proliferator-activated receptor
PTH	parathyroid hormone
PUD's	peptic ulcer disorder
PUFAs	polyunsaturated fatty acids
QR	quinone reductase
ROS	reactive oxygen species
RSA	radical scavenging activity
RSV	respiration syncytial virus
SAD	schizoaffective disorder
SAMe	S-adenosylmethionine
SIPD	substance-induced psychotic disorder
SNPs	single nucleotide polymorphisms
SOD	superoxide dismutase
SPD	shared psychotic disorder
SPI	soy protein isolate
SPMs	specialized pro-resolving mediators
STAT1	sign transducer and activator of transcription 1
T2DM	type 2 diabetes mellitus
TC	total cholesterol
TG	triglycerides
TRAMP	transgenic adenocarcinoma of the mouse prostate
UC	ulcerative colitis
USDA	United States Department of Agriculture
UTI	urinary tract infections
WCVD	wine and cardiovascular disease
WHO	World Health Organization

PREFACE

Recently, functional food sources have gained popularity for well-being within health circles. Also called nutraceuticals, functional food sources are exceptionally nutritious and provide various amazing medical advantages. For instance, they might ensure protection against infection, prevent supplement deficiencies, and enhance proper development and growth.

Functional food varieties are ingredients that offer medical advantages beyond their apparent health benefits. A few types contain supplements and/or other extra ingredients that assist in improving well-being. A few incorporate food varieties braced with nutrients, minerals, probiotics, and/or fiber. Supplement-rich ingredients like organic products, vegetables, nuts, seeds, and grains are regularly viewed as functional food varieties as well. Oats, for example, contain a kind of fiber called beta-glucan, which has been found to diminish aggravation, upgrade insusceptible capacity, and further develop heart health. Essentially, leafy foods are loaded with oxidation prevention agents, which are gainful mixtures that help secure against sickness.

This new book, *Applications of Functional Foods in Disease Prevention*, comprises 15 chapters that deal with the role of functional foods applications in the prevention and management of various diseases like diabetes, autism, obesity, microbial infection, ulcers, mental disorder, cardiac health, bone health, etc.

Chapter 1, Overview of Functional Foods, written by Shubham J. Khairnar et al., provides introduction, overview, classifications, and details on functional foods markets.

Chapter 2, Functional Foods in Autism, written by Khemchand R. Surana and his coworker, discusses using bioactive compounds to prevent, manage, or treat ASD (autism spectrum disorder).

Chapter 3, "Functional Foods in Cancer Management and Prevention," written by Khemchand R. Surana and Sunil K. Mahajan, discusses foods that have been studied for their role in cancer risks. These foods include grapes, tomatoes, pink peppers, berries, chestnuts, ginseng, tea, soy, garlic, cruciferous vegetables, etc.

Khemchand R. Surana and his associates provide a summary of the role of functional foods (diet) in brain development, signaling networks, and neurotransmitters within the brain and cognition memory in Chapter 4, Novel Applications of Functional Foods in Brain Health.

Khemchand R. Surana et al. discusses the nutritive elements which are responsible for weight loss in addition to proper nutrition to reduce or control obesity in Chapter 5, Functional Foods in Management of Weight Loss and Obesity.

Afsar S. Pathan and his colleagues summarize the different types of functional foods which are helpful for different types of functional diseases in Chapter 6, Functional Foods in Health and Diseases.

Food products may help to reduce the risk of heart disease by lowering blood lipid levels, improving arterial compliance, lowering low-density lipid oxidation, decreasing plaque formation, free radical scavengers, and inhibiting platelet aggregation when consumed in sufficient amounts on a regular basis, described by Shubham J. Khairnar in Chapter 7, Role of Functional Foods in Cardiovascular Diseases.

Afsar S. Pathan and his associates focus on the different types of diabetes mellitus (DM) and different types of functional foods and their role in diabetes mellitus (DM) in Chapter 8, Functional Foods in Prevention of Diabetes Mellitus.

Khemchand R. Surana et al. reported the beneficial effects of honey, carbohydrate, vitamins, minerals, probiotics, etc., against bacterial infection in Chapter 9, Functional Foods in Microbial Infection.

Chapter 10, Strategies of Functional Foods in Ulcers, written by Khemchand R. Surana and a coworker, reports that diet is the simplest element of a complete technique to avoid ulcer problems. In the chapter, they look at a formulation (PHF-1) containing decided functional foods which act through means of numerous mechanisms to prevent and treat advanced ulcers, after being evaluated for their efficacy against ethanol-caused gastric ulcers described in Chapter 11, Functional Foods in Acidity Management and Preventions, written by Khemchand R. Surana and Sunil K. Mahajan.

Afsar S. Pathan et al. focuses on the role of different functional foods in psychotic disorders and different types of psychotic disorders in Chapter 12, Role of Functional Foods in Psychotic Disorders. The use of certain elements for anorexia, conventional behaviors, reduction of intestinal absorption, metabolic damage, sickness metabolism of lipids and carbohydrates, and deficiency of vitamins, iron, zinc, and copper are discussed in Chapter 13, Functional Foods in Viral Infections, written by Khemchand R. Surana and coworker.

Chapter 14, Role of Functional Foods in Growth and Development, illustrates insight related to the various categories of functional foods involved in the growth and development of human beings written by Shanti Bhushan Mishra and his associates.

Trishna Das and her associates provide an overview of current opinions on various nutritional aspects of functional foods and their implications in the management of osteoporosis in Chapter 15, Overview of Functional Foods and Their Impact on Osteoporosis.

CHAPTER 1

OVERVIEW OF FUNCTIONAL FOODS

SHUBHAM J. KHAIRNAR,[1] MITHUN RUDRAPAL,[2] EKNATH D. AHIRE,[1] MOHINI R. JAGTAP,[3] and SANJAY J. KSHIRSAGAR[1]

[1]*MET's Institute of Pharmacy, Bhujbal Knowledge City, Adagoan, Nashik, Maharashtra, India*

[2]*Rasiklal M. Dhariwal College of Pharmacy, Chinchwad, Pune, Maharashtra, India*

[3]*JMCTE Institute of Pharmacy, Nashik, Maharashtra, India*

ABSTRACT

Because there is a direct link between food and health, different scientific research has been conducted to determine the impact of foods or dietary substances on key bodily functions. The phrase "functional food," which refers to foods that have distinct health benefits, was first used in Japan. The current research examines the variables that have influenced the creation of functional foods, as well as the many definitions and classifications provided by various writers. It also includes an overview of the various functional ingredients found in various foods, as well as their health benefits, state of the art in commodity-based functional food development and research, the key factors needed for mass appeal, and some highlights on global market share growth for such products. It was also mentioned that many countries still lack specific criteria for validating health claims of functional foods by completing longitudinal studies before launching on the market.

Applications of Functional Foods in Disease Prevention. Raj K. Keservani & Eknath D. Ahire, PhD (Eds.)

1.1 INTRODUCTION

Food is a term that refers to a component that is required for multiple life-sustaining tasks such as energy production, nutrition delivery, and support of various metabolic activities, as well as body growth and maintenance. Nutrition research in the early 20th century was preoccupied with preventing deficits and promoting body growth (Doyon et al., 2008). During the last two decades, our understanding of the impact of diet on health and well-being has greatly expanded, leading to the development of new and better meals that reduce the risk of various chronic diseases. Functional foods are traditional foods that have been modified to provide health benefits over non-modified items. A number of factors have converged to propel the development of functional foods, including transitional health, urbanization, and its effects, changing demography with an aging population, food security, loss of traditional food culture, and awareness of personal health deterioration caused by busy lifestyles with poor convenience food choices and a competitive food market. Insufficient exercise, greater self-medication, increased levels of knowledge from health authorities, nutrition media, the link between diet and health, and scientific breakthroughs in nutrition research are all elements that have influenced (German et al., 1999).

The phrase "functional food" was coined in Japan in the 1980s to describe manufactured foods that, in addition to being nutritious, contain substances that enhance various physiological functions. Japan is currently the only country that has developed a regulatory approval process for functional foods. Foods for Specified Health Use, or FOSHU, is a word established by the Japanese to describe foods to which a functional component has been added for a specific health benefit (Bailey& Berry et al., 2002). The Japanese Ministry of Health and Welfare has given FOSHU its seal of approval. Various definitions of functional food, ranging from very simple to highly complicated (Table 1.1), have been offered by different agencies, academic organizations, and companies around the world. As a result, there is no legal definition of functional foods because all meals consumed serve some function in some fashion (Surana et al., 2021).

Though there are several definitions, the general consensus is that functional food is any healthy food that looks similar to conventional foods, is consumed as part of a regular diet, and is claimed to have physiological benefits such as health-promoting or disease-preventing properties in addition to the basic function of supplying nutrients. Aside from the definitions above, a number of other terminologies (Table 1.2) have emerged, which

TABLE 1.1 Various Definitions of Functional Foods

Source/Reference(s)	Definition of Functional Food
International Food Information Council (IFIC, 1999), Washington, USA (Day et al., 2009; Gray et al., 2003)	Foods that may be beneficial to one's health in addition to providing basic nourishment.
Health Canada, Ontario, Canada/ Shahidi (Shahidi, 2006)	Functional foods are defined as foods that look and taste like traditional foods but have been shown to provide physiological benefits.
Doyon & Labrecque (2008)	Gave a conceptual definition based on four key factors: health benefits, food nature, level of function, and consumption pattern, all of which describe functional food.
US General Accounting Office (GAO), Washington, USA/Noonan & Noonan (2004)	Food that purports to provide health benefits in addition to those provided by basic nutrition.
Saher (2004)	"If it has been successfully proved to affect beneficially one or more target functions in the body, beyond adequate nutritional effects," food can be considered functional.
Food and Nutrition Board (FNB) of the National Academy of Sciences, Washington, USA/Thomas and Earl (Earl, 1994), Kruger & Mann (2003)	Functional food is defined as "any modified food or food ingredient that may provide a health advantage beyond that of the traditional nutrients it contains" and includes "any modified food or food ingredient that may provide a health benefit beyond that of the traditional nutrients in contains."
Weststrate et al. (2002)	Functional meals should primarily be directed at improving function or lowering (longer-term) illness risk in 'healthy' individuals, rather than disease therapy in 'sick' people, which is regarded scientifically relevant. The improvement in function could be linked to: (i) growth, development, and differentiation; (ii) substrate metabolism; (iii) defense against reactive oxidant species; (iv) the cardiovascular system; (v) gastrointestinal physiology and function; and (vi) behavior and psychological functions.
Hugget et al. (1996) & Charalampopoulos et al. (2002)	Beyond fundamental nutritional requirements, functional foods or food ingredients have a positive effect on the host's health and/or minimize the risk of chronic disease.
The European Commission Concerted Action Group on Functional Food Science in Europe (FUSOSE), The International Life Sciences Institute/ Blades (2000)	Food that has been sufficiently demonstrated to benefit one or more functions in the body, beyond adequate nutritional impacts, in a way that is relevant to either a better state of health and well-being and/or a reduction of risk, is referred to as 'functional' food.
Korver (1997)	Functional foods must meet the following minimum criteria: they are foods (not pharmaceuticals), and they have an additional function "above and beyond normal nutrition." They influence a specific nutritional function, and there is at least a reasonable hypothesis that this change enhances health.
Goldberg (1994)	Any food or food item that, in addition to its nutritional worth, has a favorable impact on an individual's health, physical performance, or mental condition.

are occasionally related or interchanged with functional foods (Ahire et al., 2021).

1.2 CLASSIFICATION OF FUNCTIONAL FOODS

According to Holm (2003), Siró et al. (2008), Klimas et al. (2008), it is possible to classify functional foods in different classes depending on origin or modification:

- Calcium fortified bread, omega-3 fortified breads, and phytosterol fortified spreads are examples of food products fortified with a variety of components that have a good impact on diseases and health.
- Foods that have been liberated to combat anti-nutritional chemicals created during processing, such as a poisonous component or a food allergy.
- Food raw materials (fruits and vegetables) in which a health contributing factor has been increased by some postharvest treatment, e.g., functional grapes with higher antioxidant developed by ultraviolet irradiation, or other nutrients/food raw materials (fruits and vegetables) in which a health contributing factor has been increased by some postharvest treatment, e.g., eggs or meats high in omega-3 fatty acids, beef high in conjugated linoleic acid or some other nutrients
- Novel foods with improved health benefits, such as rice with high iron or B vitamin content, vegetable oils with improved fatty acid composition, foods without food allergens, high-carotene canola oil, wheat with increased lutein levels, or berries with higher antioxidant content, are produced by genetic manipulation or selection of new varieties not previously consumed.

As a result, natural-traditional goods containing components that have a favorable impact on health are strictly not functional foods. For example, cranberry juice, which has a positive impact on urinary tract infections, is not a functional food when ingested as such. The created food, on the other hand, is a functional food if the juice or its health-promoting element is added to another food in isolated form to improve health. Probiotic and prebiotic foods are another type of useful food. Probiotics are defined as "live bacteria that impart a health benefit on the host when eaten in sufficient numbers." Lactic acid bacteria (LAB) and bifidobacterial, the most researched and

TABLE 1.2 Different Terms Linked or Interchanged with Functional Foods

Bioactive Compounds	Bioactive substances are naturally occurring chemical compounds found in or obtained from plants, animals, or marine sources that provide the intended health/wellness effect (Hardy, 2000).
Dietary Supplements	Dietary supplements, as the name implies, are a dietary supplement in which the active component is added to food or ingested in the form of tablets, powders, or liquids, and they do not replace a complete meal or food (Hardy, 2000).
Functional Ingredients	Manufacturers in the food industry utilize functional ingredients, which are standardized and described preparations, fractions, or extracts containing bioactive substances of varied purity (Hardy, 2000).
Medical Foods	Medical foods are defined as "formulated to be consumed or administered eternally under the supervision of a physician, and which is intended for the specific dietary management of a disease or condition for which distinctive nutritional requirements, on the basis of recognized scientific principles, are established by medical evaluation," according to the US Food and Drug Administration (FDA) (Hardy, 2000).
Natural Health Products	Homeopathic preparations, substances used in traditional remedies, minerals or trace elements, vitamins, amino acids, essential fatty acids, or other plant, animal, or microbe-derived substances are all examples of natural health products (NHP). These items are typically sold in medicinal or 'dosage' form in order to diagnose, cure, or prevent disease, restore or correct function, or maintain or promote health. Nutraceuticals are also included in NHP (Hardy, 2000).
Nutraceutical	The term nutraceutical refers to any substance that has medicinal or health benefits, such as illness prevention and treatment, and is either a food or a portion of food. These are derived from foods and can be taken in pill, capsule, potion, or liquid form (Hardy, 2000; Keservani et al., 2010a, b; Shahidi, 2006).

commonly used bacteria in the probiotic area, are natural components of the gut microbiota. Foods contain several strains of helpful bacteria, making them naturally probiotic. Prebiotics, as defined by Gibson and Roberfroid (Gibson, 1995), are non-digestible food ingredients such as starches, dietary fibers (Keservani et al., 2020; Rane et al., 2020), other non-absorbable sugars, sugar alcohols, and oligosaccharides that benefit the host by stimulating the growth and/or activity of one or a small number of bacteria in the colon, thereby improving the host's health (Charalampopoulos et al., 2003; Stanton, et al., 2005). Foods fortified with probiotics and prebiotics, such as probiotic yogurts and prebiotic bread, are classed as functional foods and provide health benefits.

1.3 FOOD–HEALTH RELATIONSHIPS FOR NEED-BASED

Understanding the need for specific food qualities in addressing various health problems, as well as the contribution of specific food ingredients to such benefit, would undoubtedly aid in the development of functional foods. Table 1.3 categorizes such relationships, whilst Table 1.4 provides several functional components along with their health benefits.

TABLE 1.3 Food–Health Relationships Categorized for Food Product Development (Asia et al., 2005)

Health Category	Food Characteristics
Diseases linked to environmental deterioration and food production practices	Foods that are environmentally conscious (e.g., those that are manufactured in a sustainable manner; biodegradable or edible packaging; recognizable biosecurity for animal-derived foods; genetic material).
Protein quality, fat quality, and vitamin status all play a role in disease.	Fish or its plant or microbial food surrogates; nutrient-dense foods.
Physical inactivity and health (particularly obesity; loss of lean mass, especially muscle).	Low-energy-density, high-nutrient-density foods.
Menopause, macular degeneration, and osteopenia are examples of phytochemical deficient illnesses.	Plant-based cuisine and their variety are being emphasized more.
Diseases associated with changing demographics. Aging causes a rapid loss of conventional food cultures and the adoption of new ones.	Anti-aging foods, particularly those that help to delay changes in body composition, sensory function, immunological function, cardio-respiratory function, and cognitive performance, as well as anti-inflammatory foods. Traditional meals are kept in a handy, inexpensive, and identifiable form.
Changes in mood and new psychosocial factors.	Food that has a positive impact on one's mood.
Foodborne disease and food safety from a microbiological standpoint.	• Pre- and probiotic foods; • Foods that boost the immune system.
Illnesses associated with food chemical safety (e.g., pesticide residues).	Food certification and regional origin.

Source: Adapted from Asia et al., 2005

TABLE 1.4 List of Some Functional Ingredients, Sources, and Benefits (Ferrari, 2007; Holm, 2003; Shahidi, 2009)

Functional Ingredient	Source/Food	Beneficial Effects
Alginic acid and xylofucans	Sargassum vulgare (algae)	Antiviral activity
Anthocyanins, catechins, cyanidins, flavonols, myricetin, and quercetin	Grape peels, juice, and wine (from red grapes); berries and eggplant	Antioxidant activity, inhibits LDL oxidation, superoxide scavengers
Calcium	Milk and derivatives	Decreases colon cancer, decreases proliferative and irritative effects on intestine
Capsaicin	Red and green peppers	Superoxide anion scavenger
Carnosol and carnosoic acid	Rosemary extracts	Inhibits peroxidation of oils
Carotenoids and lycopene	Tomatoes, carrots, bell peppers, green leafy vegetables, fruits, and palm oils	Antioxidant, anti-inflammatory, anticarcinogenic activity, inhibits LDL oxidation, decreases the risk of cardiovascular disease
Conjugated linoleic acids (CLA)	Lamb, turkey, beef	Reduce breast cancer, atherosclerosis, diabetes type 2
Curcuminoids: diferuloyl methane, phydroxycinnamoyl methane, and bisphydroxycinnamoyl methane	Curcuma longa (turmeric)	Decreases lipid oxidation, avoid ischemichypoxic
Daidzein, genistein, and isoflavones	Soybean and derivatives	Inhibits lipid peroxidation, decreases cardiovascular and cancer risks
Fiber (soluble and insoluble) phytochemicals	Oat cereal, fortified juice, Chondruscrispus (algae)	Reduces total and LDL cholesterol
Flavonoids (flavones, flavanones, flavonols, and flavan-3-ols)	Celery, parsley, citrus, onions, tea, green beans, tomatoes, cocoa, apples, berries, certain beans, and chocolate	Antioxidant, antiproliferative, antihypertensive, anticarcinogenic, antithrombotic, cell cycle arrest, induction of phase-2 enzymes, inhibition of phase-1 enzymes, inhibition of LDL-oxidation, improvement of vascular tone
Gallic acid, sulfides and thiols, quercetin, and organosulfur compounds	Garlic, onion, and broccoli	Decreases blood pressure
Lignans	Linseed, fruits, and vegetables	Estrogenic

TABLE 1.4 *(Continued)*

Functional Ingredient	Source/Food	Beneficial Effects
Phenolic acids and monounsaturated fatty acids	Coffee, green tea, black tea, cereal bran, fruits, extra-virgin olive oil	Anti-inflammatory
Prebiotic	Banana, garlic, onions, milk, honey, artichokes	Stimulate growth and/or activity helpful bacteria in the colon, thus improves host health
Probiotics	Fermented dairy products	Support intestinal tract health, boost immunity
Resveratrol, other grape polyphenols, ginsenosides, and ascorbic acid	Grapes, red wines, Panax Ginseng (ginseng), citrus fruits	Decreases blood pressure, prevents advanced malignance
Sesaminol	Grapes, peanuts	Antioxidant, cardioprotective, lifespan extension
Tocopherols, ubiquinol, tocotrienols, ω-3 fatty acids, phytosterols, and others	Almonds, nuts wheat, fish oil, oils (from rice, soy, olive), fat, and margarine	Decreases blood cholesterol, inhibits lipid peroxidation

1.4 STATE OF THE ART IN RESEARCH AND DEVELOPMENT OF FUNCTIONAL FOOD

Dairy, bakery, confectionery, soft drinks, and baby foods are some of the most popular food industry segments where functional food products have been introduced (Kotilainen, 2006; Menrad, 2003). The most common forms of functional products with substances ranging from plant to animal origin that are aimed to lower high blood pressure, cholesterol, blood sugar, and osteoporosis have been brought into the market (Charalampopoulos, et al., 2002; Sanders et al., 1998), and are briefly discussed in the following subsections.

1.4.1 DAIRY-BASED

Milk is a good source of calcium, an important vitamin that can help prevent osteoporosis and colon cancer. Beyond proteins, minerals, and vitamins, it contains bioactive minor elements. Immunoglobulins, hormones, growth factors, cytokines, nucleotides, polyamines, enzymes, and peptides are examples of these elements.

Because of their amino acid sequences, peptides express the majority of bioactivities. These peptides are derived from native milk proteins through proteolysis and lower blood pressure, thereby aiding in the prevention of cardiovascular diseases (CVD), such as myocardial infarction, stroke, and heart failure, as well as improving immunity and providing a variety of other health benefits. Bioactive peptides can be included into functional foods as ingredients. At the worldwide level, such personalized dietary formulas are currently being created to maximize health through nutrition (Anonymous, 2009; Gry et al., 2007). According to Prabhakar and Mallika (Prabhakar et al., 2010), the market is flooded with dairy-based health-promoting products. A few examples are: (i) Calpis Co. (Japan) has produced sour milk with brand name 'Calpis' containing blood pressure lowering functional ingredients, Valine-Proline-Proline and Isoleucine-Proline-Proline; (ii) Davisco (USA) has developed hydrolyzed whey protein isolate having β-lactoglobulin fragments for controlling blood pressure with brand name 'BioZate,' and whey protein isolate with glycomacropeptide for controlling blood pressure, dental caries, as well as infection from viruses and bacteria, with brand name 'BioPure-GMP'; (iii) DMV International (The Netherlands) produces two hydrolysates in the brand name 'C12' and 'Cysteine peptide,' both containing milk protein derived peptide as bioactive component which reduces blood pressure, and helps in raising energy level, respectively; and (iv) 'Clarinol' is the brand name which delivers conjugated linoleic acid from yogurt to reduce body weight is manufactured by lipid nutrition (an International Company). Probiotic dairy products make up the majority of dairy-based functional foods. In some situations, probiotic products also contain prebiotics, according to Siró et al. (Alvarez-León et al., 2006; Siró, et al., 2008) (Table 1.5).

1.4.2 CEREAL-BASED

Insoluble fibers from cereals serve as roughage and a natural functional component (Rodríguez, et al., 2006; Vitaglione et al., 2008). Cereals also contain water-soluble fibers like glucan and arabinoxylan, as well as oilgosaccharides like galacto- and fructo-oligosaccharides, which, in addition to promoting a variety of physiological benefits, can also selectively stimulate the growth of lactobacilli and bifidobacterial in the colon and act as prebiotics. Glucan is found in abundance in barley and oats, and it is an active factor in decreasing cholesterol and lowering the glycemic index. Nutrition bars, chews, breakfast cereals, beverages, baked goods, yogurt, ice cream, pasta, and dietary supplements have all successfully integrated extracted-glucan components. Nutraceuticals Inc., Ceapro Inc., and Parrheim Inc., for example, provide glucan concentrates and foods fortified with it (Anonymous, 2009). Izydorczyk et al. discovered that three barley cultivars with different starch properties increased the total and soluble dietary fiber content of wheat flour flat bread considerably. Aside from separating the components from cereals, processed whole cereal is a rich source of bioactives that has been underutilized (phytochemicals).

Antioxidant (protects cells from oxidative damage), antiproliferative (prevents undesirable cancerous cells from replicating), carcinogen detoxifier, hypocholesterolemic, enzyme and hormone stimulant, antibacterial and antiviral, anti-inflammatory, ligand to cell wall (some phytochemicals bind physically to human cell walls, preventing pathogen adhesion), and potential inhibitor of various actions affecting the immune system are all possible functions of phytochemicals. (Kuhnen et al., 2009) Maize seed was discovered to be an interesting source of xanthophylls, which efficiently block vessel formation, implying that such compounds could play a role in the prevention of disorders linked with vascular dysfunction. Magiform (Sweden) has produced müsli, biscuits, pasta, and crackers for patients with chronic gastrointestinal (GI) tract problems, all of which contain specially processed grains (Mark-Herbert, 2004). Aside from glucan, oats and barley are rich in tocopherols, phenolics, phytosterols, and insoluble fibers, all of which are helpful to one's health. All cereals are typically ideal substrates for the growth of human-derived probiotic strains, however, they are influenced by a variety of parameters such as composition, processing, probiotic strain growth substrate creation, and so on. Encapsulation of cereal fractions improves their usefulness. For example, Wang et al. discovered that encapsulated starch granules can operate as a delivery method for probiotic bacteria (Wang et al., 1999) and Charalampopoulos et al. reviewed it (Charalampopoulos, et al., 2002). Furthermore, resistant starch, which can be found in heated starch and starch-containing foods, is a functional fiber that helps colonic bacteria process carbs, lowering the risk of bowel disease (Charalampopoulos et al., 2002). Several epidemiological studies have found a negative relationship between whole grains and their dietary fibers and cardiovascular disease and cancer (Kushi et al., 1999 &Bingham, S 1990). Recent research has shown that eating wheat aleurone-rich meals increases fasting plasma betaine, which helps to lower fasting homocysteine and LDL cholesterol in adults (Price et al., 2010).

TABLE 1.5 List of Market Available Probiotic Products

Brand/Trade Name	Description	Producer
Actimel	Probiotic drinking yogurt with *Lactobacillus caseiimunitass*® cultures	Danone (France)
Activia	Creamy yogurt containing BifidusActiRegularis®	Danone (France)
Jovita Probiotisch	Blend of cereals, fruit, and probiotic yogurt	H&J Bruggen (Germany)
ProViva	Refreshing natural fruit drink and yogurt in many different flavors containing *Lactobacillus plantarum*	Skànemejerier (Sweden)
Rela	Yogurts, cultured milks, and juices with *Lactobacillus reuteri*	Ingman Foods (Finland)
Revital active	Yogurt and drink yogurt with probiotics	Olma (Czech Republic),
Yakult	Milk drink containing *Lactobacillus casei* Shirota	Yakult (Japan)
Vitality	Yogurt with pre-and probiotics and ω-3	Müller (Germany)
Vifit	Drink yogurts with LGG (lactobacillus strain), vitamins, and minerals	Campina (The Netherlands)
Snack Fibra	Snacks and bars with natural fibers and extra minerals and vitamins	Celigüets (Spain)

Potential inhibitor of a variety of actions that affect the onset and course of a variety of pathogenic processes. Kuhnen et al. (2006) discovered that maize seed contains xanthophylls that efficiently block vessel development, implying that such substances may have a role in the prevention of disorders linked with vascular dysfunction. As researched by Wang et al. (1999) and reviewed by Charalampopoulos et al. encapsulation of cereal fractions improves their functioning. For example, encapsulated starch granules can work as a delivery method for probiotic bacteria (Siró et al., 2008). Furthermore, resistant starch, which can be found in heated starch and starch-containing foods, is a functional fiber that helps colonic bacteria process carbs, lowering the risk of bowel disease (Siró et al., 2008). Several epidemiological studies have found a negative relationship between whole grains and their dietary fibers and cardiovascular disease and cancer (Bingham, 1990; Kushi, et al., 1999). Recent research has shown that eating wheat aleurone-rich meals increases fasting plasma betaine, which helps to lower fasting homocysteine and LDL cholesterol in adults (Price, et al., 2010). Flaxseed is the main ingredient in this dish. Flaxseed (linseed), a member of the Linaceae family of oilseeds, is gaining popularity as a versatile nutritional ingredient that can be easily incorporated into a variety of foods because it contains proteins, dietary fiber, polysaccharides, polyphenolics, and essential fatty acids that promote health and may prevent disease. Flaxseed is said to contain 51–55% – linolenic acid (ALA), which helps to prevent heart disease, inflammation, high blood pressure, cancer, skin illnesses, and immunological problems like renal failure, rheumatoid arthritis, and multiple sclerosis (Kelley, et al., 1991; Vijaimohan et al., 2006). A human clinical investigation reveals that flaxseed phytoestrogens provide health benefits connected to menopausal symptoms such as osteoporosis (Tham et al., 1998). Flaxseed elevates certain omega-3 fatty acids in plasma, according to long-term placebo-controlled research in healthy menopausal women (Dodin, et al., 2008; Lin et al., 2002) looked at whether a diet supplemented with 5% flaxseed inhibited prostate cancer growth (Hasler et al., 2010) analyzed the evidence that flaxseed eating lowers total and LDL cholesterol, as well as platelet aggregation. These findings have prompted Oroweat Foods (USA), Dimpflmeier Bakery Ltd. (Canada), and others to incorporate flaxseed into a variety of cereal-based foods such as breads, cereals, pastas, muffins, and other baked goods. According to the Flax Council of Canada, daily consumption of 5 g (1 teaspoon) flax oil or 8 g (1 tablespoon) milled flax offers enough ALA to meet dietary requirements (Flax Council of Canada, 2010). In 1998, the US Food and Drug Administration (FDA) designated it as a generally recognized as safe (GRAS) product (Anonymous, USFDA. 2011). Linola™, a low-linolenic-acid flax type, is utilized as a source of oil for cooking, salads, margarine, and other food applications. As a result, whole and milled flaxseed, as well as cold pressed flax oil, have grown in popularity (Anonymous, USFDA. 2011).

based on tea. According to Shi et al. (2005), tea is a significant source of polyphenols, particularly catechins, such as epicatechins (EC), epicatechingallate (ECG), epigallocatechin (EGC), and epigallocatechin gallate (EGCG), which contribute for 25–30% of the weight obtained after drying fresh tea leaf (Balentine et al., 2019; Ho et al., 1992). Green tea polyphenols are anticarcinogenic, antibacterial, antifungal, antiviral, anticariogenic, antimutagenic, thermogenic, kidney-protecting, deodorizing, and blood glucose and blood lipid-controlling. 'Sunphenon,' a green tea product from Taiyo (Japan), is a purified polyphenol series rich in natural green tea catechins derived from the green tea leaf (Anonymous. Taiyo Green Tea Nutritional Center, 2011). CCA Industries, Inc., NJ, USA, produces the 'Mega-T' brand of green tea-based weight loss products (Anonymous, CCA Online Industries Inc. N.J., 2011). Green tea drinking enhances basal metabolism, resulting in weight loss, according to (Mermel, 2014; Zemel et al., 2002). After oxidation, catechins are said to create theaflavins and thearubigins, which provide black tea its antioxidant, antibacterial, anti-inflammatory, and anticancer properties (Pasha et al., 2005; Shahidi, 2009). Zemel et al. (2002) analyzed a number of epidemiological research that linked tea and its flavonoids (Keservani and Sharma, 2014) to a reduced risk of lung cancer. Based on fruits and veggies because different types of phytochemicals are present in great amounts in fruits and vegetables, they provide a wide range of advantages (Shi et al., 2005). Scientists are attempting to evaluate the health benefits of various phytochemicals found in fruits and vegetables in order to create various functional foods (Hunter et al., 2008). Berry fruits have the highest antioxidant content among fruits. Flavonoids like anthocyanins, flavonols, and cathecins, phenolic acids like hydroxycinnamic acids, and tannins like proanthoscyanidins and ellagitannins are among the polyphenols found in berries. Berry fruits are also utilized as a sweetener in aromatized cereals and as a colorant in breakfast cereals. Berry seed oils, berry oils, berry powders, berry extracts, and berry fiber powders are all new functional food additives (Keservani et al., 2016b, 2017, 2020). It can aid in the prevention of heart disease, cancer, diabetes, and premature aging, as well as improving eyesight and brain function (Eyduran et al., 2009). It was shown that by consuming a specific fruit-juice mixture made mostly of berry fruits, blood antioxidant capacity increased significantly, whereas lipid peroxydase activity dropped (Eyduran et al., 2009). The polyphenol resveratol is abundant in red grapes and the alcoholic beverage made from them, red wine (Jang et al., 1997; Zemel et al., 2002). Resveratrol has been shown to have cardioprotective, cancer-fighting, anti-inflammatory, antioxidant, and lipid-metabolizing properties (Dell'Agli, 2004; Zemel et al., 2002). This range of activity has recently been expanded due to its ability to manage obesity and diabetes (Jarouliya and Keservani 2019; Szkudelska, et al., 2010). Murcia's Center of Edafology and Applied Biology has devised a post-harvest treatment that uses UV irradiation pulses to produce 200-fold resveratrol-enriched functioning table grapes. In the last few years, a number of in vitro, clinical, and epidemiological research have been published that show red wine has health-promoting qualities in the prevention of cardiovascular disease and certain types of cancer. According to Holm (2003) of the European Commission, a wine and cardiovascular disease (WCVD) project was carried out to assess the effects of red wine polyphenolic components and alcohol on indicators of arteriosclerosis and thrombosis. Anthocyanins, flavonols (quercetin), and resveratrol were found as antioxidants. Lycopene is a carotenoid antioxidant found in red fruits and vegetables, particularly tomatoes and tomato-based products. Lycopene has been shown to have antioxidant, anti-hypertensive, anti-atherosclerotic, and anti-diabetic properties (Day et al., 2009; Kohlmeier et al., 1997). An examination of epidemiological data found a link between tomato consumption and the risk of prostate, lung, and stomach cancer, as well as the prevalence of coronary heart disease (Bruno et al., 2007; Rao et al., 2003). The possibility of incorporating lycopene-rich byproducts from tomato processing businesses into functional foods is being explored. Adding dried tomato peel to meat to fortify dry fermented sausage, enriching low-quality edible oils with lycopene from tomato peels or tomato puree to induce thermal stability, feeding hens lycopene-rich byproducts to enrich the egg yolk with lycopene, encapsulation of supercritical fluid extract of lycopene-rich tomato pulp waste, enrichment of extruded snacks with barley-tomato pomace blends, and mixture of Kong et al. (2010). According to Liu et al. (2008), the antioxidant properties of lycopene, vitamin E, vitamin C, and -carotene combinations were much better than the sum of the separate antioxidant effects. Sun-Waterhouse et al. (2008) devised an aqueous method for extracting high pectic polysaccharides and good phenolic antioxidants from food grade apple fiber for use in functional foods. Shui and Leong (2006) looked for antioxidant nutraceuticals in star fruit (Averrhoa carambola) residue, which is usually abandoned after juice extraction. Gruenwald (2006) examined the presence of bioactive natural ingredients in fruits such as sea buckthorn (*Hippophae rhamnoides*), baobab (*Adansonia digitata*), sausage tree (*Kigelia pinnata*), noni (*Morinda citrifolia*), mangosteen (*Garcinia mangostana*), and mabola plum (*Parinari curatellifolia*), all of which have endless applications as functional ingredients in beverages and foods to provide various nutrients (Cederroth et al., 2009). Genistein and daidzein are the two main isoflavones found in soy. Consumption of soybeans and soy products appears to protect against hormone-dependent malignancies such as breast and prostate cancer, as well as having positive effects on cardiovascular disease, osteoporosis, and menopausal symptoms (Hasler, 2010). Total phenolic content and total isoflavones were found to have a strong relationship with antioxidant activity in soybeans and soy products (Devi et al., 2009).

In addition to these bioactive components, soy is a top-rated source of high-quality protein and fat (Jaroulia and Keservani, 2019a; Keservani et al., 2015; Singh et al., 2017b), with 15% saturated fat, 61% polyunsaturated fat, and 24% monounsaturated fat. Epidemiological and clinical studies strongly suggest that people who eat soy protein on a regular basis have a lower risk of age-related diseases, lower total and LDL-cholesterol levels in their blood (Cederroth et al., 2009; Jooyandeh, 2011) and a lower risk of glycosuria, which is a strong predictor of diabetes if they eat more than 12.6 g/day. Consumption of isoflavones by postmenopausal women was also linked to a reduced BMI and increased HDL values. The oil is separated from the soy protein by crushing the beans. Isoflavones are still present in soy protein. Soy protein isolate (SPI) is more refined than soy protein but contains fewer isoflavones; both of these products are typically considered acceptable to use in foods with no limits. Food makers can now make health claims like "25 grams of soy protein per day, as part of a low-fat, low-cholesterol diet, may reduce the risk of heart disease." Soybeans provides around 31 gm of soy protein per serving (Brown, 2011). Several studies have found an inverse relationship between the consumption of vegetables high in carotene and glucosinolates (from brassica vegetables) and the risk of cancer (Stoll et al., 2003; Ovesen 1999; Verhoeven et al., 1996) used enzymes to extract total carotene from carrot pomace, which they then put into a functional drink. Carrot pomace has been used in a variety of cuisines, including bread, cake, dressing, and pickles. Bombardelli and Morazzoni created a carotenoid-based antioxidant that can be employed as a functional food additive (Bombardelli et al., 1995; Singh et al., 2017a). Black carrots are high in water soluble phenolic chemicals including hydroxycinnamic acid and its derivatives, which have strong radical scavenging activity (RSA) and can help to prevent cancer and heart disease. Its extract is utilized as a food colorant and antioxidant (Day et al., 2009 & Lambert et al., 2005). Pasta made with black carrot concentrate has been found to be popular among consumers (Day et al., 2009). Garlic supplements such as garlic powder, garlic oil, and garlic extract (which contain a sulfur compound for improved cardiovascular functions) and saw palmetto extract (which contains fatty acid and phytosterol that reduce benign prostate hyperplasia) are two plant-derived products that are expected to do well in the global marketplace (Anonymous, World Nutraceutical Ingredients, 2009; Roldán et al., 2008) classified onion byproducts, such as bagasse, as antioxidant and antibrowning food ingredients (Plaza et al., 2008; Rodrguez et al., 2006) Both revealed antioxidant properties of vegetable fibers that act as roughage and natural functional component (Fuentes-Alventosa et al., 2009; Izydorczyk et al., 2008). The functional features of high dietary fiber powders made from asparagus byproducts, which can be used to make fiber-enriched functional meals, were investigated. Total LDL cholesterol levels are known to be reduced by plant sterols and stanols (saturated sterols), lowering the risk of coronary heart disease (CHD). Holm (2003) found that consuming 1.3 g of plant sterol esters or 3.4 g of plant stanol esters per day considerably lowers total cholesterol, whereas consuming 2–2.5 g per day lowers LDL cholesterol by roughly 10%. Various foods and dietary supplements enriched with sterols or stanol esters are available, such as the 'Bayer Heart Health AdvantageTM,' 'Benecol®' spread, and 'Centrum Cardio®' dietary supplement, which contain 0.4, 0.85, and 0.8 g sterol per serving, respectively (94).According to an epidemiological study, daily consumption of spreads enhanced with plant sterols efficiently decreases total and LDL cholesterol levels in the blood. Isanga and Zhang (2007) and Yang et al. (2009) investigated the total and soluble free phytochemical components, including phenolic and flavonoid content, as well as the antioxidant potential of various nuts. Walnuts had the most soluble free phenolic content, pecans had the highest soluble free flavonoid content, but walnuts had the best antioxidant activity, according to the researchers. More than 30 studies have demonstrated that adding nuts in one's diet helps lessen the risk of heart disease, according to the International Tree Nut Council Nutrition Research and Education Foundation. According to the studies, there are a large number of functional substances that can be used for human benefit.

Scientists at the University of Georgia recently produced a functional peanut with a higher resveratrol content based on seafood. Apart from providing high quality protein, seafoods are rich in omega-3 fatty acids, antioxidants, and other bioactive compounds (Prabhasankar et al., 2009). Various components of marine algae, such as its oil (high in docosahexaenoic acid, DHA, an omega-3 fatty acid) and polysaccharides (dietary fiber), are natural sources of functional ingredients that have a variety of health benefits, including preventing cardiovascular disease (Keservani et al., 2016a) and acting as an antihypertensive, antioxidant, antiviral, and anticancer agent (Plaza et al., 2008; Rupérez, et al., 2002). Prabhasankar et al. (2009) examined the chemical, functional, and structural features of pasta with edible Japanese seaweed, wakame (*Undaria pinnatifida*), and found it to be sensorially satisfactory. Holm (2003) described the invention of (i) multiple triacylglycerol preparations from fish oil with varied EPA: DHA ratios in combination with a stable form of folate, and (ii) a microencapsulating procedure for the synthesis of powered fish oils. Both of these ingredients were used in a variety of food products as part of nutritional research to see how they affected CVD, infant development, and neural tube abnormalities in newborns. According to Careche (2011), functional fish have only been developed by husbandry with high levels of polyunsaturated fatty acids. However, either through aquaculture or the development of filet-based synthetic food, there is obviously a possibility for expanding the range of useful seafoods (products). Fish oils have also been used to modify cow feed in order to produce egg and beef with high DHA and EPA content (Siró et al., 2008). This has prompted the change of products at the most fundamental level, such as feeding rather than injecting animals. According to Jeon et al. (2000), dietary fibers such as chitin and chitosan (structural elements in the exoskeleton of crustaceans such as crabs, shrimp, and others) can be used in functional foods as antimicrobial preservatives, antitumor, immuno-potentiating, hypocholesterolemic, and antihypertensive agents. Epidemiological studies have found an inverse relationship between fish and fish oil consumption and the incidence of colorectal cancer, prostate cancer, and stroke (Anonymous, Fish Oil-Evidence from Epidemiological Studies, 2011). A double-blind, placebo-controlled trial suggests that taking fish oil supplements can help with rheumatoid arthritis. A meat-based dish is a good source of vital nutrients like taurine, L-carnitine, creatine, conjugated linoleic acid (CLA), and endogenous antioxidants, which can be improved with a few tweaks. According to Holm, adding 1% CLA to animal feed decreases the risk and severity of atherosclerosis and type 2 diabetes. CLA may also have antithrombotic and anti-cancer properties, however human intervention studies are required before any conclusions can be drawn. Olmedilla-Alonso and colleagues and Jiménez-Colmenero et al. created functional meat-based products with nuts and discovered that they are linked to a lower risk of cardiovascular disease. Fernández-Ginés et al. reviewed the addition of fish oil, vegetables, fiber, soy, natural extracts with antioxidant qualities, and sodium chloride reduction to produce functional beef. To promote joint health, chondroitin sulfate (CS) and glycosamine extracted from bovine skin are being tested as functional ingredients in food and beverages. However, in other countries, these ingredients are prohibited. Eggs containing omega-3 polyunsaturated fatty acids have been created. Egg-based blades are significant in the prevention of heart disease, atherosclerosis, thrombosis, and blood pressure; Paliyath reported their incorporation into breads, biscuits, and soups. Surai and Sparks also stated that after adding antioxidants, vitamin D, E, B12, and folic acid, as well as carotenoids, selenium, and DHA, eggs can be used as functional foods. 'Columbus' eggs, fortified with omega-3 and vitamin E, were initially introduced in 1997 by Belovo, Belgium, and have since been sold in the United Kingdom (since 1998), the Netherlands (since 1999), India, Japan, and South Africa (since 2000). In Europe, the output of 'Columbus' eggs now surpasses 50 million per year. Pilgrim's Pride Company, Gold Circle Farms, OmegaTech (USA), and Freshlay Foods (UK) have all recently released enriched eggs of this type. Mark-Herbert examined if an egg carrying a large amount of immunoglobulin (antibodies), often known as 'IgY-eggs,' can be used as an oral vaccine for consumers. The addition of omega-3 enhanced eggs to the diet reduced plasma triglycerides, blood platelet aggregation, and the risk of fatal ischemic heart disease, according to epidemiological studies and double-blinded cross-over trials.

Based on herbs include a variety of functional compounds that are useful when used in food, either whole or extracted. Curcumin, a yellow polyphenolic pigment found in the Indian spice turmeric, has been shown to have therapeutic potential in Alzheimer's disease, Parkinson's disease, multiple sclerosis, epilepsy, cerebral injury, cardiovascular disease, cancer, allergy, asthma, bronchitis, colitis, rheumatoid arthritis, renal ischemia, psoriasis, diabetes, obesity, depression, fatigue, and HIV/AIDS.Although epidemiological evidence is ambiguous, epidemiological evidence suggests that turmeric use may lessen the risk of some malignancies and provide other beneficial biological effects in humans. Curcumin may also have a therapeutic role in many of the conditions listed above, according to multiple clinical trials. Curcumin antioxidant component has just been granted GRAS certification, allowing it to be used in baked goods, milk products, fats and oils, jams and jellies, gelatins and puddings, soups, snacks, candies, frozen dairy, imitation dairy, non-alcoholic beverages, as well as spices and flavors. The addition of turmeric powder to bread significantly boosted the curcumin concentration and antioxidant activity.

Cinnamon's polyphenol component, methylhydroxychalcone, assists in glucose and lipid metabolism. Integrity Neutraceuticals International (INI) offers a water soluble cinnamon extract that was produced in collaboration with the United States Department of Agriculture (USDA). Clinical trials with cinnamon and its extract have proven that it contributes to good health. The soluble dietary fiber galactomannan is extracted from fenugreek seeds by Parry Nutraceuticals in India. The separated fiber has no harsh taste or odor, making it excellent for usage as a blood sugar management and other health benefits element. Gingerols and other antioxidant-active chemicals (flavonoids) are among the chemical components found in ginger root. Shakib and Gabrial suggested that barley-wheat fenugreek-ginger bread might be used as a substitute for white wheat bread since it provides a balanced diet that can assist healthy subjects maintain appropriate postprandial blood glucose levels. According to Franck, chicory inulin-type fructans are a non-digestible carbohydrate that serves as a prebiotic dietary fiber, a low-glycemic and low-calorie sugar alternative, and a fat substitute. Ramrez et al. employed preparative-supercritical fluid chromatography to extract antioxidant and antibacterial components from rosemary for application as functional food additives. Ginkgo (*Ginkgo biloba*), St. John's wort (*Hypericum perforatum*), ginseng (Panax ginseng), kava (*Piper methysticum*), black cohosh (*Actaea racemosa*), and Echinacea (*Echinacea purpura*, *Echinacea pallida*, and *Echinacea augustifolia*) are some of the other botanicals often employed. Essential oil (lemon grass oil) from *Cymbopogon citratus* has antioxidant and antibacterial properties, according to (Sacchetti et al., 2004), because it reduces unsaturated fatty acid oxidation. Isoflavones, which have estrogenic properties, are abundant in red clover. It contains substantial levels of polyphenolic compounds, which are bioactive antioxidants and radical scavengers, and hence could be a useful functional dietary ingredient or supplement. Because of the risk of cardiovascular instability and hemorrhage, the American Society of Anesthesiologists has advised consumers not to take herbal medicines for two weeks before getting anesthesia (Benkouider, 2011; Hasler, 2001; Rane et al., 2020a–c; Sharma et al., 2020). Hasler et al. (2001) also stated that excessive intake produces disease or creates an imbalance in the body, such as when iron is consumed in excess, hemochromatosis develops, and when folic acid is used in excess, the symptoms of pernicious anemia in vitamin B12 deficient individuals are masking. As a result, prior to their integration into foods, long-term research of such botanicals should be conducted. The amount to be used, when it should be taken, and any precautions that must be followed should all be indicated.

According to Holm (2003), Vitamin B6, B12, D, and K reduce the incidence of CVD and osteoporosis, but vitamins C and E may aid in the prevention of colorectal cancer. The most prominent vitamin manufacturers, Roche Vitamins and BASF AG, have introduced specialized 'bioactive ingredients' to the market, which are utilized in functional foods (Kotilainen et al., 2006) such as calcium-enriched yogurt and milk fortified with omega-3 fatty acids, folic acid, and vitamins (Ahire et al., 2020).

1.5 KEY FACTORS TO SUCCESS OF FUNCTIONAL FOOD

A variety of factors influence the success of functional foods, including:

- Concentrate on your overall well-being;
- Health advantage for a common ailment;
- Market positioning and mass distribution;
- Effective health benefit communication;
- Expansion of a current brand/food business; and
- Put a premium on taste, convenience, and fair pricing.

In all of these areas, a successful functional food, in addition to its health benefits, must be competitive. Consumers must be better informed about the active ingredients in food products, as well as their health advantages, in order to solve general well-being issues.

1.6 FUNCTIONAL FOOD MARKET

Consumers worldwide are becoming more interested in and accepting of functional foods, resulting in thriving marketplaces in Asia, North America, Western Europe, Latin America, Australia, and New Zealand, with an estimated market size of $7–63 billion US dollars by 2010 (Shakib et al., 2010). By 2013, it is estimated to be worth at least $90.5 billion US dollars. The United States is the most important market for functional foods, followed by Japan and Europe, which together account for almost 90% of worldwide sales. From 200 in 2006 to over 800 in 2008, the number of functional food debuts in the North American market has surged. New products with fiber, probiotics, antioxidants, and herbal sweeteners such as stevia are hitting the market. Harmonium International Inc. (probiotics), CV Technologies Inc. (natural health products), Lassonde Industries Inc. (functional juices), and Ocean Nutrition Canada Ltd. (omega-3 fatty acids), among others, have experienced significant growth in this field, as evidenced by a 32% increase in the number of functional foods producing companies. This has prompted research institutes such as Agriculture and Agri-Food Canada (AAFC), the University of Manitoba, and St. Boniface General Hospital to work together to conduct further research into the health claims made by various foods. According to Starling, the US market would expand by 21% by 2015, owing to high demand for energy and fortified dairy products such as probiotic yogurts. Japan is the world's second-largest market. Between 1988 and 1998, more than 1,700 functional food products were launched in Japan, according to Menrad (Menrad, 2003). In 2005, Japan has the world's largest and most inventive range of functional foods, with sales of 17 billion US dollars, with FOSHU accounting for 6 billion US dollars. Table 1.6 shows a few examples of FOSHU with a low recognition rate. Germany, France, the United Kingdom, and the Netherlands are the most prominent countries in the functional food market in Europe, with 305 items launched in Germany between 1999 and 2000. Functional dairy products are the leading product category in Europe, with sales of roughly 1.35 billion US dollars in 1999. Unilever (UK), Raisio Food Group (Finland), Nestlé (Switzerland), Danone (France), Iparlat (Spain), and Corporacion Alimentaria Penasanta, SA have dominated the European functional food market over the previous two decades (CAPSA, Spain). Innovative products such as Unilever's 'Flora Pro-Active' and Nestle's 'Petit' sipping yogurt have added a new dimension to the market and increased competitiveness. Manufacturers must now show scientific evidence to back up their product claims under new European laws, which has hampered the launch of functional food products. Holm (Holm, 2003) stated that the functional food markets have developed differently in the three main market segments in Europe, namely dairy products (46%) and cereal products (28%), while functional drinks (58%) are the primary market in the United States and Japan, respectively, followed by cereals (17% in the United States) and confectionery (15% in Japan). The Indian functional food market is expected to develop at a rate of 12–15% annually. In 2009, the whole market for nutraceuticals, which includes food supplements, vitamins, and minerals, was valued at 44 billion INR, and it is expected to grow to 95 billion INR by 2013. Pharmaceutical companies, including GlaxoSmithKline, Dabur, Cadila Healthcare, and Novartis are paying attention to the developing Indian nutraceuticals sector. Abbott, a US pharmaceutical company, has reportedly invested in the Indian functional foods business, particularly in newborn and adult nutrition. India's probiotics sector is a "significant growth market of the future," according to a report, with yearly growth of 22.6% through 2015. Food supplements, which account for 49% of probiotic sales in India, outnumber foods and beverages, which account for only 4%.GlaxoSmithKline Consumer Healthcare has a healthy malt drink named 'Viva,' British Biologicals has a nutritional supplement including whey protein concentrate as one of the ingredients, and Parry Nutraceuticals has a carotenoids variety. According to reports, the Indian government is working to set probiotic product rules. Furthermore, an econometric study is being conducted that examines the current demand prognosis for functional foods and beverages in India's states, union territories, and cities. In contrast to a few decades ago, when consumers were confused about functional food and doubtful about its acceptability, the increased interest and acceptance of functional food has had a significant impact on its market. Although legal standards have yet to be enforced, they have alleviated concern, resulting in an upward trend and intense rivalry in this industry. Because of rising medical costs and a population with a longer life expectancy in developed countries, certain nutritional recommendations have been developed, prompting the food industry to develop products that assist consumers in adhering to these recommendations, such as low-fat products, high-fiber products, and so on (Korver, 1997).

1.7 CONCLUSIONS

Functional food is any healthy food that resembles conventional foods in appearance, is consumed as part of a regular diet, and is said to offer physiological benefits such as health-promoting or disease-preventing properties in addition to its basic function of giving nutrients. Functional food can be a raw product created by genetic manipulation, a feeding study, or a processed product fortified with a variety of components that have a good impact on illnesses and health. Efforts are being made around the world to develop newer functional foods. The global market for functional foods is predicted to reach a value of at least 90.5 billion US dollars by 2013.

ACKNOWLEDGMENT

We thank the MET's, Institute of Pharmacy, BKC, affiliated under Savitribai Phule Pune University, Nashik. EDA wants to thank the Ministry of Tribal Affairs Govt. of India for providing financial assistance through a fellowship (Award No.: 202021-NFST-MAH-01235).

KEYWORDS

- **fortified food**
- **functional food**
- **healthy food**
- **physiological benefits**

REFERENCES

Aggarwal, B. B., & Harikumar, K. B., (2009). Potential therapeutic effects of curcumin, the anti-inflammatory agent, against neurodegenerative, cardiovascular, pulmonary, metabolic, autoimmune and neoplastic diseases. *The International Journal of Biochemistry & Cell Biology, 41*(1), 40–59.

Ahire, E. D., Surana, K. R., Patil, C. D., Shah, H. S., Sonwane, G. B., & Talele, S. G., (2020). Role of omega-3 fatty acids in different neurodegenerative disorders. *Applied Pharmaceutical Science and Microbiology: Novel Green Chemistry Methods and Natural Products*, 173.

Ahire, E., Sonawane, V., Surana, K., & Talele, G., (2021). Drug discovery, drug-likeness screening, and bioavailability: Development of drug-likeness rule for natural products. *Applied Pharmaceutical Practice and Nutraceuticals.* Apple Academic Press.

Alvarez-León, E. E., Román-Vinas, B., & Serra-Majem, L., (2006). Dairy products and health: A review of the epidemiological evidence. *British Journal of Nutrition, 96*(S1), S94–S99.

Anonymous, (2009). *What Are Functional Foods and Nutraceuticals?* Available from: http://www4.agr.gc.ca/AAFC-AAC/display-afficher.do?id=1171305207040 (accessed on 22 February 2023).

Anonymous, (2010). *Flax Council of Canada.* Available from: http://www.flaxcouncil.ca (accessed on 22 February 2023).

Anonymous, (2011). *United States Food and Drug Administration.* Agency Response Letter. GRAS Notice No. GRN 00002. Available from: http://www.cfsan.fda.gov/~rdb/opa-g002.html (accessed on 22 February 2023).

Anonymous, CCA Online Industries Inc. N.J. Available from: https://www.ccainvestor.com/press-releases/detail/16/cca-industries-inc-introduces-a-new-diet-aid-supplement (accessed on 22 February 2023).

Anonymous, *Functional Food: Enhancing Resveratrol Content of Peanuts.* University of Georgia Research Foundation. Available from: University of Georgia Innovation Gateway | iBridge.Network (ibridgenetwork.org) (accessed on 22 February 2023).

Anonymous, *Global Market Review of Functional Foods-Forecasts (2010).* Available from: http://www.marketresearch.com/product/display.asp?productid=1080567 (accessed on 22 February 2023).

Anonymous, *GRAS Opens Functional Food Road for Curcumin Antioxidant*. Available from: http://www.nutraingredients-usa.com/Industry/GRAS-opens-functional-food-road-for-curcuminantioxidant (accessed on 22 February 2023).

Anonymous, *Phytosterols: Sterols & Stanols*. Available from: http://my.clevelandclinic.org/heart/women/sterolstanol.aspx (accessed on 22 February 2023).

Anonymous, *Taiyo Green Tea Nutritional* Center. Available from: https://www.taiyointernational.com/press-releases/taiyo-publishes-new-book-green-tea-polyphenols-nutraceuticals-of-modern-life/ (accessed on 22 February 2023).

Anonymous, *World Nutraceutical Ingredients*. Available from: http://www.marketresearch.com/product/display.asp?productid=2496625&SID=66174803-464432607-408883852 (accessed on 22 February 2023).

Anonymous. *Meat and Meat Products as Functional Foods*. Available from: https://medicalxpress.com/news/2009-06-meat-products-functional-foods.html (accessed on 22 February 2023).

Arts, I. C., (2008). A review of the epidemiological evidence on tea, flavonoids, and lung cancer. *The Journal of Nutrition*, *138*(8), 1561S–1566S.

ASA, (2000). *What You Should Know About Herbal Use and Anesthesia* (pp. 1, 2). American Society of Anesthesiologists, Park Ridge, IL, USA.

Asia, P. I., (2005). Functional foods and their impact on nutrition and health: Opportunities in the Asia Pacific. *Asian Functional Foods*, 1.

AT, D., & Action, E. C., (1999). Scientific concepts of functional foods in Europe: Consensus document. *British Journal of Nutrition, 81*(1).

Bailey, R. *Foods for Specified Health Use (FOSHU) as Functional Foods in Japan: Japan has a Regulatory Framework for the Growing Area of "Functional Foods."* Available from: https://go.gale.com/ps/i.do?id=GALE%7CA54829177&sid=sitemap&v=2.1&it=r&p=HRCA&sw=w&userGroupName=anon%7E96345582 (accessed on 22 February 2023).

Balentine, D. A., Harbowy, M. E., & Graham, H. N., (2019). Tea: The plant and its manufacture; chemistry and consumption of the beverage. In: *Caffeine* (pp. 35–72). CRC Press.

Benkouider, C. *Functional Food-Fad or Future? Euromonitor International*. Available from: https://www.just-food.com/analysis/functional-food-fad-or-future/ (accessed on 22 February 2023).

Berry, C., (2002). Biologic: Functional foods. *QJM-Int. J. Med., 95*, 639, 640.

Bertram, J. S., (1993). Cancer prevention by carotenoids. Mechanistic studies in cultured cells. *Annals of the New York Academy of Sciences, 691*, 177–191.

Bingham, S., (1990). Mechanisms and experimental evidence relating dietary fiber and starch to protection against large bowel cancer. *Proc. Nutr. Soc., 49*(2), 153–171.

Blades, M., (2000). Functional foods or nutraceuticals. *Nutrition & Food Science*.

Bombardelli, E., & Morazzoni, P., (1996). Carotenoid-based 'functional food' antioxidant ingredients. *Trends in Food Science & Technology, 11*(7), 377.

Brown, J. L., (2011). *Soy Proteins and Soy Isoflavones*. College of Agricultural Sciences and Cooperative Extension. The Pennsylvania State University.

Bruno, R. S., Wildman, R. E., & Schwartz, S. J., (2007). Lycopene: Food sources, properties, and health. *Handbook of Nutraceuticals and Functional Foods, 2*.

Careche, M., Borderias, A. J., Sanchez-Alonso, I., & Lund, E. K., (2011). Functional seafood products. In: *Functional Foods* (pp. 557–581). Woodhead Publishing.

Cederroth, C. R., & Nef, S., (2009). Soy, phytoestrogens and metabolism: A review. *Molecular and Cellular Endocrinology, 304*(1, 2), 30–42.

Charalampopoulos, D., Pandiella, S. S., & Webb, C., (2003). Evaluation of the effect of malt, wheat and barley extracts on the viability of potentially probiotic lactic acid bacteria under acidic conditions. *International Journal of Food Microbiology, 82*(2), 133–141.

Charalampopoulos, D., Wang, R., Pandiella, S. S., & Webb, C., (2002). Application of cereals and cereal components in functional foods: A review. *International Journal of Food Microbiology, 79*(1, 2), 131–141.

Cohen, S. L., Moore, A. M., & Ward, W. E., (2005). Flaxseed oil and inflammation-associated bone abnormalities in interleukin-10 knockout mice. *The Journal of Nutritional Biochemistry, 16*(6), 368–374.

Day, L., Seymour, R. B., Pitts, K. F., Konczak, I., & Lundin, L., (2009). Incorporation of functional ingredients into foods. *Trends in Food Science & Technology, 20*(9), 388–395.

Dell'Agli, M., Buscialà, A., & Bosisio, E., (2004). Vascular effects of wine polyphenols. *Cardiovascular Research, 63*(4), 593–602.

Devi, M. A., Gondi, M., Sakthivelu, G., Giridhar, P., Rajasekaran, T., & Ravishankar, G. A., (2009). Functional attributes of soybean seeds and products, with reference to isoflavone content and antioxidant activity. *Food Chemistry, 114*(3), 771–776.

Dodin, S., Cunnane, S. C., Mâsse, B., Lemay, A., Jacques, H., Asselin, G., Tremblay-Mercier, J., et al., (2008). Flaxseed on cardiovascular disease markers in healthy menopausal women: A randomized, double-blind, placebo-controlled trial. *Nutrition, 24*(1), 23–30.

Doyon, M., & Labrecque, J., (2008). Functional foods: A conceptual definition. *British Food Journal.*

Earl, R., & Thomas, P. R., (1994). *Opportunities in the Nutrition and Food Sciences: Research Challenges and the Next Generation of Investigators*. National Academies Press.

Ernst, E., (2002). The risk-benefit profile of commonly used herbal therapies: Ginkgo, St. John's Wort, Ginseng, Echinacea, Saw Palmetto, and Kava. *Annals of Internal Medicine, 136*(1), 42–53.

eservani, R. K., Sharma, A. K., & Jarouliya, U., (2015). Protein and peptide in drug targeting and its therapeutic approach. *Ars Pharmaceutica (Internet), 56*(3), 165–177.

Eyduran, S. P., (2009). Functional components of berry fruits and their usage in food technologies. *African Journal of Agricultural Research, 4*(5), 422–426.

Fernández-Ginés, J. M., Fernández-López, J., Sayas-Barberá, E., & Pérez-Alvarez, J. A., (2005). Meat products as functional foods: A review. *Journal of Food Science, 70*(2), R37–R43.

Ferrari, C. K., (2007). Functional foods and physical activities in health promotion of aging people. *Maturitas, 58*(4), 327–339.

Franck, A., (2006). Inulin-type fructans: Functional food ingredients. *Trends Food Sci. Tech., 17*, 39–41.

Fuentes-Alventosa, J. M., Rodríguez-Gutiérrez, G., Jaramillo-Carmona, S., Espejo-Calvo, J. A., Rodríguez-Arcos, R., Fernández-Bolaños, J., Guillén-Bejarano, R., & Jiménez-Araujo, A., (2009). Effect of extraction method on chemical composition and functional characteristics of high dietary fiber powders obtained from asparagus by-products. *Food Chemistry, 113*(2), 665–671.

Galland-Irmouli, A. V., Fleurence, J., Lamghari, R., Luçon, M., Rouxel, C., Barbaroux, O., Bronowicki, J. P., et al., (1999). Nutritional value of proteins from edible seaweed *Palmaria palmata* (Dulse). *The Journal of Nutritional Biochemistry, 10*(6), 353–359.

German, B., Schiffrin, E. J., Reniero, R., Mollet, B., Pfeifer, A., & Neeser, J. R., (1999). The development of functional foods: Lessons from the gut. *Trends in Biotechnology, 17*(12), 492–499.

Gibson, G. R., & Roberfroid, M. B., (1995). Dietary modulation of the human colonic microbiota: Introducing the concept of prebiotics. *The Journal of Nutrition, 125*(6), 1401–1412.

Gillingham, L. G., Caston, L., Leeson, S., Hourtovenko, K., & Holub, B. J., (2005). The effects of consuming docosahexaenoic acid (DHA)-enriched eggs on serum lipids and fatty acid compositions in statin-treated hypercholesterolemic male patients. *Food Research International, 38*(10), 1117–1123.

Goel, A., Kunnumakkara, A. B., & Aggarwal, B. B., (2008). Curcumin as "curecumin": From kitchen to clinic. *Biochemical Pharmacology, 75*(4), 787–809.

Goldberg, I., (1994). *Functional Foods: Designer Foods, Pharmafoods, and Nutraceuticals* (pp. 3, 4). An Aspen Publication, Chapman and Hall, London, UK.

Gray, J., Armstrong, G., & Farley, H., (2003). Opportunities and constraints in the functional food market. *Nutrition & Food Science*.

Gruenwald, J., (2009). Novel botanical ingredients for beverages. *Clinics in Dermatology, 27*(2), 210–216.

Gry, J., Black, L., Eriksen, F. D., Pilegaard, K., Plumb, J., Rhodes, M., Sheehan, D., et al., (2007). EuroFIR-BASIS – a combined composition and biological activity database for bioactive compounds in plant-based foods. *Trends in Food Science & Technology, 18*(8), 434–444.

Hardy, G., (2000). Nutraceuticals and functional foods: Introduction and meaning. *Nutrition (Burbank, Los Angeles County, Calif.), 16*(7, 8), 688, 689.

Hasler, C. M., (2010). *Functional Foods: Their Role in Disease Prevention and Health Promotion*. Available from: http://www.nutriwatch.org/04Foods/ff.html (accessed on 22 February 2023).

Hasler, C., Moag-Stahlberg, A., Webb, D., & Hudnall, M., (2001). How to evaluate the safety, efficacy, and quality of functional foods and their ingredients. *Journal of the American Dietetic Association, 101*(7), 733–739.

Ho, C. -T., (1992). "Phenolic compounds in food: An overview." *Phenolic Compounds in Food and Their Effects on Health II, ACS, 1*, 2–7.

Holm, F., (2003). *New Functional Food Ingredients: Cardiovascular Health*. Institut national de la recherche agronomique.

Huggett, A. C., & Schilter, B., (1996). Research needs for establishing the safety of functional foods. *Nutrition Reviews, 54*(11), S143.

Hunter, D. C., Zhang, J., Stevenson, L. M., & Skinner, M. A., (2008). Fruit-based functional foods II: The process for identifying potential ingredients. *International Journal of Food Science & Technology, 43*(12), 2123–2129.

Isanga, J., & Zhang, G. N., (2007). Biologically active components and nutraceuticals in peanuts and related products. *Food Reviews International, 23*(2), 123–140.

Izydorczyk, M. S., Chornick, T. L., Paulley, F. G., Edwards, N. M., & Dexter, J. E., (2008). Physicochemical properties of hull-less barley fiber-rich fractions varying in particle size and their potential as functional ingredients in two-layer flat bread. *Food Chemistry, 108*(2), 561–570.

Jang, M., Cai, L., Udeani, G. O., Slowing, K. V., Thomas, C. F., Beecher, C. W., Fong, H. H., et al., (1997). Cancer chemopreventive activity of resveratrol, a natural product derived from grapes. *Science, 275*(5297), 218–220.

Jarouliya, U., & Keservani, R. K., (2019). Pathways leading to child obesity. In: Debasis, B., (ed.), *Global Perspectives on Childhood Obesity: Current Status, Consequences and Prevention* (2nd edn., pp. 137–146). Academic Press, Elsevier, chapter 12. ISBN: 9780128128404.

Jarouliya, U., & Keservani, R. K., (2019a). Protein function as cell surface and nuclear receptor in human diseases. In: Sanjay, B., & Debarshi, K. M., (eds.), *Medicinal Chemistry with Pharmaceutical Product Development* (pp. 1–32). Apple Academic Press, CRC Press, Taylor & Francis Group. Chapter 1. ISBN: 9781771887106.

Jeon, Y. J., Shahidi, F., & Kim, S. K., (2000). Preparation of chitin and chitosan oligomers and their applications in physiological functional foods. *Food Reviews International, 16*(2), 159–176.

Jiménez-Colmenero, F., Sánchez-Muniz, F. J., & Olmedilla-Alonso, B., (2010). Design and development of meat-based functional foods with walnut: Technological, nutritional and health impact. *Food Chemistry, 123*(4), 959–967.

Jooyandeh, H., (2011). Soy products as healthy and functional foods. *Middle-East Journal of Scientific Research, 7*(1), 71–80.

Kelley, D. S., Branch, L. B., Love, J. E., Taylor, P. C., Rivera, Y. M., & Iacono, J. M., (1991). Dietary α-linolenic acid and immunocompetence in humans. *The American Journal of Clinical Nutrition, 53*(1), 40–46.

Keservani, R. K., & Sharma, A. K., (2014). Flavonoids: Emerging trends and potential health benefits. *Journal of Chinese Pharmaceutical Sciences, 23*(12), 815.

Keservani, R. K., Kesharwani, R. K., Sharma, A. K., Vyas, N., & Chadoker, A., (2010b). Nutritional supplements: An overview. *International Journal of Current Pharmaceutical Review and Research, 1*(1), 59–75.

Keservani, R. K., Kesharwani, R. K., Vyas, N., Jain, S., Raghuvanshi, R., & Sharma, A. K., (2010a). Nutraceutical and functional food as future food: A review. *Der Pharmacia Lettre, 2*(1), 106–116.

Keservani, R. K., Sharma, A. K., & Kesharwani, R. K., (2016a). Nutraceutical and functional foods for cardiovascular health. In: *Food Process Engineering* (pp. 257–278, 291–312). Apple Academic Press, CRC Press. ISBN: 9781771884020.

Keservani, R. K., Sharma, A. K., & Kesharwani, R. K., (2016b). Medicinal effect of nutraceutical fruits for the cognition and brain health. *Scientifica*, 10. Article ID 3109254. doi: 10.1155/2016/3109254.

Keservani, R. K., Sharma, A. K., & Kesharwani, R. K., (2017). An overview and therapeutic applications of nutraceutical and functional foods. *Recent Advances in Drug Delivery Technology*, 160–201. ISBN:9781522507543.

Keservani, R. K., Sharma, A. K., & Kesharwani, R. K., (2020). *Nutraceuticals and Dietary Supplements: Applications in Health Improvement and Disease Management*. CRC Press. ISBN: 9781771888738.

Khan, A., Safdar, M., Ali, K. M. M., Khattak, K. N., & Anderson, R. A., (2003). Cinnamon improves glucose and lipids of people with type 2 diabetes. *Diabetes Care, 26*(12), 3215–3218.

Khorshidian, N., Yousefi Asli, M., Arab, M., Adeli, M. A., & Mortazavian, A. M., (2016). Fenugreek: Potential applications as a functional food and nutraceutical. *Nutrition and Food Sciences Research, 10, 3*(1), 5–16.

Klimas, M., Brethour, C., & Bucknell, D., (2008). *Introduction and Background* (p. 3). In international market trends analysis for the functional foods and natural health products industry in the United States, Australia, the United Kingdom, and Japan. Nutri-Net Canada, Ontario, Canada.

Kohlmeier, L., Kark, J. D., Gomez-Gracia, E., Martin, B. C., Steck, S. E., Kardinaal, A. F., Ringstad, J., et al., (1997). Lycopene and myocardial infarction risk in the EURAMIC study. *American Journal of Epidemiology, 146*(8), 618–626.

Kong, K. W., Khoo, H. E., Prasad, K. N., Ismail, A., Tan, C. P., & Rajab, N. F., (2010). Revealing the power of the natural red pigment lycopene. *Molecules, 15*(2), 959–987.

Korver, O., (1997). Healthy' developments in the food industry. *Cancer Letters, 114*(1, 2), 19–23.

Kotilainen, L., Riikka, R., Catherine, R., & Eija, P., (2006). *Health Enhancing Foods: Opportunities for Strengthening Developing Countries, 1*, 30–38.

Kroyer, G. T., (2004). Red clover extract as antioxidant active and functional food ingredient. *Innovative Food Science & Emerging Technologies, 5*(1), 101–105.

Kruger, C. L., & Mann, S. W., (2003). Safety evaluation of functional ingredients. *Food and Chemical Toxicology, 41*(6), 793–805.

Kuhnen, S., Lemos, P. M. M., Campestrini, L. H., Ogliari, J. B., Dias, P. F., & Maraschin, M., (2009). Antiangiogenic properties of carotenoids: A potential role of maize as functional food. *Journal of Functional Foods, 1*(3), 284–290.

Kushi, L. H., Meyer, K. A., & Jacobs, Jr. D. R., (1999). Cereals, legumes, and chronic disease risk reduction: Evidence from epidemiologic studies. *The American Journal of Clinical Nutrition, 70*(3), 451s–458s.

Lambert, J. D., Hong, J., Yang, G. Y., Liao, J., & Yang, C. S., (2005). Inhibition of carcinogenesis by polyphenols: Evidence from laboratory investigations. *The American Journal of Clinical Nutrition, 81*(1), 284S–291S.

Larsson, S. C., Virtamo, J., & Wolk, A., (2011). Fish consumption and risk of stroke in Swedish women. *The American Journal of Clinical Nutrition, 93*(3), 487–493.

Lewis, N. M., Seburg, S., & Flanagan, N. L., (2000). Enriched eggs as a source of n-3 polyunsaturated fatty acids for humans. *Poultry Science, 79*(7), 971–974.

Lim, H. S., Park, S. H., Ghafoor, K., Hwang, S. Y., & Park, J., (2011). Quality and antioxidant properties of bread containing turmeric (*Curcuma longa* L.) cultivated in South Korea. *Food Chemistry, 124*(4), 1577–1582.

Lin, X., Gingrich, J. R., Bao, W., Li, J., Haroon, Z. A., & Demark-Wahnefried, W., (2002). Effect of flaxseed supplementation on prostatic carcinoma in transgenic mice. *Urology, 60*(5), 919–924.

Liu, D., Shi, J., Ibarra, A. C., Kakuda, Y., & Xue, S. J., (2008). The scavenging capacity and synergistic effects of lycopene, vitamin E, vitamin C, and β-carotene mixtures on the DPPH free radical. *LWT-Food Science and Technology, 41*(7), 1344–1349.

Maheshwari, R. K., Singh, A. K., Gaddipati, J., & Srimal, R. C., (2006). Multiple biological activities of curcumin: A short review. *Life Sciences, 78*(18), 2081–2087.

Mark-Herbert, C., (2004). Innovation of a new product category—Functional foods. *Technovation, 24*(9), 713–719.

Menrad, K., (2003). Market and marketing of functional food in Europe. *Journal of Food Engineering, 56*(2, 3), 181–188.

Mermel, V. L., (2004). Old paths new directions: The use of functional foods in the treatment of obesity. *Trends in Food Science & Technology, 15*(11), 532–540.

Noonan, W. P., & Noonan, C., (2004). Legal requirements for "functional food" claims. *Toxicology Letters, 150*(1), 19–24.

O'Donnell, C. D., (2006). Nutritionals at the tipping point-global functional ingredients market. *Prepared Foods, 175*, NS11–NS17.

Olmedilla-Alonso, B., Granado-Lorencio, F., Herrero-Barbudo, C., & Blanco-Navarro, I., (2006). Nutritional approach for designing meat-based functional food products with nuts. *Critical Reviews in Food Science and Nutrition, 46*(7), 537–542.

Ovesen, L., (1999). Functional foods: Some relevant considerations?. *British Food Journal.*

Paliyath, G., (2001). Guide to functional food ingredients. *Trends Food Sci. Tech., 12*, 302–303.

Pasha, C., & Reddy, G., (2005). Nutritional and medicinal improvement of black tea by yeast fermentation. *Food Chemistry, 89*(3), 449–453.

Patil, B. S., Jayaprakasha, G. K., Chidambara, M. K. N., & Vikram, A., (2009). Bioactive compounds: Historical perspectives, opportunities, and challenges. *Journal of Agricultural and Food Chemistry, 57*(18), 8142–8160.

Patterson, C. A., (2008). *Oat and Barley β-Glucans Unique Soluble Fibers* (pp. 1–4). Agr. Agri-Food Canada, Ontario, Canada.

Plaza, M., Cifuentes, A., & Ibáñez, E., (2008). In the search of new functional food ingredients from algae. *Trends in Food Science & Technology, 19*(1), 31–39.

Prabhakar, K., & Mallika, E. N., (2010). *Functional Foods of Milk Origin* (pp. 106–110). SMVS Dairy Year Book 2010, Serva Manav Samiti, Uttar Pradesh, India.

Prabhasankar, P., Ganesan, P., Bhaskar, N., Hirose, A., Stephen, N., Gowda, L. R., Hosokawa, M., & Miyashita, K. J. F. C., (2009). Edible Japanese seaweed, wakame (Undaria pinnatifida) as an ingredient in pasta: Chemical, functional and structural evaluation. *Food Chemistry, 115*(2), 501–508.

Prasad, K., (2000). Flaxseed: A source of hypocholesterolemic and antiatherogenic agents. *Drug News & Perspectives, 13*(2), 99–104.

Price, R. K., Keaveney, E. M., Hamill, L. L., Wallace, J. M., Ward, M., Ueland, P. M., McNulty, H., et al., (2010). Consumption of wheat aleurone-rich foods increases fasting plasma betaine and modestly decreases fasting homocysteine and LDL-cholesterol in adults. *The Journal of Nutrition, 140*(12), 2153–2157.

Ramírez, P., García-Risco, M. R., Santoyo, S., Señoráns, F. J., Ibáñez, E., & Reglero, G., (2006). Isolation of functional ingredients from rosemary by preparative-supercritical fluid chromatography (Prep-SFC). *Journal of Pharmaceutical and Biomedical Analysis, 41*(5), 1606–1613.

Rane, B. R., & Keservani, R., (2020). Introduction to nutraceuticals and dietary supplements. In: *Nutraceuticals and Dietary Supplements* (pp. 35–56). Apple Academic Press. ISBN: 9781771888738.

Rane, B. R., Bharath, M. S., Patil, R. R., Keservani, R. K., & Jain, A. S., (2020b). Novel approaches in nutraceuticals. In: Rajesh, K. K., Raj, K. K., & Anil, K. S., (eds.), *Enhancing the Therapeutic Efficacy of Herbal Formulations Through Novel Drug Delivery Systems* (pp. 241–266). IGI Global International Publisher, Pennsylvania, USA, Chapter 11. ISBN13: 9781799844532.

Rane, B. R., Patil, A. S., Keservani, R. K., & Jain, A. S., (2020a). Novel approaches in herbal formulation. In: Rajesh, K. K., Raj, K. K., & Anil, K. S., (eds.), *Enhancing the Therapeutic Efficacy of Herbal Formulations Through Novel Drug Delivery Systems* (pp. 43–68). IGI Global International Publisher, Pennsylvania, USA, Chapter 2. ISBN13: 9781799844532.

Rane, B. R., Tadavi, S. A., & Keservani, R. K., (2020c). Naturopathy. In: Anil, K. S., Raj, K. K., & Surya, P. G., (eds.), *Herbal Product Development* (pp. 321–347). Apple Academic Press, CRC Press, Taylor & Francis Group, Chapter 12. ISBN: 9781771888776.

Rao, L. G., Guns, E., & Rao, A. V., (2003). Lycopene: Its role in human health and disease. *Agro. Food, 7*, 25–30.

Rodríguez, R., Jimenez, A., Fernández-Bolanos, J., Guillen, R., & Heredia, A., (2006). Dietary fiber from vegetable products as source of functional ingredients. *Trends in Food Science & Technology, 17*(1), 3–15.

Roldán, E., Sanchez-Moreno, C., De Ancos, B., & Cano, M. P., (2008). Characterization of onion (*Allium cepa* L.) by-products as food ingredients with antioxidant and anti-browning properties. *Food Chemistry, 108*(3), 907–916.

Rupérez, P., (2002). Mineral content of edible marine seaweeds. *Food Chemistry, 79*(1), 23–26.

Sacchetti, G., Maietti, S., Muzzoli, M., Scaglianti, M., Manfredini, S., Radice, M., & Bruni, R., (2005). Comparative evaluation of 11 essential oils of different origin as functional antioxidants, antiradicals and antimicrobials in foods. *Food Chemistry, 91*(4), 621–632.

Saher, M., Arvola, A., Lindeman, M., & Lähteenmäki, L., (2004). Impressions of functional food consumers. *Appetite, 42*(1), 79–89.

Sanders, M. E., (1998). Overview of functional foods: Emphasis on probiotic bacteria. *International Dairy Journal, 8*(5, 6), 341–347.

Shahidi, F., (2009). Nutraceuticals and functional foods: Whole versus processed foods. *Trends in Food Science & Technology, 20*(9), 376–387.

Shakib, M. C., & Gabrial, S. G., (2010). Post-prandial responses to different bread products based on wheat, barley and fenugreek or ginger or both in healthy volunteers and their effect on the glycemic index of such products. *J. Am. Sci., 6*(10), 89–96.

Sharma, A. K., Keservani, R. K., & Gautam, S. P., (2020). *Herbal Product Development*. Apple Academic Press, CRC Press, Taylor & Francis Group. ISBN: 9781771888776.

Shi, J., Nawaz, H., Pohorly, J., Mittal, G., Kakuda, Y., & Jiang, Y., (2005). Extraction of polyphenolics from plant material for functional foods—Engineering and technology. *Food Reviews International, 21*(1), 139–166.

Shui, G., & Leong, L. P., (2006). Residue from star fruit as valuable source for functional food ingredients and antioxidant nutraceuticals. *Food Chemistry, 97*(2), 277–284.

Sierksma, A., Weststrate, J. A., & Meijer, G. W., (1999). Spreads enriched with plant sterols, either esterified 4, 4-dimethylsterols or free 4-desmethylsterols, and plasma total-and LDL-cholesterol concentrations. *British Journal of Nutrition, 82*(4), 273–282.

Singh, P., Kesharwani, R. K., & Keservani, R. K., (2017a). Antioxidants and vitamins: Roles in cellular function and metabolism. In: *Sustained Energy for Enhanced Human Functions and Activity* (pp. 385–407). Academic Press. ISBN:9780128054130.

Singh, P., Kesharwani, R. K., & Keservani, R. K., (2017b). Protein, carbohydrates, and fats: Energy metabolism. In: *Sustained Energy for Enhanced Human Functions and Activity* (pp. 103–115). Academic Press. ISBN:9780128054130.

Siró, I., Kápolna, E., Kápolna, B., & Lugasi, A., (2008). Functional food. Product development, marketing and consumer acceptance—A review. *Appetite, 51*(3), 456–467.

South, J. *Cinnamon Lowers Blood Sugar and Blood Lipids*. Available from: https://www.diabetes.co.uk/natural-therapies/cinnamon.html (accessed on 22 February 2023).

Stanton, C., Ross, R. P., Fitzgerald, G. F., & Van, S. D., (2005). Fermented functional foods based on probiotics and their biogenic metabolites. *Current Opinion in Biotechnology, 16*(2), 198–203.

Stoll, T., Schweiggert, U., Schieber, A., & Carle, R., (2003). Process for the recovery of a carotene-rich functional food ingredient from carrot pomace by enzymatic liquefaction. *Innovative Food Science & Emerging Technologies, 4*(4), 415–423.

Sun-Waterhouse, D., Farr, J., Wibisono, R., & Saleh, Z., (2008). Fruit-based functional foods I: Production of food-grade apple fiber ingredients. *International Journal of Food Science & Technology, 43*(12), 2113–2122.

Surai, P. F., & Sparks, N. H. C., (2001). Designer eggs: From improvement of egg composition to functional food. *Trends In Food Science & Technology, 12*(1), 7–16.

Surana, K. R., Ahire, E. D., Sonawane, V. N., Talele, S. G., & Talele, G., (2021). Informatics and methods in nutrition design and development. *Natural Food Products and Waste Recovery: Healthy Foods, Nutrition Design, and Extraction of Valuable Compounds*, 33.

Szkudelska, K., & Szkudelski, T., (2010). Resveratrol, obesity and diabetes. *Alternative Medicine Review; Eur. J. Pharmacol., 15*(2), 168, 169.

Tham, D. M., Gardner, C. D., & Haskell, W. L., (1998). Potential health benefits of dietary phytoestrogens: A review of the clinical, epidemiological, and mechanistic evidence. *The Journal of Clinical Endocrinology & Metabolism*, *83*(7), 2223–2235.

Verhoeven, D. T., Goldbohm, R. A., Van, P. G., Verhagen, H., & Van, D. B. P. A., (1996). Epidemiological studies on brassica vegetables and cancer risk. *Cancer Epidemiology and Prevention Biomarkers, 5*(9), 733–748.

Verhoeven, D. T., Verhagen, H., Goldbohm, R. A., Van, D. B. P. A., & Van, P. G., (1997). A review of mechanisms underlying anti-carcinogenicity by brassica vegetables. *Chemico-Biological Interactions, 103*(2), 79–129.

Vijaimohan, K., Jainu, M., Sabitha, K. E., Subramaniyam, S., Anandhan, C., & Devi, C. S., (2006). Beneficial effects of alpha linolenic acid rich flaxseed oil on growth performance and hepatic cholesterol metabolism in high fat diet fed rats. *Life Sciences*, *79*(5), 448–454.

Vitaglione, P., Napolitano, A., & Fogliano, V., (2008). Cereal dietary fiber: A natural functional ingredient to deliver phenolic compounds into the gut. *Trends in Food Science & Technology*, *19*(9), 451–463.

Wang, X., Brown, I. L., Evans, A. J., & Conway, P. L., (1999). The protective effects of high amylose maize (amylomaize) starch granules on the survival of *Bifidobacterium* spp. in the mouse intestinal tract. *Journal of Applied Microbiology, 87*(5), 631–639.

Weststrate, J. A., Van, P. G., & Verschuren, P. M., (2002). Functional foods, trends and future. *British Journal of Nutrition, 88*(S2), S233–S235.

Yang, J., Liu, R. H., & Halim, L., (2009). Antioxidant and antiproliferative activities of common edible nut seeds. *LWT-Food Science and Technology, 42*(1), 1–8.

Zemel, M. B., Thompson, W., Zemel, P., Nocton, A. M., Milstead, A., Morris, K., & Campbell, P., (2002). Dietary calcium and dairy products accelerate weight and fat loss during energy restriction in obese adults. In: *American Journal of Clinical Nutrition* (Vol. 75, No. 2, pp. 342S, 343S). 9650 Rockville Pike, Subscriptions, RM L-3300, Bethesda, MD 20814-3998 USA: Amer Soc Clinical Nutrition.

CHAPTER 2

FUNCTIONAL FOODS FOR AUTISM

PANKAJ R. AHER,[1] KHEMCHAND R. SURANA,[2] EKNATH D. AHIRE,[3] DHANANJAY M. PATIL,[4] and SANJAY J. KSHIRSAGAR[3]

[1]*Department of Pharmaceutical Chemistry, Loknete Dr. J. D. Pawar College of Pharmacy, Manur, Kalwan, Nashik, Maharashtra, India*

[2]*Department of Pharmaceutical Chemistry, Shreeshakti Shaikshanik Sanstha, Divine College of Pharmacy, Satana, Nashik, Maharashtra, India*

[3]*Department of Pharmaceutics, MET's Institute of Pharmacy, Adgaon, Nashik, Maharashtra, India*

[4]*Department of Pharmaceutics, Shreeshakti Shaikshanik Sanstha, Divine College of Pharmacy, Satana, Nashik, Maharashtra, India*

ABSTRACT

Autism is specified by impairments in social connections, communication deficits, and repetitive or stereotyped behavior. Stress the identical qualities of nutritional deficiencies in these children and the modification to evaluation and treatment that are needed to tailor these methods to the child, thus resulting in a more effective intervention. Keeping that factor in mind, it has to be cited that dietary lacking is a whole lot extra common than what's discovered with other children with disabilities, which have a few similarities to ASD along with language impairments. A traditional feeding control or feeding issues are diagnosed whilst children are not able to consume or drink enough portions to keep dietary wellbeing. The mode of such behavior that brought about an analysis of a feeding disorder can contain refusal of foods, selectivity of kind and texture of foods and disruptive behavior related to feeding or meal time.

Applications of Functional Foods in Disease Prevention. Raj K. Keservani & Eknath D. Ahire, PhD (Eds.)

Moderately selective eaters choose especially from one group and once in a while from one or other groups. The "bioactive compounds" represent the effectors of useful food effectiveness. Samples of functional foods are grapes wealthy in resveratrol, blueberries rich in anthocyanins, flavonoids which have urged to be a robust cluster of antioxidants, and walnuts rich in minerals and polyunsaturated fatty acid fatty acids which are shown to possess medicament properties. In recent years, many clinical trials have been conducted exploitation bioactive compounds within the prevention, management, or treatment of each autism spectrum disorder and schizophrenia.

2.1 INTRODUCTION

The latest definition of functional food, developed via the Functional Food Center (FFC) in 2012 at the 10th FFC International Conference in Santa Barbara, California, is: "Natural or processed foods that comprise identified or unknown biologically active compounds; which, in defined, effective, non-poisonous amounts, offer a clinically examined and documented fitness advantage for the prevention, management, or remedy of chronic disease." Food product employing a new definition through Danik M. Martirosyan and Jaishree Singh is unique. It highlights the function of foods in an extremely normal eating regimen for nutrient optimization and emphasizes the significance of "bioactive compounds" that represent the effectors of functional food efficaciousness (Pangrazzi, et al., 2020; Peterson, et al., 2019) examples of functional foods are grapes rich in resveratrol, blueberries rich in anthocyanins, flavonoids which might be proven to be an efficient inhibitor institution, and nuts wealthy in minerals and omega-3 fatty acids that have been proven to have anti-inflammatory properties. In recent years, many scientific research were completed at the usage of bioactive compounds inside the prevention, control or remedy of ASD and psychosis (Casey et al., 2008). They have furthermore been accomplished as supplementary compounds to potentiate traditional medications. This study has focused on the feature of various varieties of diets, which incorporates the ketogenic diet regime or gluten-unfastened diets, further to looking at how bioactive compounds can play a feature in schizophrenia management. These studied bioactive additives embody omega-3 fatty acids, the antioxidant vitamins C, D and E, folate supplementation and Curcumin. Many more ongoing clinical trials are investigating the function of particular diets and the supplementation of fatty acids (DHA, omega-3 FAs), antioxidants or minerals (www.clinicaltrials.gov: autism spectrum problems and diet) withinside the manipulate of ASD (Serra et al., 2019). It has been advocated that nutritional supplementation cannot maximum successfully addresses the neuropsychiatric pathology of ASD and schizophrenia, however additionally the scientific situations often related to the issues together with cardiovascular diseases, obesity, and metabolic syndrome (Joshi et al., 2018). These situations are due to food selectivity, horrific dietary habits, or complications related to the common region medicinal drugs used for treatment. Functional food is strongly counseled for the general prevention and control of cardio-metabolic problems due to horrific dietary habits. The onset of systemic medical conditions along with weight issues and metabolic syndrome in patients with ASD or schizophrenia is essentially because of nutritional deficiencies and the adoption of diets rich in pro-inflammatory food at the side of diffused sugars and processed food (Ghalichi et al., 2016). Chronic low-grade systemic contamination has been suggested to play an essential function in a now no longer unusual place pathogenic mechanism observed in each ASD and schizophrenia, important to the activation of inflammatory cells (microglia cells) withinside the brain (neuroinflammation). Neuroinflammation reasons adjustments in neuronal mobileular communication and synaptic plasticity. It truly is considered to be the primary mechanism underlying analyzing and memory processes. In addition, it has moreover been noted that neuroleptic tablets used to deal with schizophrenia can purpose oxidative stress and in turn neuronal damage (Volkert, 2010; Vaz, 2010). Recently, extra emphasis has been positioned on analyzing the nutritional recognition of humans with ASD and schizophrenia. In this financial disaster, we are able to illustrate a number of the modern-day scientific evidence and research that address the link among unique beneficial food and bioactive compounds and the improvement and manipulation of each ASD and schizophrenia. As there may be a more fast growth withinside the medical literature and studies in this field, the writer would like to awareness on that this financial ruin isn't always prepurported to be an entire look at protective all of the scientific facts available regarding all bioactive compounds and their correlation with neuropsychiatric problems (Chistol et al., 2018). It is as an alternative supposed to interest on the complexity of the pathogenesis hypothesis of these persistent and debilitating problems and to expose the capacity feature of a few beneficial foods in dealing with those illnesses (Siniscalco et al., 2018). The primary and critical feature of the food is to provide nutrients or vitamins [i.e., carbohydrates, proteins (Singh et al., 2017b), fats, nutrients, and minerals] to expand properly, which are essential for growth, maintenance, and wellness. In addition, non-vitamins at the side of fibers, probiotics, antioxidants, and bioactive compounds are covered withinside the meals. Although they're now no longer at as quickly as required to assist the easy features of organisms, their presence is useful to promote fitness and wellness. When a sort of meals can provide fitness advantages further to its nutritional value, it's far some distance described as functional food.

Generally, a regular hallmark of these aliments is the presence of antioxidant molecules at the factor of flavonoids, carotenoids, and phenols. Typical examples of functional foods are many varieties of fruits, vegetables, nuts, seeds, complete grains, legumes, and seafood (Sharma et al., 2020). When beneficial compounds are isolated from the food matrix, they're known as nutraceuticals. This time period changed into coined with the useful resource of the usage of De Felice in 1989, combining the words "nutrients" and "pharmaceutics," to indicate "food or components of food that provide scientific or fitness advantages, which include the prevention and treatment of illnesses" (Keservani et al., 2010a, b, 2020; Johnson et al., 2013).

Hence, when curious compounds are found in nutritious foods, they may be in a position to be enhanced and offered as nutraceuticals in the market. Despite this, it is essential to remember that bioactive compounds can be much less powerful when extracted from useful meals, and, in some cases, toxic effects may additionally appear (Binnendyk, 2004; Lucyshyn, 2005). Nevertheless, the ones molecules can be useful in counteracting pathological situations and helping body health. Recently, many studies have commenced exploring the position of nutraceuticals withinside the context of persistent problems along with cancer, diabetes, atherosclerosis, cardiovascular illnesses (Keservani et al., 2016a), and neurological issues (Keservani et al., 2016b, 2017), and exciting consequences had been achieved. In addition, proof exists that nutraceutical compounds may play a in human beings with ASD (Allen-Vercoe, 2015; Toh, 2015). In that studies, the antioxidant ability of nutraceutical compounds played a main role in enhancing the disorder situation. Thus, functional food may constitute an effective tool for counteracting a huge spectrum of illnesses manner to the aggregate of several nutraceuticals, the maximum of which is probably antioxidant compounds (Singh et al., 2017a). As formerly discussed, nutritional factors may be important to counteract ASD-like symptoms and symptoms in man or woman with ASD (Fodstad, 2009; Matson, 2009).

2.2 FUNCTIONAL FOODS FOR THE MANAGEMENT OF AUTISM

Activation is the primary underlying biological approaches of ASD and schizophrenia. Figure 2.1 illustrates the molecular mechanisms concerned in neuronal development (Essa, 2020 and Qoronfleh, 2020). Functional food is defined as natural or processed foods that include recognized or unknown biologically-active compounds; which, in defined, effective, and non-toxic amounts, offer a clinically proven and documented health advantage for the prevention, management, or remedy of chronic disease. Importantly, the

Functional Food Center defines functional food as "components of an ordinary diet for optimized nutrition." This definition includes traditional whole foods, such as oranges or bran flakes, instead of the using pills, capsules, or supplements (Farhangi, et al., 2018). The noted bioactive compounds encompass flavonoids (Keservani and Sharma, 2014), essential fatty acids, minerals, vitamins (Singh et al., 2017a), and phytonutrients, whilst the continual illnesses they may help cope with embody diabetes, cardiovascular illnesses, hypertension, and illnesses affecting the crucial frightened system, along with ASD. ASD is the neurodevelopmental disorders, with the age of onset of ASD taking place in the early segment of life (commonly the analysis takes area round age of 2–3 years old, with a better occurrence for males), counting on precipitating elements (Hetzroni and Tannous, 2004). In fact, it's far well-established that the end result of an interaction among the individual's genetic constitution and environmental elements which act as "2nd hit." These environmental of mental stressors and trauma, immunological stressors, infections, chemical substances exposure, substance abuse, and dietary deficiencies (Singh, 2018).

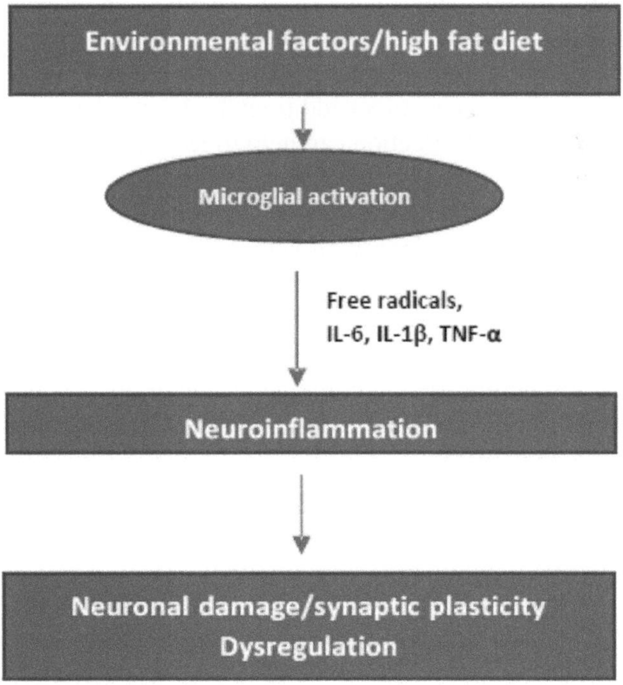

FIGURE 2.1 Brain injury and high-fat diet activate microglia and kill neurons.

It has been pronounced that prenatal and perinatal exposure to environmental risk factors which include dietary habits, maternal diabetes, stress, medications, or infections are associated with a greater risk of developing ASD (Mirenda, 2003). This chapter will interest on particular functional foods which have been proven through clinical evidence to help brain functioning and the general fitness of people affected via way of means of ASD. Glutamate is also involved in the synthesis of the effective antioxidant glutathione that is synthesized from glutamate, cysteine, and glycine the usage of enzymes. One of these enzymes is the glutathione synthetase that wishes magnesium molecules to carry out its feature (Manea et al., 2009). Deficiencies in vitamin B6 or magnesium can result in abnormalities in neurotransmitter synthesis and alterations in the physiological feature of glutamate or GABA. Foods that are rich in magnesium include dark leafy greens, nuts and seeds, fish, soybeans, avocados, bananas, darkish chocolate, and fats-free or low-fat yogurt. Magnesium deficiency induces an inflammatory reaction that impacts in leukocyte and macrophage activation, the discharge of inflammatory cytokines and acute-section proteins which includes CRP and excessive production of free radicals, which cause cell damage (Lasheras et al., 2020; Jarouliya and Keservani, 2019b). Neuroinflammation represents a crucial mechanism involved in the pathogenesis of ASD, even though it is probably secondary to altered pathways of the so-called intestine-microbiome-brain communication. Systemic infection can affect additionally the mind and its development. Targeting the reasons of infection can help to lessen the signs and symptoms of ASD through addressing the abnormalities in the gut system and consequently one's nutritional patterns. Several mechanisms are proposed for the gut-microbiota-mind axis along with neural, hormonal, immunological, and metabolic (Moes and Frea, 2002). The hypothesis is based on the statement that the infection of the gastrointestinal (GI) tract reasons a growth withinside the cytokines release that could move the blood-brain barrier and cause an immune reaction withinside the brain. Neuroinflammation in early stages of lifestyles can adjust the maturation of neurons and synaptic plasticity through converting the glutamatergic system. During development, microglia, the quiescent immune cell type in the brain, regulates neurogenesis, synaptic plasticity and synaptic striping (Neely et al., 2016) (Table 2.1).

Flavonoids, then, might also additionally lower the levels of every oxidation and infection therefore inflicting ameliorative effects in normal brain functioning. Children with ASD were defined in numerous research as picky eaters and as more at risk of consuming an imbalanced diet (Allison, et al., 2012). An example of an imbalanced food plan is one that lacks of dietary

TABLE 2.1 Specific Flavonoids Related for ASD Pathogenesis (Oxidative Stress and Inflammation)

Sl. No.	Phytochemical	Bioactive Compounds	Food Source	Physiological Effect
1.	Anthocyanidins	Malvidin	Pomegranates, blueberries, black grapes	Anti-inflammatory activity and Antioxidant
2.	Flavonols	Kaempferol, myricetin, quercetin	Berries, tomatoes, kale, grapes, bell peppers, cocoa, broccoli, spinach, sweet potatoes	Mitigation of microglia-mediated neuroinflammation Anti-carcinogenic, anti-inflammatory, antioxidant, and antiviral activities.
3.	Flavanones	Naringenin, Hesperetin,	Grapes, citrus fruits (lemons and oranges)	Anti-inflammatory antioxidant,
4.	Flavones	Apigenin, luteolin	Oregano, celery, fresh parsley, olives, peppers, and rosemary	Suppression of oxidative stress via anti-inflammatory effects on NF-kB; offers brain support, protection, and increases in memory
5.	Flavanols	Epicatechin-gallates, catechin procyanidins,	Tea, grapes, lentils, cocoa, black grapes, apples with peel on, apricots, cherries, peaches, strawberries, blackberries, blueberries, and raspberries.	Decrease of the hypothalamic inflammation and microglia over activation. Antioxidant, free radicals scavenging properties. Improve cognition.
6.	Isoflavones	Daidzein	Soy products grape seeds,	Improves adipose insulin resistance, inflammation, and cognitive function.
7.	Flavans	Genistein	Soybeans	Improves glucose metabolism and cognitive function. Antioxidant and neuroprotective activities.
8.	Flavonoid glycoside	Rutin	Buckwheat, unpeeled apples, orange juice, asparagus (specially the bottom part), grapefruit, lemons, oranges	Neuroprotective, nephroprotective, Anti-inflammatory, antioxidant, hepatoprotective effects.
9.	Phenolic acids	Caffeic acid, Ferulic acid	Apples, oranges, coffee beans, blueberries, peaches, potatoes, pears	Improve in cognition and neurodegeneration Antioxidant and anti-inflammatory properties.
10.	Hydroxy-benzoic juice, longan seeds	Gallic acids	Strawberries, grape, and raspberry grape	Anti-inflammatory properties. NF-κB inhibitors
11.	Diferuloyl methane	Curcumin	Turmeric plants	Antioxidant activities and anti-inflammatory

Source: Theoharides, Asadi, & Panagiotidou (2012); Theoharides, Asadi, & Patel (2013); and Usui, Kobayashi, & Shimada (2023).

fibers. Fibers are essential to preventing cardiovascular disease, diabetes, and cancer, and selling ordinary gastrointestinal health and mineral absorption withinside the intestinal tract (Navarro et al., 2016). The FDA lately up to date the respectable list of fibers and posted guidance files on dietary fiber. The fiber observed in entire grains, fruits, vegetables, nuts, and seeds in conjunction with fibers in meals additives that have been automatically processed, chopped or ground (Sigafoos and Saggers, 1995). Only seven isolated or synthetic fibers had been identified as having physiological advantages withinside the up-to-date vitamins labeling guidelines of 2016:

- Beta-glucan soluble fiber;
- Psyllium husk;
- Cellulose;
- Guar gum;
- Pectin;
- Locust bean gum;
- Hydroxypropyl methylcellulose.

Regular consumption of foods rich in nutritional fibers can enhance typical health and prevent different conditions along with diabetes, atherosclerosis, diverticulitis, and obesity (Jarouliya and Keservani 2019a). A diet rich in fibers is suggested for people with the above health conditions, because it has been proven to lessen inflammatory biomarkers along with C-reactive protein (CRP) and IL-6 (Avery et al., 2018). Foods excessive in fiber consist of grains (barley, oats, and rice), vegetables (potatoes, egg plants, uncooked spinach), fruits (grapes, kiwi, oranges,), legumes (peas, beans, inexperienced beans), nuts and seeds (almonds, flaxseed, sesame seeds). Dietary fiber may be additionally classified into many distinct families based on their bioactive compounds (Gale et al., 2011). These households encompass arabinoxylan, inulin, lignin, psyllium, pectin, bran, cellulose, β-glucan and resistant starch, and guar gum (McDowell et al., 2007) (Table 2.2).

2.3 FUNCTIONAL FOOD AND NUTRACEUTICAL COMPOUNDS

The primary and fundamental function of food is to offer vitamins (i.e., carbohydrates, proteins, lipids, vitamins and minerals) to the structure, which may be vital for growth, preservation and well-being (Brondino et al., 2015.). Additionally, non-nutrients consisting of fiber, probiotics, antioxidants, and bioactive compounds are protected in meals. Although they're now not without delay had to guide the number one competencies

TABLE 2.2 Dietary Fibers and Different Food Sources

Food Group	Bioactive Compounds	Physiological Effects
• **Grains:** Whole Barley, oats, rice, wheat (whole grain), wheat germ.	Cellulose, B-glucan, phytochemicals (phenolic acids, flavonoids, lignin's, tocols, sterols, folate).	Boosting the immune system, immunostimulation, reducing the risk of colonic cancer, high blood pressure and gallstones.
• **Legumes:** Peas, beans, green beans, celery, lentils.	Cellulose, pectin's, gums,	Improved blood sugar control and insulin sensitivity, laxation.
• **Vegetables:** Potatoes, beetroot, raw celery, raw broccoli, raw spinach, cucumber, green onions, eggplants, raw carrots.	Lignin, inulin, gums.	Bulking capacity and stimulate the growth of intestinal bifidobacterial.
• **Fruits:** Apples with skin, grapes, strawberries, kiwi, oranges, pineapple, pears, peach, pears, bananas, peach.	Pectin	Lipid reduction, GI benefits, blood sugar reduction.
• **Nuts and Seeds:** Almonds, sesame seeds, cashew, flaxseed.	Lignans	Inflammatory diseases including intestinal inflammation, brain function, immune system, cancer.

Source: Dhingra, Michael, Rajput, & Patil (2012); López-Marcos et al. (2015); and Turner & Lupton (2011).

of organisms, their presence is useful in promoting health and well-being. Form of food can offer health advantages further to its dietary value; it's far extensively defined as a functional food (Gallup et al., 2021). Generally, a standard feature of those meals is the presence of antioxidant molecules consisting of carotenoids, flavonoids and phenols. Typical examples of functional foods are many forms of fruits, vegetables, nuts, seeds, complete grains, legumes, and seafood. When functional compounds are isolated from the food matrix, they're known as nutraceuticals (Cascio et al., 2012). It needs to be taken into consideration that bioactive compounds can be a whole lot much less amazing even as extracted from centered elements and in a few instances, poisonous outcomes may also appear. However, those molecules may be beneficial to counter pathological conditions and assist the body's fitness. Recently, many researches have commenced to discover the position of nutraceuticals withinside the context of persistent illnesses consisting of cancer, diabetes, atherosclerosis, cardiovascular ailment and neurological problems, with exciting results. Additionally, there's evidence that nutraceuticals may additionally play a beneficial position in human beings with ASD (Vissoker et al., 2015).

2.3.1 *OMEGA-3 FATTY ACIDS*

A developing body of evidence has proven that human beings with psychosis eat fewer fruits and vegetables. Additionally, ingesting fruits and veggies has been proven to assist lessen the risk of cardiovascular disease, diabetes, and ordinary mortality (James, et al., 2010). Fruits and vegetables are rich in antioxidants, minerals, and nutrients that promote widespread health through lowering cell harm related to age and inflammation. Green tea, broccoli, onions and berries incorporate a large quantity of plant-primarily based totally phytonutrients (Conciatore & Andrea, 2015). Examples of nutrients include carotenoids, ellagic acid, flavonoids, resveratrol, glucosinolates, and phytoestrogens. Pumpkins and carrots are desirable sources of alpha and beta carotene. Lycopene, the only antioxidant, is observed in tomatoes, watermelon and grapefruit. Spinach, kale and cabbage are desirable sources of lutein and zeaxanthin. Ellagic acid is observed in lots of berries and numerous vegetable flours, inclusive of strawberries, raspberries, and pomegranates (Peterson & Ibañez, 2018). Catechins, hesperidin, and flavonols, inclusive of quercetin, are excellent bioactive compounds that may assist in restricting oxidation and inflammation. These components may be observed in inexperienced tea, citrus fruits, apples, berries, cabbage and onions. Resveratrol is observed in grapes, crimson wine and red grape juice. Glucosinolates are observed in cruciferous greens, inclusive of Brussels sprouts, kale, and broccoli. Sesame seeds and flax seeds are desirable sources of lignans, a form of phytoestrogen (Peterson et al., 2016).

2.3.2 *MINERALS*

Minerals, particularly zinc, have also been thought to play a role in psychosis. Zinc is important for all physiological systems, which include neuronal functioning. It is an essential trace mineral necessary for the proper functioning of cells, including DNA replication, transcription, protein synthesis, cell membrane maintenance and cell transport (Rhine and Tarbox, 2009; Keservani et al., 2015). Found specifically in red meat, poultry, fish and dairy products, zinc also plays a role in the functioning of the endocrine, immunological and neuronal systems. Zinc dysregulation is linked to a decrease in immunological functioning. Additionally, minerals, especially zinc, had been concept to play a function in psychosis (Silbaugh et al., 2017). Zinc is essential for all physiological systems, which includes neuronal functioning. It is an essential trace mineral necessary for the proper functioning of cells, including DNA replication, transcription, protein synthesis, maintenance of cell membranes and cell transport. Present in particular in red meat, poultry, fish and dairy products, zinc also plays a role in the functioning of the endocrine, immunological and neuronal systems. Zinc dysregulation is associated with impaired immunological functioning, impaired cognitive function, and gastrointestinal disturbances. Zinc deficiency is also associated with neuropsychiatric manifestations (McKenzie, 2012).

2.3.3 VITAMINS

Vitamins even have a potential protecting function in psychosis, in particular for vitamins B6, B12 and folate. These nutrients are concerned in different mechanisms such as homocysteine metabolism, neurotransmitter synthesis and oxidative strain levels (Gorrindo et al., 2012). The significance of the adequacy of vitamins A, C and E to play a function in psychosis become additionally considered. Foods containing vitamin B6 are bananas, pistachios, salmon, enriched tofu and candy potatoes. Foods wealthy in diet B12 are seafood, eggs, soy, meat, cheese, and fortified dairy products additionally continually function dietary yeasts and fungi (Sharp et al., 2018). Conversely, the ingredients to choose are complete end result and vegetables—in particular, broccoli, onions, berries and grapes; fiber-wealthy ingredients, together with vegetables, end result, complete grains, bran, legumes, nuts, and seeds; ingredients excessive in omega-3 fatty acids which include fish, shellfish, nuts, and flax seeds; ingredients excessive in folate (critical for methionine metabolism) which include edamame, spinach, okra, artichokes, asparagus, broccoli, brussels sprouts, lettuce, and avocado; and zinc-rich ingredients together with oysters, wheat germ, liver, pumpkin seeds, baked beans, soy products, beef, and pork (Gray et al., 2017).

2.3.4 CAROTENOIDS

Carotenoids are a class of fat-soluble, red, orange, and yellow natural pigments synthesized through plants, bacteria, and fungi. Carrots, tomatoes, and pumpkins are given the characteristic shadeation by these chemicals.

Importantly, many carotenoids show antioxidant and anti-inflammatory properties. Several research showed that human beings consuming greater carotenoids of their diets had decreased risk of persistent diseases, collectively with cardiovascular diseases, photosensitivity diseases, cataracts, and age-associated macular degeneration. In addition, carotenoids had been examined to decorate cognitive standard overall performance in middle-aged people, suggesting that those compounds may be useful in helping thoughts health (Li et al., 2018).

2.3.5 POLYPHENOLS

In addition to carotenoids, polyphenols were proven to play a useful function in ASD. Phenols are a category of nutraceutical compounds that constitute the maximum ample organization of secondary metabolites produced through plants. These molecules are located in lots of foods, along with vegetables, end result and seeds (Cosbey and Muldoon, 2017). Since they incorporate one or greater hydroxyl companies connected to a benzene ring, they have got the capacity to donate electrons to oxidized molecules. For this reason, phenols are taken into consideration effective antioxidant molecules. Phenolic compounds are a totally heterogeneous group and showcase wonderful chemical properties (Peterson et al., 2019). Several polyphenols were proven to modulate the sirtuin 1 signaling pathway and alter key cell processes which include the cell cycle, DNA repair, protein aggregation, infection, and mitochondrial function. Furthermore, this beauty of molecules may additionally intrude with numerous key cell signaling pathways (which include the ones mediated through activator protein 1 (AP1), nuclear aspect kappa light chain enhancer of activated B cells (NFkB), and sign transducer and activator of transcription 1 (STAT1), in the long run serve to defend in opposition to oxidative stress, infection and mobile proliferation (Begum et al., 2017) are promising molecules for assuaging the signs of ASD (Silbaugh, et al., 2016).

2.4 CONCLUSION

The antioxidant potential of nutraceutical compounds has played an important role in improving the condition of the disease. Therefore, functional foods can be a powerful tool in fighting a wide range of diseases. It should be noted that polyphenols may have low bioavailability in the brain, as not all compounds can cross the blood-brain barrier. Carotenoids, astaxanthin, a carotenoid belonging to the terpene family, have shown promise in the context of neurodegenerative diseases. Like many carotenoids, astaxanthin is a fat-soluble pigment that is orange-red in color and has a strong ability to reduce reactive oxidant molecules. Due to the functional management of food, it is easy to stress or deal with stereotypical behaviors. Functional foods rich in resveratrol, anthocyanins, flavonoids which have been shown to be a potential group of antioxidants and nuts rich in minerals and omega-3 fatty acids which are said to have anti-inflammatory properties. Nowadays, bioactive compounds are used in the prevention, management, or treatment of ASD (autism spectrum disorder) and schizophrenia. There are several types of functional food, and its role in brain diseases also evoked its good management tool to use satisfactorily. Demonstrated the importance of functional foods in the prevention, treatment, and management of autism.

ACKNOWLEDGMENT

We thank the METs, Institute of Pharmacy, BKC, affiliated under Savitribai Phule Pune University, Nashik, Divine College of Pharmacy Satana, for their constant support and for providing all facilities to complete this work. EDA thanks the Ministry of Tribal Affairs Govt. of India for providing financial assistance in the form of a fellowship (Award No.: 202021-NFST-MAH-01235).

KEYWORDS

- **autism**
- **carotenoids**
- **functional food**
- **omega-3 fatty acids**
- **polyphenols**
- **vegetables**

REFERENCES

Allison, J., Wilder, D. A., Chong, I. V. Y., Lugo, A., Pike, J., & Rudy, N., (2012). A comparison of differential reinforcement and noncontingent reinforcement to treat food selectivity in a child with autism. *Journal of Applied Behavior Analysis, 45*(3), 613–617.

Avery, J. A., Ingeholm, J. E., Wohltjen, S., Collins, M., Riddell, C. D., Gotts, S. J., Kenworthy, L., et al., (2018). Neural correlates of taste reactivity in autism spectrum disorder. *Neuro Image: Clinical, 19*, 38–46.

Begum, P. S., Madhavi, G., Rajagopal, S., Viswanath, B., Razak, M. A., & Venkataratnamma, V., (2017). Probiotics as functional foods: Potential effects on human health and its impact on neurological diseases. *International Journal of Nutrition, Pharmacology, Neurological Diseases, 7*(2), 23.

Binnendyk, L., & Lucyshyn, J., (2005). Clinical issues: Family centered positive behavior support for a child with autism and severe food refusal behavior. *Perspectives on Language Learning and Education, 12*(1), 10–13.

Brondino, N., Fusar-Poli, L., Rocchetti, M., Provenzani, U., Barale, F., & Politi, P., (2015). Complementary and alternative therapies for autism spectrum disorder. *Evidence-Based Complementary and Alternative Medicine, 2015.*

Cascio, C. J., Foss-Feig, J. H., Heacock, J. L., Newsom, C. R., Cowan, R. L., Benningfield, M. M., Rogers, B. P., & Cao, A., (2012). Response of neural reward regions to food cues in autism spectrum disorders. *Journal of Neurodevelopmental Disorders, 4*(1), 1–11.

Casey, S. D., Perrin, C. J., Merical, C. L., Lecomte, J. M., Milligan, J., & Walsh-Czekalski, M., (2008). Increasing bite acceptance and reducing food refusal in a child with autism: Moving beyond the clinic. *Journal of Behavior Analysis in Health, Sports, Fitness and Medicine, 1*(1), 34.

Chistol, L. T., Bandini, L. G., Must, A., Phillips, S., Cermak, S. A., & Curtin, C., (2018). Sensory sensitivity and food selectivity in children with autism spectrum disorder. *Journal of Autism and Developmental Disorders, 48*(2), 583–591.

Cosbey, J., & Muldoon, D., (2017). EAT-UP™ family-centered feeding intervention to promote food acceptance and decrease challenging behaviors: A single-case experimental design replicated across three families of children with autism spectrum disorder. *Journal of Autism and Developmental Disorders, 47*(3), 564–578.

Dhingra, D., Michael, M., Rajput, H., & Patil, R. T., (2012). Dietary fiber in foods: A review. *Journal of Food Science and Technology, 49*, 255–266.

Essa, M. M., & Qoronfleh, M. W., (2020). *Personalized Food Intervention and Therapy for Autism Spectrum Disorder Management* (Vol. 24). Springer Nature.

Farhangi, M. A., Javid, A. Z., Sarmadi, B., Karimi, P., & Dehghan, P., (2018). A randomized controlled trial on the efficacy of resistant dextrin, as functional food, in women with type 2 diabetes: Targeting the hypothalamic–pituitary–adrenal axis and immune system. *Clinical Nutrition, 37*(4), 1216–1223.

Gale, C. M., Eikeseth, S., & Rudrud, E., (2011). Functional assessment and behavioral intervention for eating difficulties in children with autism: A study conducted in the natural environment using parents and ABA tutors as therapists. *Journal of Autism and Developmental Disorders, 41*(10), 1383–1396.

Gallup, J., Perihan, C., Tatsuma, Y., & Ikuta, S., (2021). Creating inclusive functional content using dot-codes: An exploration of multistep recipes for individuals with autism in post-secondary settings. In: *Education and Technology Support for Children and Young Adults with ASD and Learning Disabilities* (pp. 149–166). IGI Global.

Ghalichi, F., Ghaemmaghami, J., Malek, A., & Ostadrahimi, A., (2016). Effect of gluten free diet on gastrointestinal and behavioral indices for children with autism spectrum disorders: A randomized clinical trial. *World Journal of Pediatrics, 12*(4), 436–442.

Gorrindo, P., Williams, K. C., Lee, E. B., Walker, L. S., McGrew, S. G., & Levitt, P., (2012). Gastrointestinal dysfunction in autism: Parental report, clinical evaluation, and associated factors. *Autism Research, 5*(2), 101–108.

Grey, I. M., Lydon, H., Healy, O., & Thomas, J., (2017). Using a novel variant of stimulus fading in the treatment of food refusal and co-morbid challenging behavior in a child with autism. *Global Journal of Intellectual & Developmental Disabilities, 1*(5), 79–85.

Hetzroni, O. E., & Tannous, J., (2004). Effects of a computer-based intervention program on the communicative functions of children with autism. *Journal of Autism and Developmental Disorders, 34*(2), 95–113.

James, S. J., Melnyk, S., Jernigan, S., Pavliv, O., Trusty, T., Lehman, S., Seidel, L., et al., (2010). A functional polymorphism in the reduced folate carrier gene and DNA hypomethylation in mothers of children with autism. *American Journal of Medical Genetics Part B: Neuropsychiatric Genetics, 153*(6), 1209–1220.

Jarouliya, U., & Keservani, R. K., (2019a). Pathways leading to child obesity. In: Debasis, B., (ed.), *Global Perspectives on Childhood Obesity: Current Status, Consequences and Prevention* (2nd edn., pp. 137–146). Academic Press, Elsevier, chapter 12. ISBN: 9780128128404.

Jarouliya, U., & Keservani, R. K., (2019b). Protein function as cell surface and nuclear receptor in human diseases. In: Sanjay, B., & Debarshi, K. M., (eds.), *Medicinal Chemistry with Pharmaceutical Product Development* (pp. 1–32). Apple Academic Press, CRC Press, Taylor & Francis Group. Chapter 1. ISBN: 9781771887106.

Johnson, J. W., Blood, E., Freeman, A., & Simmons, K., (2013). Evaluating the effectiveness of teacher-implemented video prompting on an iPod touch to teach food-preparation skills to high school students with autism spectrum disorders. *Focus on Autism and Other Developmental Disabilities, 28*(3), 147–158.

Joshi, D., Roy, S., & Banerjee, S., (2018). Prebiotics: A functional food in health and disease. In: *Natural Products and Drug Discovery* (pp. 507–523). Elsevier.

Keservani, R. K., & Sharma, A. K., (2014). Flavonoids: Emerging trends and potential health benefits. *Journal of Chinese Pharmaceutical Sciences, 23*(12), 815.

Keservani, R. K., Kesharwani, R. K., Sharma, A. K., Vyas, N., & Chadoker, A., (2010b). Nutritional supplements: An overview. *International Journal of Current Pharmaceutical Review and Research, 1*(1), 59–75.

Keservani, R. K., Kesharwani, R. K., Vyas, N., Jain, S., Raghuvanshi, R., & Sharma, A. K., (2010a). Nutraceutical and functional food as future food: A review. *Der Pharmacia Lettre, 2*(1), 106–116.

Keservani, R. K., Sharma, A. K., & Jarouliya, U., (2015). Protein and peptide in drug targeting and its therapeutic approach. *Ars Pharmaceutica (Internet), 56*(3), 165–177.

Keservani, R. K., Sharma, A. K., & Kesharwani, R. K., (2016a). Nutraceutical and functional foods for cardiovascular health. In: *Food Process Engineering* (pp. 257–278, 291–312). Apple Academic Press, CRC Press. ISBN: 9781771884020.

Keservani, R. K., Sharma, A. K., & Kesharwani, R. K., (2016b). Medicinal effect of nutraceutical fruits for the cognition and brain health. *Scientifica*, 10. Article ID 3109254. doi: 10.1155/2016/3109254.

Keservani, R. K., Sharma, A. K., & Kesharwani, R. K., (2017). An overview and therapeutic applications of nutraceutical and functional foods. *Recent Advances in Drug Delivery Technology*, 160–201. ISBN:9781522507543.

Keservani, R. K., Sharma, A. K., & Kesharwani, R. K., (2020). *Nutraceuticals and Dietary Supplements: Applications in Health Improvement and Disease Management*. CRC Press. ISBN: 9781771888738.

Lasheras, I., Seral, P., Latorre, E., Barroso, E., Gracia-García, P., & Santabárbara, J., (2020). Microbiota and gut-brain axis dysfunction in autism spectrum disorder: Evidence for functional gastrointestinal disorders. *Asian Journal of Psychiatry, 47*, 101874.

Li, X. S., Pinto-Martin, J. A., Thompson, A., Chittams, J., & Kral, T. V., (2018). Weight status, diet quality, perceived stress, and functional health of caregivers of children with autism spectrum disorder. *Journal for Specialists in Pediatric Nursing, 23*(1), e12205.

López-Marcos, M. C., Bailina, C., Viuda-Martos, M., Pérez-Alvarez, J. A., & Fernández-López, J., (2015). Properties of dietary fibers from agroindustrial coproducts as source for fiber-enriched foods. *Food and Bioprocess Technology*, *8*, 2400–2408.

Manea, I., Buruleanu, L., Avram, D., & Bratu, M., (2009). Researches concerning the obtaining of lacto fermented cabbage juice–functional food. *Scientific Study & Research, 10*(2), 157.

Matson, J. L., & Fodstad, J. C., (2009). The treatment of food selectivity and other feeding problems in children with autism spectrum disorders. *Research in Autism Spectrum Disorders, 3*(2), 455–461.

McDowell, C., Duffy, K., & Kerr, K. P., (2007). Increasing food acceptance and reducing challenging behavior in a four-year-old girl with autism. *European Journal of Behavior Analysis, 8*(2), 267–276.

McKenzie, E. L., (2012). *Evaluation of a Parent Implemented Intervention for Food Selectivity in Children with Autism* (Vol. 1, pp. 1–100). University of Canterbury, New Zealand.

Mirenda, P., (2003). Toward functional augmentative and alternative communication for students with autism. *Language, Speech, and Hearing Services in Schools, 34*(1), 203–216.

Moes, D. R., & Frea, W. D., (2002). Contextualized behavioral support in early intervention for children with autism and their families. *Journal of Autism and Developmental Disorders, 32*(6), 519–533.

Navarro, F., Liu, Y., & Rhoads, J. M., (2016). Can probiotics benefit children with autism spectrum disorders? *World Journal of Gastroenterology, 22*(46), 10093.

Neely, L. C., Ganz, J. B., Davis, J. L., Boles, M. B., Hong, E. R., Ninci, J., & Gilliland, W. D., (2016). Generalization and maintenance of functional living skills for individuals with autism spectrum disorder: A review and meta-analysis. *Review Journal of Autism and Developmental Disorders, 3*(1), 37–47.

Pangrazzi, L., Balasco, L., & Bozzi, Y., (2020). Natural antioxidants: A novel therapeutic approach to autism spectrum disorders? *Antioxidants, 9*(12), 1186.

Peterson, K. M., Piazza, C. C., & Volkert, V. M., (2016). A comparison of a modified sequential oral sensory approach to an applied behavior-analytic approach in the treatment of food selectivity in children with autism spectrum disorder. *Journal of Applied Behavior Analysis, 49*(3), 485–511.

Peterson, K. M., Piazza, C. C., Ibañez, V. F., & Fisher, W. W., (2019). Randomized controlled trial of an applied behavior analytic intervention for food selectivity in children with autism spectrum disorder. *Journal of Applied Behavior Analysis, 52*(4), 895–917.

Peterson, K., & Ibañez, V., (2018). Food selectivity and autism spectrum disorder. *Teaching Exceptional Children, 20*(10), 3–11.

Peterson, K., & Ibañez, V., (2018). Food selectivity and autism spectrum disorder: Guidelines for assessment and treatment. *Teaching Exceptional Children, 50*(6), 322–332.

Rhine, D., & Tarbox, J., (2009). Chewing gum as a treatment for rumination in a child with autism. *Journal of Applied Behavior Analysis, 42*(2), 381–385.

Serra, D., Almeida, L. M., & Dinis, T. C., (2019). Polyphenols as food bioactive compounds in the context of autism spectrum disorders: A critical mini-review. *Neuroscience & Biobehavioral Reviews, 102*, 290–298.

Sharma, R., Ghimire, S., & Dhungel, K. U., (2020). Autism and food selectivity. *Janaki Medical College Journal of Medical Science, 8*(1), 64–74.

Sharp, W. G., Postorino, V., McCracken, C. E., Berry, R. C., Criado, K. K., Burrell, T. L., & Scahill, L., (2018). Dietary intake, nutrient status, and growth parameters in children with autism spectrum disorder and severe food selectivity: An electronic medical record review. *Journal of the Academy of Nutrition and Dietetics, 118*(10), 1943–1950.

Sigafoos, J., & Saggers, E., (1995). A discrete-trial approach to the functional analysis of aggressive behavior in two boys with autism. *Australia and New Zealand Journal of Developmental Disabilities, 20*(4), 287–297.

Silbaugh, B. C., Penrod, B., Whelan, C. M., Hernandez, D. A., Wingate, H. V., Falcomata, T. S., & Lang, R., (2016). A systematic synthesis of behavioral interventions for food selectivity of children with autism spectrum disorders. *Review Journal of Autism and Developmental Disorders, 3*(4), 345–357.

Silbaugh, B. C., Wingate, H. V., & Falcomata, T. S., (2017). Effects of lag schedules and response blocking on variant food consumption by a girl with autism. *Behavioral Interventions, 32*(1), 21–34.

Singh, A. U., (2018). Dysbiosis, small intestinal bacterial overgrowth and biofilms in autism and chronic illness. *AIMS Molecular Science, 5*(2), 160–165.

Singh, P., Kesharwani, R. K., & Keservani, R. K., (2017a). Antioxidants and vitamins: Roles in cellular function and metabolism. In: *Sustained Energy for Enhanced Human Functions and Activity* (pp. 385–407). Academic Press. ISBN:9780128054130.

Singh, P., Kesharwani, R. K., & Keservani, R. K., (2017b). Protein, carbohydrates, and fats: Energy metabolism. In: *Sustained Energy for Enhanced Human Functions and Activity* (pp. 103–115). Academic Press. ISBN:9780128054130.

Siniscalco, D., Brigida, A. L., & Antonucci, N., (2018). Autism and neuro-immune-gut link. *AIMS Mol. Sci., 5*, 166–172.

Tanner, A., & Andreone, B. E., (2015). Using graduated exposure and differential reinforcement to increase food repertoire in a child with autism. *Behavior Analysis in Practice, 8*(2), 233–240.

Theoharides, T. C., Asadi, S., & Panagiotidou, S., (2012). A case series of a luteolin formulation (NeuroProtek®) in children with autism spectrum disorders. *International Journal of Immunopathology and Pharmacology, 25*(2), 317–323.

Theoharides, T. C., Asadi, S., & Patel, A. B., (2013). Focal brain inflammation and autism. *Journal of Neuroinflammation, 10*, 1–7.

Toh, M. C., & Allen-Vercoe, E., (2015). The human gut microbiota with reference to autism spectrum disorder: Considering the whole as more than a sum of its parts. *Microbial Ecology in Health and Disease, 26*(1), 26309.

Turner, N. D., & Lupton, J. R., (2011). Dietary fiber. *Advances in Nutrition, 2*(2), 151–152.

Usui, N., Kobayashi, H., & Shimada, S., (2023). Neuroinflammation and oxidative stress in the pathogenesis of autism spectrum disorder. *International Journal of Molecular Sciences, 24*(6), 5487.

Vissoker, R. E., Latzer, Y., & Gal, E., (2015). Eating and feeding problems and gastrointestinal dysfunction in autism spectrum disorders. *Research in Autism Spectrum Disorders, 12*, 10–21.

Volkert, V. M., & Vaz, P. C., (2010). Recent studies on feeding problems in children with autism. *Journal of Applied Behavior Analysis, 43*(1), 155–159.

CHAPTER 3

FUNCTIONAL FOODS IN CANCER MANAGEMENT AND PREVENTION

KHEMCHAND R. SURANA[1] and SUNIL K. MAHAJAN[2]

[1]*Department of Pharmaceutical Chemistry, Shreeshakti Shaikshanik Sanstha, Divine College of Pharmacy, Satana, Nashik, Maharashtra, India*

[2]*Department of Pharmaceutical Chemistry, MGV's Pharmacy College, Panchvati, Nashik, Maharashtra, India*

ABSTRACT

Cancer is the main purpose of death in the world. Common cancers encompass prostate cancer, breast cancer, bladder cancer, colorectal cancer, and pores and skin cancers. Treatment of most cancers may be very costly. Therefore, they want to prevent the circumstance primarily or maybe secondarily. Foods, like chemicals, have cumulative consequences and include materials, each important and non-important, that have been proven to be healthy. This brought about the acclamation and reputation of certain ingredients as presenting fitness benefits and purposeful ingredients. Functional ingredients and nutraceuticals are chemopreventive to cancers as evidenced via way of means of unique research performed over the years. This does now no longer conclude that purposeful ingredients and nutraceuticals have an all-preventive impact due to the fact a few studies findings implicate them in cancer risks and progression. These evaluation goals to talk about a number of those ingredients and the chemoprotective compounds therein to useful resources in addition to scientific studies. Foods evaluated in cancer risks are grapes, tomatoes, pink peppers, berries, chestnuts, ginseng, *Rhodiola*, tea, soy, garlic, and cruciferous vegetables. This evaluation additionally

Applications of Functional Foods in Disease Prevention. Raj K. Keservani & Eknath D. Ahire, PhD (Eds.)

takes into consideration the so-referred to as antioxidant nutrients – selenium, nutrients A, B, C, D, and E. Others encompass glucosamine, chondroitin, melatonin, MSM, which have been taken as supplements. Anyway, researches are nevertheless ongoing referring to cancers and foods as this listing is supposedly now no longer exhaustive.

3.1 INTRODUCTION

Functional foods have received excellent significance nowadays for their foreseen visualized function in the remedy of diverse pathologies including most cancers. Functional foods because the name suggests encompass foods which can be of dietary significance and also are concerned in resolving the diseased state of the body. They had been stated for his or her useful position in cancer prevention and treatment. Several purposeful foods which have been received from diverse sources including plants, animals and microbial sources are stated to own significant anticancer results with various mechanisms of action viz. anti-metastatic, induction of apoptosis, anti-proliferative, anti-angiogenic, scavenging of free radicals, inhibition of matrix metalloproteinases, etc. Examples of functional foods that evinced anticancer capacity encompass cereals, vegetables, beverages, dairy products, fish oil, beef, mushroom, probiotics, etc. (Kabir, 2020; Islam and Siddiqua, 2020).

The entire grain cereals which might be fed on global consist of wheat (*Triticum aestivum* L.), rice (*Oryza sativa* L.), maize (*Zea mays* ssp. mays L.), rye (*Secale cereale* L.), oats (*Avena sativa* L.), and barley (*Hordeum vulgare* L.). Whole grain cereals include various components with anticancer potential which can be placed commonly in the germ and bran layers which might be eliminated from the endosperm at some point of refining. Bran is the difficult outer layer of cereal grains this is profoundly wealthy in a myriad of healthful phytochemicals together with phenolics, flavonoids, glucans, and pigments and has been found to evince major anticancer activity. Bitter melon (*Momordica charantia* L.) is an extensively fed on vegetable all through the sector that contains numerous bioactive additives together with polyphenols, flavonoids, and saponins and has been said to own anticancer capacity (Kumar, 2015; Aghajanpour et al., 2017).

Thus, there are numerous examples of functional foods displaying anticancer activity that can be exhibited through various modes of action, e.g., sour melon inhibits proliferation of cancer cells, fish oil reveals anticancer activity through inhibition of metastasis, numerous species of mushrooms in particular *Tricholoma matsutake* act through inducing apoptosis and has proven useful outcomes in the treatment of oral cancer. Some functional foods have proven anticancer efficacy through inhibition of pathological angiogenesis together with green tea, purple grapes, etc. (Kumar and Barmanray, 2007; Gul et al., 2016).

Several functional foods evinced anticancer activity through scavenging free radicals, e.g., cauliflower, cabbage, garlic, tomato, broccoli, green tea, red beets, soybean, carrot, onion, cranberries, blueberry, cocoa, citrus fruits, blackberry, red grapes and prunes. Inhibition of matrix metalloproteinases through functional foods is but every other mechanism developed for cancer treatment, e.g., amla (*Emblica officinalis*) inhibit MMP-1, turmeric (*Curcuma longa* L.) inhibit MMP-2 and MMP14, Camellia sinensis inhibit MMP-1, MMP-2, MMP-3, MMP-7 and MMP-9 thereby exerting their anticancer effect. Thus, functional foods provide amazing ability for the prevention and treatment of cancer and consequently might also additionally act as an opportunity for cancer therapy (Asgari et al., 2011, Kabir, 2020).

Some plant foods lessen the risk of certain cancers. "Nutraceutical" became devised via way of means of Stephen De Felice in 1989 from the combination of terms, "nutrition" and "pharmaceutical." He described nutraceutical as, "a food (or part of food) that gives clinical or health benefits, inclusive of the prevention and or treatment of a disease." Thus functional foods may be seemed as nutraceuticals in the event that they function as such (Keservani et al., 2010a, b; Maddi et al., 2007). Plant foods and fitness-primarily based totally researches have surely demonstrated plant bioactive compounds to include antioxidants and different healthy properties. For fantastic consequences, large doses of a few antioxidants need to be taken, and this could present an unwanted condition (Maddi et al., 2007). Nutraceuticals have been deployed in lots of conditions requiring therapeutic needs, cancers prevention inclusive. This became established via way of means of a look at to analyze the function of nutraceuticals in prostate cancer prevention (Balsano and Alisi, 2009). Antioxidants are so-known as due to the fact they prevent oxidations that arise in living tissues of the body. The oxidations are because of a group of materials known as reactive oxygen species. Reactive oxygen species consist of superoxide radicals, hydroxyl radicals, and hydrogen peroxide and are involved in oxidative damage to DNA, cells and proteins alike. Evidence is available that oxidative stress is involved with inside the pathogenesis of many chronic diseases, and cancer isn't always an exception. The interest that has been given functional foods and nutraceuticals is because of their acclaimed antioxidant properties (Chikwere, 2017; Milner, 1999; Ahire et al., 2021).

3.2 APPLICATIONS OF FUNCTIONAL FOODS

There is an extensive variety of the primary use of functional foods:

1. **Prevent Disease:** The function of functional foods in combating in opposition to diseases.
2. **Prevent Aging:** Functional foods strategies to slow down the aging technique via way of means of tonifying the kidneys and spleen, nourishing essential substances of the body, and roving the disease-causing factors.
3. **Maintain and Enhance Appearance:** Functional foods enhance the luster of facial pores and skin, lowering facial wrinkles, putting off facial pigments and enhancing facial pores and skin edema, enhance blood and to bolster the characteristic of the spleen and lung.
4. **Control Body Weight:** Functional foods has the ability to take away immoderate body fats and lowering body weight via way of means of putting off dampness and facilitate diuresis.

Overall, the functionality of foods are carried out for a range of applications during fitness protection and medication, aging prevention, and cosmetological purposes (Lai, 2012; Riezzo et al., 2005).

3.3 FUNCTIONAL FOODS USED IN TREATMENT OF CANCER

Europe scientific institute described functional food as "a food product can most effective be taken into consideration as functional, if collectively with the simple nutritional effect it has an useful impact on one or extra feature of the human organism consequently both through enhancing popular and physical fitness or lowering the risk of evolution of sickness." Functional food often known as nutrients imparting materials suggests probably energetic conduct withinside the treatment and prevention of cancer. Whey protein suggests up ground-breaking and the suppositions had been made with confirmations that it segregates has a cytotoxic effect on melanoma. A lesser number of tumor foci became visible whilst whey protein hydrolysate become sorted through colon malignant increase bearing rodents. Table 3.1 represents the listing of functional foods together with their biological applications. Cruciferous veggies inclusive of broccoli, cauliflower, and cabbage incorporate sure non-nutritional phytochemicals which might be useful for cancer prevention, while onion and garlic incorporate certain biologically active compounds which have anticarcinogenic and anticholinergic effects, different functional foods encompass functional milk, dairy products, sea buckthorns, and a few functional seafood lipid and proteins (Jarouliya and Keservani 2019b; Aghajanpour et al., 2017). Whey protein or milk protein suggests anticancer activity, as a functional food facilitates withinside the dealing with of chronic sickness. Folate, a water solvable completely oxidized form of Vitamin B, it performs an extensive proportion withinside the growth of certain sicknesses like cancer (Parodi, 2007). Linola seed holds oil that is amusing in linoleic acid and yield conjugated linoleic acid beneficially diminishes the jeopardy of breast cancer in women. Functional food is used as CMDs medicinal drugs due to its latent houses inclusive of anti-inflammatory, antioxidants (Keservani et al., 2015c), antiestrogenic, and immunomodulatory (Keservani et al., 2015d); different purified bioactive complexes are advanced to pharmaceutical capsules for the dealing of unembellished persistent symptoms (Lyu et al., 2017). Phenolic and polyphenolic (Keservani and Sharma, 2014) compounds act as a particular a part of secondary metabolites of plant life that flip as free radical scavengers and inhibitors of LDL fatty acid oxidation and DNA cracking. As a result, nutritional polyphenols are used to prevent cardiovascular diseases (CVDs) (Keservani et al., 2016a) and certain tumors. Prunes (Prunus domestica L.), dried plums fruits are very wealthy in bioactive compounds, their phenolic compounds inhibit human LDL oxidation in vitro, and performs a shielding function in opposition to chronic disease cancer (Shahidi, 2004; Sanders, 1998; Keservani et al., 2016b).

TABLE 3.1 List of Functional Foods with Their Biological Uses

Sl. No.	Functional Foods	Biological Use
1.	Ginger, tea, onion, berry, mushroom, garlic, carrot	Anti-carcinogenic, decrease lipid
2.	Ginger, apple, beetroot, spinach, broccoli	Anti-carcinogenic, enhanced drug detoxification
3.	Green tea, garlic, onion	Anti-carcinogenic, antibacterial
4.	Soybean cabbage, fennel	Anti-carcinogenic, anti-estrogenic
5.	Green tea, garlic, orange	Anti-carcinogenic

3.3.1 APPLE

Apples are a prime wellspring of nutritional phytoconstituents like flavonoids. They moreover comprise enormous stages of polyphenols additionally, and unique phytochemicals. The usage of one apple every day became accounted for to reduce the chance of colorectal malignancy. The have a look at additionally prophesied that the chance of colorectal tumor condensed round 50% upon feeding of supplementary according to day an apple (Tu et al., 2017). Apple phloretin (Ph), which may be obtained from apples and juices, is a famous inhibitor of the type II glucose carrier (GLUT2). Expanded articulation of GLUT2 has been discovered in several human neoplasm tissues, which include liver, breast, and gastric malignancy cells. Apple extricates basically constrained the TNF-R-initiated NF-KB enactment at a part of 5 mg/ml. In malignant increase cells, NF-KB instigates safety from anticancer chemotherapeutic operators with the aid of using increasing cell enlargement and restraining apoptosis. Along those lines, restraint of NF-KB actuation in malignant increase cells is worthwhile in disorder treatment with the aid of using lowing the obstruction to chemotherapy. Subsequently, apple usage has been related to a delivered down chance of malignant increase (Yang et al., 2009, Kaur and Das, 2011).

3.3.2 BEETROOT

Beetroot (*Beta vulgaris* L.) is an usually expended vegetable, which furthermore has strong antioxidant properties. Beetroots and their juices from biological invention have an excessive content material of bioactive complexes. These safe plants, while applied and combined with extreme anticancer medications, for example, doxorubicin (*Adriamycin*), can act synergistically and relieve treatment-associated medicine poisonousness. Betanin, the betacyanin constituent basically accountable for red beet shading, is a cancer prevention agent with a mainly fairly free radical-looking action and is a modulator of oxidative pressure. Red beetroot go away is a not unusual place Mediterranean weight-reduction plan vegetable, that is an ample and less expensive supply of herbal antioxidant and anticancer vitamins (Singh et al., 2017a) and may be delivered to juice because it improves the juice's dietary value and does now no longer have a widespread impact on its sensory properties (Yoon and Liu, 2007).

3.3.3 BLUEBERRY

Anthocyanins are reddish herbal pigments, a class of flavonoids, broadly circulated in fruit and especially in blueberries (*Vaccinium* spp.). Epidemiological research has related their consumption favorably with reduced mortality because of CVDs and a few cancer types. One of the herbal and safe to eat items containing a wealthy supply of phenolic nutritional compounds is Blueberry. Quercetin was formerly shown to have cytotoxic outcomes on cancer in vitro as one of the most plentiful nutritional phenolic compounds found in blueberry extracts. Kaempferol is a blueberry extract flavonol much like quercetin that has formerly been shown to be effective for its anticarcinogenic outcomes. In vivo and in vitro evaluation of berries as antioxidants and anticancer causing agents has improved our knowledge of the complicated mechanism and their effect on populace health and illnesses. Blueberries act through pathways through which carcinogenesis has been averted by inhibiting the production of pro-inflammatory molecules and oxidizing pressure and products, along with DNA damage, inhibiting cell growth, and improved apoptosis (Kazimierczak et al., 2014).

3.3.4 GINGER

Ginger (*Zingiber officinale*) is and the world over identified plant utilized in cooking and has long been taken into consideration to have therapeutic properties. Ginger (*Zingiber officinale*), an individual from the family Zingiberaceae, consists of sugars (50–70%), phenol, lipids (3–8%) and terpenes, mixes as a constituent of ginger rhizome. Other than these, amino acids, crude fiber, debris, protein (Keservani et al., 2015; Singh et al., 2017b), phytosterols, nutrients, and minerals are moreover present. The ginger has a massive component in treating some infections, inclusive of gastrointestinal complexities, deal with belly upset, and looseness of the bowels, rheumatic problems, queasiness, fundamental colds, fever, and tipsiness. Furthermore, ginger has anti-neoplastic and chemopreventive properties. Ginger root and its essential phenolic mixes, for example, gingerols and zerumbone have anti-carcinogenic action, cancer prevention agents, and calming movement through prompting apoptosis in malignant increase cells (Zein et al., 2015). The *Zingiberaceae* flowers comprise essential oils which encompass terpenoids, phenylpropanoids, flavonoids (Keservani and Sharma, 2014), and sesquiterpenes which have noteworthy antitumor action. Steamed ginger gives a brand new supply of chemical variant in its structure as compared to unprocessed ginger. Recently, an observe observed that ginger steam treatment significantly advanced its anti-carcinogenic outcomes on colon cancer (Faria et al., 2010).

3.3.5 GRAPES

Grapes are wealthy in polyphenols which are useful for the treatment of breast cancer it indicates pro-apoptotic, anti-invasive, anti-proliferative, anti-angiogenic, and anti-metastatic action in breast cancer. Resveratrol, a polyphenolic compound in grapes is one of the robust anticancer agents; those compounds show the progression-inhibition impact on cancer cells. Resveratrol derivatives gradual the boom of cancer cells and manipulate CV results additionally a neuroprotective and prevent metabolic diseases. Pterostilbene and piceid show a comparable impact and inhibit cellular proliferation as resveratrol on cell antioxidation structures with the aid of using improving cell resistance. Resveratrol moreover contains some degrees to detoxing or cellular reinforcement compounds by actuating a redox sensitive record factor. Apoptosis is precipitated with the aid of using resveratrol with the initiation of P53 and with the aid of using suppression of protein translation in T47D human breast cancer (Johnson and Arjmandi, 2013; Prasad and Tyagi, 2015).

3.3.6 MUSHROOM

Mushroom (*Inonotus obliquus*) has been utilized as a beneficial food and traditional Chinese spices for a protracted time, having anti-inflammatory, calming, and anticancer constituents. Homopolysaccharides are common to be the important thing active mushrooms' polysaccharides, special varieties of polysaccharides of mushroom have herbal houses along with anti-obesity, anti-diabetic, antiviral, antimicrobial, and anti-carcinogenic effects that exhibit the numerous wellness advancing potential (Ramakrishnan, 2013). Therapeutic mushrooms with anticancer activity contain polysaccharides, polysaccharide-protein edifices, nutritional fibers, and particular varieties of proteins, terpenoids, steroids, phenols, and so forth. In traditional Chinese medicine, Cordyceps has been applied to deal with some situations inclusive of cancer (Stuart and Robb, 2013).

3.3.7 ONIONS

Onion is one of the most extensively and to a first rate extent wolfed vegetables. From epidemiologic considers, onion utilization is popularly related to reduce paces of cardiac contamination. Onions are stuffed with adversarial to malignancy sulfur mixes which might be profoundly gainful for disorder chemoprevention. Progressed contemplates having indicated that onion is probably avert the liver, colon, prostate, cell breakdown withinside the lungs, and so on with excessive cancer prevention properties (Ma et al.,

2013). Allium usage has been contrarily recommending a chemoprotective function, for example, onions. Figure 3.1 suggests special useful ingredients alongside in which they act in special degrees of tumor progression. Various herbal carotenoids (herbal fat-soluble pigment) inclusive of carotene, lutein, zeaxanthin, lycopene, and astaxanthin except beta carotene had been verified to have anticholinergic activities; they inhibit cancer cellular proliferation (Ivanova et al., 2014).

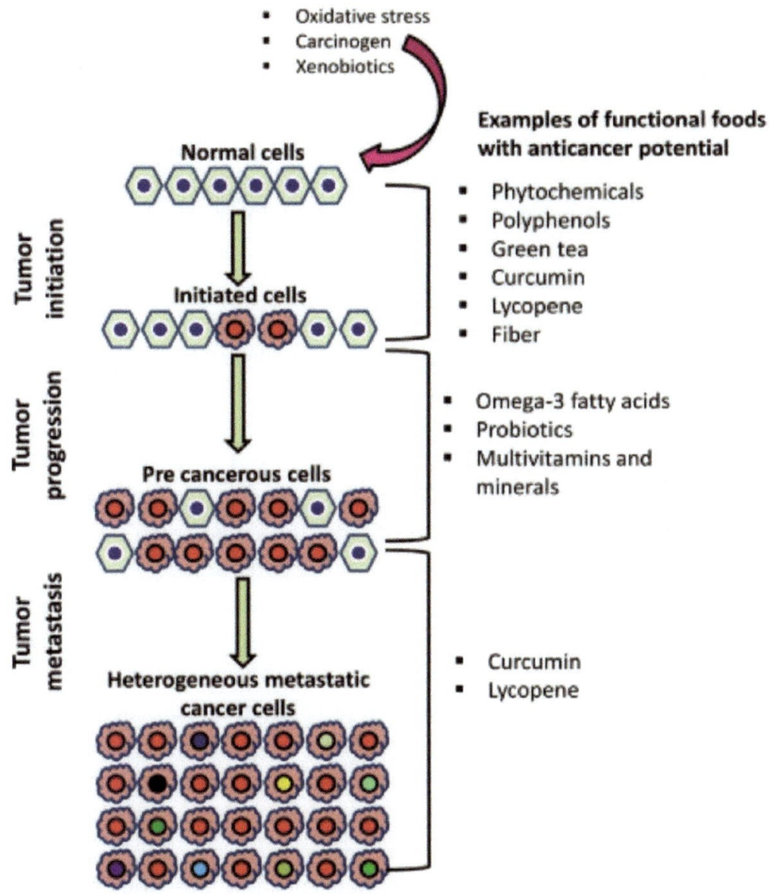

FIGURE 3.1 Role of functional foods in cancer prevention and treatment.

Moringa in any other case referred to as horseradish tree or drumstick tree. It is the maximum examine species for its anticancer interest and are applied in countries in which malnutrition is a significant problem. Due to the fact, its flowers, cases, leaves, and seeds comprise first rate nutritional gain and used to stop loss of healthy sustenance in individuals. And due to the nearness of positive bioactive compounds, it suggests pharmacological properties that show anticancer exercises and is probable applied as a functional food (Alpsoy et al., 2013; Keservani et al., 2017, 2020). Tea polyphenol suggests preventive measures in opposition to osteoporosis, cancer, and CVDs for those who consume three cups of tea per day. Artichoke may be used as a supply of functional food that stops cancer in individuals. Furthermore, those paleolithic consuming regimens inclusive of lean meats, fish, natural products, vegetables, nuts, and seeds display promising final results withinside the counteraction of malignant increase cancer and all different CVDs as they comprise ω-3 unsaturated fat, polyphenols, filaments, and plan sterols those bioactive mixes reduce the prevalence of contamination like cancer (Xiao and Parkin, 2007). Rice bran is an ironic foundation of nutritional fiber, minerals, vitamins, and crucial fatty acids. Various revisions specify that rice is excessive in γ-oryzanol, which hinders tumor development. The use of flaxseed has useful consequences on colon body structure with a discount in threat of or occurrences of colorectal cancer in each human and animal experiments. Taken separately, the threat from colonic neoplasms reduced, and cellular growth control through numerous feasible pathways is likewise correlated with every of its predominant bioactive additives which include fibers, alpha-linolenic acid, lignans, and different phytochemicals (Goyal et al., 2014). Banana peel extract is listed as non-poisonous to regular cells in compliance with the requirements described by the National Cancer Standard Institute. Chronic non-transmittable illnesses inclusive of metabolic disorder, Type 2 diabetes, CVDs, and cancer show up due to helpless nutritional propensities and disenchanted manner of existence that may be handled through shifting to a greater high-quality weight-reduction plan and refreshments decision. The development of human consuming habitual for the reason that beyond several years from the paleolithic weight-reduction plan to the cutting-edge shape of food has modified a ton that brings approximately stoutness and interminable contamination inclusive of CVDs and diabetes (Gul et al., 2015).

Functional foods constructed on their rudimentary nourishing roles can reduce the chance of many persistent illnesses and display functional benefits, promoted with the privilege in their functionality to decrease cardiac conditions, focusing specially on diagnosed risk aspects, including hyperlipidemia, diabetes, obesity (Jarouliya and Keservani 2019a), and raised lipoprotein. Strategies had been made for the prevention of malignancy in people with amplified consuming of functional foods, including barley, brown rice, buckwheat, and complete grains and veggies, including broccoli, cabbage, garlic, mushrooms, and onions, or a few beverages, including coffee and green tea (Challacombe, 2021).

3.3.8 CHESTNUT

Chestnut extract possesses antioxidant activity and is defensive in opposition to gastric cancer, and moderate defensive effect in opposition to prostate cancer and breast cancer. Chestnut extract made with ethanol has a considerable negative impact on the growth of gastric cancer cells, compared to raw chestnut extract which had no impact and or maybe the water extract of chestnut which exhibited as an alternative growth selling impact on cancer cells. The flesh and internal shell of chestnuts (which include the leaves, and flowers), and raw chestnuts and processed chestnut confer antioxidant effects. The efficiency of the extracts in chemoprevention is depending on their extraction medium and dose applied (Lee et al., 2011).

3.3.9 SOY

In the studies of soy and the risk of cancer, a whole lot of interest and hobby has been given to breast cancer. Studies targeted on secondary prevention show that soy (isoflavones) lessen the risk of breast cancer, mainly amongst postmenopausal women who're on drug therapy, tamoxifen as soy did now no longer intrude with the efficacy of the drug. However, it's far advised that soy produces no sizeable impact for fast consumers, however greater useful to early consumers, as a long way as cancer chemoprevention is concerned. This suggests why soy consumption in kids is related to decreased risk of breast cancer in premenopausal women mainly however curiously any other study determined this inverse affiliation amongst postmenopausal ladies and none withinside the premenopausal. Thus those findings are nevertheless now no longer correlated and feature now no longer hooked up one common end as one study contradicts the other (Huang et al., 2012).

3.3.10 GARLIC

Garlic, which belongs to the Allium genus of vegetables, and its sulfur-containing energetic compounds are shielding in opposition to the onset and/or development of numerous cancers. Diallyl trisulfide averted the improvement of prostate cancer and the spread of cancer to the lung in a transgenic mouse version of prostate cancer. This impact turned into dose-dependent in that the experimental topics who acquired the better amounts of the treatment derived notably more benefit. The diallyl averted development of prostate cancer in doses of 1 mg and a pair of mg given to the mice three instances per week in comparison to placebo. The organosulfur compound acts with the aid of using inhibiting the expression of the androgen receptor that is actively concerned in the improvement of prostate cancers. Garlic in distinctive forms inclusive of garlic oil has particularly been shielding in opposition to liver cancer (Thoppil and Bishayee, 2011).

3.3.11 GREEN TEA

Green tea polyphenols have been proven to prevent the development of tumor cells in transgenic adenocarcinoma of the mouse prostate (TRAMP) version. Green tea has been proven to shield the bladder in opposition to mobileular death. The antioxidant efficiency of inexperienced tea extract, turned into tested as green tea decreased the oxidative pressure prompted with the aid of using H_2O_2 (oxidative agent) in ordinary or malignant human bladder cells (Jiang et al., 2011).

3.3.12 GINSENG

Ginseng, over time, has been used as a medicinal plant. Ginseng belongs to the genus Panax, including *Chinese ginseng, American ginseng*, and *Panax notoginseng*. It has obtained an extensive interest in cancer studies related to numerous cancer sorts as cancer of the blood, colon, breast, liver, and lung. Ginseng extract has been advised to be chemopreventive withinside the improvement of colon cancer in vitro and in a mouse model. It has been demonstrated that ginseng root extract has a suppressive effect on SW480 colorectal tumors, consequently preventing the multiplication of cancer cells. American ginseng extract had a suppressive impact on ulcerative colitis, an inflammation of the large gut and a risk element for colon cancer, while administered to mice, and in vitro. It is accordingly anti-inflammatory to colitis and chemopreventive to the cancer cells. The colon cancer prevention impact of ginseng has been attributed to its antioxidant property and ginsenosides had been advised to be accountable for this property, and in a few cases, particular metabolites are involved. *Panax notoginseng* saponin extract has been shown to have a scavenging impact. Ginseng has additionally been proven to defend towards breast cancer as a ginsenoside from *Panax notoginseng* substantially decreased oncogene (MDM2) expression, and the protecting effect of ginsenoside diminished while the oncogene became allowed to overexpress (Chen et al., 2013). However, heat-treated *Panax quinquefolius* root has been of a more chemoprevention towards breast most cancers in comparison to unheated ginseng. This indicates that heating ginseng earlier than intake might possibly confer to a bigger extent greater useful effect in comparison to consuming it raw. Terpenoids together with ginsenosides had been advised to have ability chemoprotective impact on liver cancer as liver cancer cells increase turned into inhibited and mice precipitated with liver cancer accompanied with the aid of using treatment with ginsenoside Rg3 had an extra survival period. Chen et al. have additionally indicated that during children, a ginsenoside metabolite, so-referred to as compound k have anticancer results on blood cancer cells (Liu et al., 2012). Most researches have actually made ginseng 'super' in cancer prevention. But very vital and excellent is the truth that the chemopreventive consequences displayed through ginseng have been all "dose-dependent" and "time-dependent." However, a cohort observe with ginseng complement confirmed no impact on risk of blood cancer incidence. This is a sign that though ginseng can be powerful as a herbal product, its complement may also don't have any affiliation with at the least positive cancers (Wang et al., 2010; Sharma et al., 2020; Rane et al., 2020a–c).

3.3.13 *RHODIOLA ROSEA*

Only some researchers have concerned *Rhodiola rosea* in cancer studies, so information regarding its involvement in the prevention of cancer is clearly limited. Nonetheless, in an in vitro study, *Rhodiola rosea* extracts averted the increase and/or the proliferation of cancer cells of the urinary bladder. Another species *Rhodiola crenulata* additionally had an inhibitory impact at the increase of breast cancer cells in an test carried out in vitro (McCormick et al., 2010).

3.3.14 TOMATO AND RED PEPPER

Tomato carries lycopene, which has been mentioned as being potent and different antioxidants. Studies recommend lycopene is chemopreventive in opposition to cancers inclusive of breast cancer, prostate cancer, gastric cancer, and lung cancer. Tomato has been proven to be powerful in opposition to liver cancer caused with the aid of using a few environmental cancer agents in a rat model. Diethylnitrosamine-initiated and nonalcoholic steatohepatitis-promoted early liver cancer turned into inhibited in rats with the aid of using lycopene-containing tomato extract. Similarly, mortality risk from prostate cancer turned into stated in studies with rats, to be decreased with tomato powder. Capsaicin (energetic thing in pepper) in a test carried out in vitro acted as an anticancer agent by inhibiting the migration of pores and skin cancer cells to different body parts demonstrating its chemopreventive effect (Chikwere, 2017).

3.3.15 POLYPHENOLS AND CAROTENOIDS

Polyphenols are a set of more than 500 phytochemicals that typically have potential in health and disorder prevention. Among them, many polyphenols which includes curcumin, quercetin, lycopene, catechins, lignans, isoflavones, and flavanones had been tested; all of them have defensive influences in opposition to cancer. This section describes with proof-based values of not unusual place polyphenols in the food regimen which might be taken into consideration perfect chemo-preventive agents through antioxidant and anti-inflammatory properties; activation of apoptosis and autophagy. Recent proof shows that polyphenols can affect epigenetic modifications through posttranslational changes and microRNA expressions related to the fate of cancer cells. Several scientific research on polyphenolic phytochemicals indicated using tea polyphenols to prevent and deal with cancer. However, the information for curcumin and soy isoflavones are inconsistent. Mechanistic proof has validated that nutritional polyphenols affect prostate cancer hazard through regulation of inflammatory genes and repairing oxidative DNA harm. In addition, there are interactions among one-of-a-kind nutritional polyphenols, that can range the prostate cancer hazard through the activation of each antioxidant and nonantioxidant mechanisms (Russo et al., 2017).

3.3.16 CURCUMIN

Curcumin, a polyphenol derived from turmeric, possesses numerous anticarcinogenic and anti-inflammatory properties. Laboratory, animal, and human studies show its preventive results on tumor invasion and metastasis through the modulation of more than one cell signaling pathway. Curcumin has been correlated with a huge variety of influences; however, reviews had been disconcerting and every now and then contradictory. Some proof shows that curcumin is an antioxidant whilst different studies show that curcumin is prooxidant, which will increase ROS and results in oxidative stress in cells. Curcumin might also additionally modify cancer growth, development, and unfold through interference with cell cycle proliferation and apoptosis in addition to suppression of angiogenesis and metastasis. New proof shows that curcumin might also additionally make chemotherapy extra powerful and defend wholesome cells from harm through radiation therapy (Hatcher et al., 2008).

3.4 DIETARY FIBER AND CRUCIFEROUS VEGETABLES

Dietary fiber is thought to prevent constipation by increasing bulk of stool, drawing water into stool. It has been proven to have a vast impact on the threat of colorectal cancer. In a cohort look at, nutritional fiber became powerful in the prevention of colorectal cancer amongst those who have been genotypically prone to the chronic disease. In the colon, the nutritional fiber has direct contact with the partitions of the intestine which in itself reduces the stress exerted at the partitions of the canal. Also, nutritional fiber-wealthy weight loss program consumption has been observed to be related to a discounted threat of breast cancer in a cohort look at with postmenopausal women. These consist of broccoli, cabbage, Brussels, cauliflower, sprouts, and greens (kale, turnip, mustard, and collard). In a look at carried out in vitro phenethyl isothiocyanate from cruciferous greens became postulated to be a chemopreventive agent because it had an inhibitory impact on human breast cancer cell lines. A case-control study that compared culmination and different veggies to cruciferous vegetables showed cruciferous veggies to be appreciably related to a discounted threat of lung cancer amongst smokers. This affiliation became even more potent in heavy smokers (people who took extra than 20 cigarettes daily) and short-term smokers (people with much less than 30 years of smoking). Raw cruciferous veggies have been observed to be more efficient than cooked ones (Islam and Siddiqua, 2020).

3.5 CONCLUSION

A balanced, well-working body culminates in a good enough diet, ensuing in the fulfillment of proper human physiology, as a result balanced living.

Dietary variables are identified as having a main effect on cancer hazard, with risks being each elevated and reduced by diverse nutritional components. Diets are related to many deaths from cancer, whilst physical inactiveness has a tendency to be related to the hazard of cancer. Researchers have made development in our information of the connection between functional foods and cancers in current decades, particularly in the region of prevention. In general, this paper addresses functional foods for cancer prevention and health benefits. Some functional foods together with green tea and red grapes have proven anticancer efficacy through inhibiting pathological angiogenesis. Through the scavenging of free radicals, a few functional foods displayed anticancer activity, together with cabbage, berries, broccoli, carrots, cauliflower, chocolate, green tea, garlic, grapes, tomatoes, onions, soybeans, prunes, and citrus fruits. The intrinsically chemopreventive lifestyles of foods and/or their additives and functional foods must now no longer be completely believed and acknowledged.

ACKNOWLEDGMENT

We thank the Divine College of Pharmacy Satana and MGVs, College of Pharmacy, Panchvati, Nashik, for their constant support and for providing all facilities to complete this work.

KEYWORDS

- **cancer**
- **nutraceuticals**
- **functional foods**
- **prevention**
- **risks**

REFERENCES

Aghajanpour, M., Nazer, M. R., Obeidavi, Z., Akbari, M., Ezati, P., & Kor, N. M., (2017). Functional foods and their role in cancer prevention and health promotion: A comprehensive review. *American Journal of Cancer Research, 7*, 740.

Ahire, E., Sonawane, V., Surana, K., & Talele, G., (2021). Drug discovery, drug-likeness screening, and bioavailability: Development of drug-likeness rule for natural products. *Applied Pharmaceutical Practice and Nutraceuticals*. Apple Academic Press.

Alpsoy, S., Uygur, R., Aktas, C., Topcu, B., Kanter, M., Erboga, M., Karakaya, O., & Gedikbasi, A., (2013). The effects of onion (*Allium cepa*) extract on doxorubicin-induced apoptosis in aortic endothelial cells. *Journal of Applied Toxicology, 33*, 364–369.

Asgari, M. M., Chren, M. M., Warton, E. M., Friedman, G. D., & White, E., (2011). Supplement use and risk of cutaneous squamous cell carcinoma. *Journal of the American Academy of Dermatology, 65*, 1145–1151.

Balsano, C., & Alisi, A., (2009). Antioxidant effects of natural bioactive compounds. *Current Pharmaceutical Design, 15*, 3063–3073.

Challacombe, J. F., (2021). *Metabolic Pathway Engineering: Analysis and Applications in the Life Sciences*. CRC Press.

Chen, Y., Xu, Y., Zhu, Y., & Li, X., (2013). Anticancer effects of ginsenoside compound k on pediatric acute myeloid leukemia cells. *Cancer Cell International, 13*, 1–6.

Chikwere, P., (2017). Functional foods and nutraceuticals, wonders in cancer risks – a review. *World Scientific News, 64*, 17–33.

Faria, A., Pestana, D., Teixeira, D., De Freitas, V., Mateus, N., & Calhau, C., (2010). Blueberry anthocyanins and pyruvic acid adducts: Anticancer properties in breast cancer cell lines. *Phytotherapy Research, 24*, 1862–1869.

Goyal, A., Sharma, V., Upadhyay, N., Gill, S., & Sihag, M., (2014). Flax and flaxseed oil: An ancient medicine & modern functional food. *Journal of Food Science and Technology, 51*, 1633–1653.

Gul, K., Singh, A., & Jabeen, R., (2016). Nutraceuticals and functional foods: The foods for the future world. *Critical Reviews in Food Science and Nutrition, 56*, 2617–2627.

Gul, K., Yousuf, B., Singh, A., Singh, P., & Wani, A. A., (2015). Rice bran: Nutritional values and its emerging potential for development of functional food—A review. *Bioactive Carbohydrates and Dietary Fiber, 6*, 24–30.

Hatcher, H., Planalp, R., Cho, J., Torti, F., & Torti, S., (2008). Curcumin: From ancient medicine to current clinical trials. *Cellular and Molecular Life Sciences, 65*, 1631–1652.

Huang, W. Y., Zhang, H. C., Liu, W. X., & Li, C. Y., (2012). Survey of antioxidant capacity and phenolic composition of blueberry, blackberry, and strawberry in Nanjing. *Journal of Zhejiang University Science B, 13*, 94–102.

Islam, S. R., & Siddiqua, T. J., (2020). Functional foods in cancer prevention and therapy: Recent epidemiological findings. *Functional Foods in Cancer Prevention and Therapy*, 405–433.

Ivanova, T., Krupodorova, T., Barshteyn, V., Artamonova, A., & Shlyakhovenko, V., (2014). Anticancer substances of mushroom origin. *Experimental Oncology.*

Jarouliya, U., & Keservani, R. K., (2019a). Pathways leading to child obesity. In: Debasis, B., (ed.), *Global Perspectives on Childhood Obesity: Current Status, Consequences and Prevention* (2nd edn., pp. 137–146). Academic Press, Elsevier, chapter 12. ISBN: 9780128128404.

Jarouliya, U., & Keservani, R. K., (2019b). Protein function as cell surface and nuclear receptor in human diseases. In: Sanjay, B., & Debarshi, K. M., (eds.), *Medicinal Chemistry with Pharmaceutical Product Development* (pp. 1–32). Apple Academic Press, CRC Press, Taylor & Francis Group. Chapter 1. ISBN: 9781771887106.

Jiang, J. W., Chen, X. M., Chen, X. H., & ZHENG, S. S., (2011). Ginsenoside Rg3 inhibits hepatocellular carcinoma growth via intrinsic apoptotic pathway. *World Journal of Gastroenterology: WJG, 17*, 3605.

Johnson, S. A., & Arjmandi, B. H., (2013). Evidence for anticancer properties of blueberries: A mini-review. *Anticancer Agents in Medicinal Chemistry (Formerly Current Medicinal Chemistry-Anti-Cancer Agents), 13*, 1142–1148.

Kabir, Y., (2020). *Functional Foods in Cancer Prevention and Therapy*. Academic Press.

Kaur, S., & Das, M., (2011). Functional foods: An overview. *Food Science and Biotechnology, 20*, 861–875.

Kazimierczak, R., Hallmann, E., Lipowski, J., Drela, N., Kowalik, A., Püssa, T., Matt, D., et al., (2014). Beetroot (*Beta vulgaris* L.) and naturally fermented beetroot juices from organic and conventional production: Metabolomics, antioxidant levels and anticancer activity. *Journal of the Science of Food and Agriculture, 94*, 2618–2629.

Keservani, R. K., & Sharma, A. K., (2014). Flavonoids: Emerging trends and potential health benefits. *Journal of Chinese Pharmaceutical Sciences, 23*(12), 815.

Keservani, R. K., Kesharwani, R. K., & Sharma, A. K., (2015c). Pulmonary and respiratory health: Antioxidants and nutraceuticals. In: Debasis, B., Harry, G. P., & Anand, S., (eds.), *Nutraceutical and Functional Foods in Human Life and Disease Prevention* (pp. 279–295). Boca Raton, Fl. CRC Press, Taylor and Francis, Chapter 17. ISBN: 9781482237214.

Keservani, R. K., Kesharwani, R. K., Sharma, A. K., & Jarouliya, U., (2015d). Dietary supplements, nutraceutical and functional foods in immune response (immunomodulators). In: Debasis, B., Harry, G. P., & Anand, S., (eds.), *Nutraceutical and Functional Foods in Human Life and Disease Prevention* (343–358). Boca Raton, FL. CRC Press, Taylor and Francis, Chapter 20. ISBN: 9781482237214.

Keservani, R. K., Kesharwani, R. K., Sharma, A. K., Vyas, N., & Chadoker, A., (2010b). Nutritional supplements: An overview. *International Journal of Current Pharmaceutical Review and Research, 1*(1), 59–75.

Keservani, R. K., Kesharwani, R. K., Vyas, N., Jain, S., Raghuvanshi, R., & Sharma, A. K., (2010a). Nutraceutical and functional food as future food: A review. *Der Pharmacia Lettre, 2*(1), 106–116.

Keservani, R. K., Sharma, A. K., & Jarouliya, U., (2015). Protein and peptide in drug targeting and its therapeutic approach. *Ars Pharmaceutica (Internet), 56*(3), 165–177.

Keservani, R. K., Sharma, A. K., & Kesharwani, R. K., (2016a). Nutraceutical and functional foods for cardiovascular health. In: *Food Process Engineering* (pp. 257–278, 291–312). Apple Academic Press, CRC Press. ISBN: 9781771884020.

Keservani, R. K., Sharma, A. K., & Kesharwani, R. K., (2016b). Medicinal effect of nutraceutical fruits for the cognition and brain health. *Scientifica*, 10. Article ID 3109254. doi: 10.1155/2016/3109254.

Keservani, R. K., Sharma, A. K., & Kesharwani, R. K., (2017). An overview and therapeutic applications of nutraceutical and functional foods. *Recent Advances in Drug Delivery Technology*, 160–201. ISBN:9781522507543.

Keservani, R. K., Sharma, A. K., & Kesharwani, R. K., (eds.). (2020). *Nutraceuticals and Dietary Supplements: Applications in Health Improvement and Disease Management*. CRC Press. ISBN: 9781771888738.

Kumar, K., & BARMANRAY, A., (2007). Nutritional evaluation and storage studies of button mushroom powder fortified biscuits. *Proteins, 96*, 325.

Kumar, K., (2015). Role of edible mushrooms as functional foods—A review. *South Asian J Food Technol Environ, 1*, 211–218.

Lai, R. M., (2012). Functional foods. *Jurnal Teknologi Pangan dan Gizi,* 1.

Lee, H. S., Kim, E. J., & Kim, S. H., (2011). Chestnut extract induces apoptosis in AGS human gastric cancer cells. *Nutrition Research and Practice, 5*, 185–191.

Liu, Z., Li, X., Simoneau, A. R., Jafari, M., & Zi, X., (2012). *Rhodiola rosea* extracts and salidroside decrease the growth of bladder cancer cell lines via inhibition of the mTOR pathway and induction of autophagy. *Molecular Carcinogenesis, 51*, 257–267.

Lyu, M., Wang, Y. F., Fan, G. W., Wang, X. Y., Xu, S. Y., & Zhu, Y., (2017). Balancing herbal medicine and functional food for prevention and treatment of cardiometabolic diseases through modulating gut microbiota. *Frontiers in Microbiology, 8*, 2146.

Ma, L., Chen, H., Dong, P., & Lu, X., (2013). Anti-inflammatory and anticancer activities of extracts and compounds from the mushroom *Inonotus obliquus*. *Food Chemistry, 139*, 503–508.

Maddi, V., Aragade, P., Digge, V., & Nitalikar, M., (2007). Phcog Rev.: Short review importance of nutraceuticals in health management. *Pharmacognosy Reviews,* 1.

Mccormick, D. L., RAO, K., Johnson, W. D., Bosland, M. C., Lubet, R. A., & Steele, V. E., (2010). Null activity of selenium and vitamin e as cancer chemopreventive agents in the rat prostate. *Cancer Prevention Research, 3*, 381–392.

Milner, J., (1999). Functional foods and health promotion. *The Journal of Nutrition, 129*, 1395S–1397S.

Parodi, P., (2007). A role for milk proteins and their peptides in cancer prevention. *Current Pharmaceutical Design, 13*, 813–828.

Prasad, S., & Tyagi, A. K., (2015). Ginger and its constituents: Role in prevention and treatment of gastrointestinal cancer. *Gastroenterology Research and Practice, 2015*.

Ramakrishnan, R., (2013). Anticancer properties of *Zingiber officinale* – Ginger: A review. *Int. J. Med. Pharm. Sci., 3*, 11–20.

Rane, B. R., Bharath, M. S., Patil, R. R., Keservani, R. K., & Jain, A. S., (2020b). Novel approaches in nutraceuticals. In: Rajesh, K. K., Raj, K. K., & Anil, K. S., (eds.), *Enhancing the Therapeutic Efficacy of Herbal Formulations Through Novel Drug Delivery Systems* (pp. 241–266). IGI Global International Publisher, Pennsylvania, USA, Chapter 11. ISBN13: 9781799844532.

Rane, B. R., Patil, A. S., Keservani, R. K., & Jain, A. S., (2020a). Novel approaches in herbal formulation. In: Rajesh, K. K., Raj, K. K., & Anil, K. S., (eds.), *Enhancing the Therapeutic Efficacy of Herbal Formulations Through Novel Drug Delivery Systems* (pp. 43–68). IGI Global International Publisher, Pennsylvania, USA, Chapter 2. ISBN13: 9781799844532.

Rane, B. R., Tadavi, S. A., & Keservani, R. K., (2020c). Naturopathy. In: Anil, K. S., Raj, K. K., & Surya, P. G., (eds.), *Herbal Product Development* (pp. 321–347). Apple Academic Press, CRC Press, Taylor & Francis Group, Chapter 12. ISBN: 9781771888776.

Riezzo, G., Chiloiro, M., & Russo, F., (2005). Functional foods: Salient features and clinical applications. *Current Drug Targets-Immune, Endocrine & Metabolic Disorders, 5*, 331–337.

Russo, g. L., Tedesco, I., Spagnuolo, C., & Russo, M., (2017). Antioxidant polyphenols in cancer treatment: Friend, foe or foil? *Seminars in Cancer Biology* (pp. 1–13). Elsevier.

Sanders, M. E., (1998). Overview of functional foods: Emphasis on probiotic bacteria. *International Dairy Journal, 8*, 341–347.

Shahidi, F., (2004). Functional foods: Their role in health promotion and disease prevention. *Journal of Food Science, 69*, R146–R149.

Singh, P., Kesharwani, R. K., & Keservani, R. K., (2017a). Antioxidants and vitamins: Roles in cellular function and metabolism. In: *Sustained Energy for Enhanced Human Functions and Activity* (pp. 385–407). Academic Press. ISBN:9780128054130.

Singh, P., Kesharwani, R. K., & Keservani, R. K., (2017b). Protein, carbohydrates, and fats: Energy metabolism. In: *Sustained Energy for Enhanced Human Functions and Activity* (pp. 103–115). Academic Press. ISBN:9780128054130.

Stuart, J. A., & Robb, E. L., (2013). *Bioactive Polyphenols from Wine Grapes*, Springer.

Thoppil, R. J., & Bishayee, A., (2011). Terpenoids as potential chemopreventive and therapeutic agents in liver cancer. *World Journal of Hepatology, 3*, 228.

Tu, S. H., Chen, L. C., & Ho, Y. S., (2017). An apple a day to prevent cancer formation: Reducing cancer risk with flavonoids. *Journal of Food and Drug Analysis, 25*, 119–124.

Wang, Y., Ausman, L. M., Greenberg, A. S., Russell, R. M., & Wang, X. D., (2010). Dietary lycopene and tomato extract supplementations inhibit nonalcoholic steatohepatitis-promoted hepatocarcinogenesis in rats. *International Journal of Cancer, 126*, 1788–1796.

Xiao, H., & Parkin, K. L., (2007). Isolation and identification of potential cancer chemopreventive agents from methanolic extracts of green onion (*Allium cepa*). *Phytochemistry, 68*, 1059–1067.

Yang, K. C., Tsai, C. Y., Wang, Y. J., Wei, P. L., Lee, C. H., Chen, J. H., Wu, C. H., & Ho, Y. S., (2009). Apple polyphenol phloretin potentiates the anticancer actions of paclitaxel through induction of apoptosis in human hep G2 cells. *Molecular Carcinogenesis* (Vol. 48, pp. 420–431). Published in cooperation with the University of Texas MD Anderson Cancer Center.

Yoon, H., & Liu, R. H., (2007). Effect of selected phytochemicals and apple extracts on NF-κB activation in human breast cancer MCF-7 cells. *Journal of Agricultural and Food Chemistry, 55*, 3167–3173.

Zein, H., Hashish, A., & Ismaiel, G., (2015). The antioxidant and anticancer activities of Swiss chard and red beetroot leaves. *Current Science International, 4*, 491–498.

CHAPTER 4

NOVEL APPLICATIONS OF FUNCTIONAL FOODS IN BRAIN HEALTH

VAISHALI N. PARDESHI,[1] TUSHAR LOKHANDE,[2] SUNIL K. MAHAJAN,[2] and KHEMCHAND R. SURANA[3]

[1]*Department of Pharmaceutical Chemistry, Abhinav Education Society's College of Pharmacy, Pune, Maharashtra, India*

[2]*Department of Pharmaceutical Chemistry, Mahatma Gandhi Vidyamandirs Pharmacy College Panchavati, Nashik, Maharashtra, India*

[3]*Department of Pharmaceutical Chemistry, Shreeshakti Shaikshaint Sanstha, Divine College of Pharmacy, Satana, Nashik, Maharashtra, India*

ABSTRACT

Functional foods are highly nutritious and related to a number of powerful health benefits. They protect against diseases prevent nutrient deficiency and proper growth and development. These are ingredients designed to enhance health. They provide health benefits that reach beyond nutritional value. Vitamins, minerals, probiotics, fiber, and nutrient-dense components like fruits, vegetables, nuts, seeds, and grains are all found in functional foods. Brain health isn't independent of the body's general health status, and every measure that causes honest and maintained overall physical health also will be useful for the brain. Normally, overeating and adhering to a high-saturated-fat Western diet, processed carbs, and high caloric density behavior, which

Applications of Functional Foods in Disease Prevention. Raj K. Keservani & Eknath D. Ahire, PhD (Eds.)

results in lifestyle diseases, is additionally dangerous for intellectuality. Functional foods play a significant role in brain function to enhance memory, mood regulation, diseases, anxiety and depression. So functional food are developed to reinforce brain function claim to stop dementia and improve cognition focus and mood. The general makeup of the human diet, as well as individual dietary components, has been demonstrated to influence brain function. Many individuals use nutraceuticals and supplements in the hopes of increasing alertness or slowing down cognitive deterioration. There has been a significant rise in the quantity of research studying the relationships between nutrition, nutraceuticals, and psychological function over the last decade or two. A variety of sources have indicated cognitive improvements as a result of this. Glucose injection, for example, increases cognitive performance, and the processes behind this benefit are becoming more understood. A food's glycemic index (GI) might potentially affect mental function. This chapter provides a summary of the role of functional foods (diet) in brain development, signaling network, and neurotransmitter within the brain and cognition memory.

4.1 INTRODUCTION

During the last decades, the demand for health-improving foods and beverages has increased in parallel with the rising costs of health care, the rise in anticipation, and therefore the desire for higher life quality. During this context, functional foods offer a replacement quite health message, promising specific effects by means of particular food components. an outsized number of definitions exist worldwide for 'functional foods,' pertaining to (processed) foods containing ingredients that beneficially influence specific body functions, additionally to being nutritious. However, we are handling an idea instead of a well-defined group of food products and an universally accepted definition is lacking (Ozen et al., 2015). The individual genotype alongside a variety of environmental factors like nutrition, parental care, social interactions, stress, and diseases, form the idea for the event and functional properties of the brain. The connection between diet and psychological state is becoming increasingly recognized and nutrition has become a crucial ancillary consideration as neighborhood of the treatment of psychiatric disorders. This has formed the idea for the concept of "brain foods"—foods that will help optimize mental performance. A functional food is often a natural food, or it can contain one or more specific components, which have a functional influence on the health and well-being of the buyer. Such components are often to food (e.g., omega-3 fatty acids), faraway from food (e.g., fat from milk), naturally enhanced through special conditions (e.g., selenium-enriched garlic), or modified to supply health benefits (e.g., phytosterol-added probiotics) (Caballero et al., 2003). Importantly, functional foods don't necessarily induces a health benefit altogether members of the population. Overall, functional foods are often confused with the subsequent related concepts.

4.1.1 NUTRACEUTICALS

Products made up of a food but sold as pill, powder, or other dosage forms not usually related to food and shown to possess beneficial or protective physiological properties against chronic diseases (Keservani et al., 2010a, b, 2020).

4.1.2 PHARMA FOODS

A universally accepted term to designate products on the food–medicine borderline, which first emerged in Japan; they're also called 'border foods' as they claim drug-like preventive and therapeutic qualities. For a long time, scientists have recognized that there is a clear link between the food we eat and our health. The current notion of functional foods arose from the progressive awareness that consuming nutritious foods results in a healthy diet, as well as the discovery of the processes by which foods affect metabolism and health. Our initial and most basic goal after eating food is to obtain nutrients for our bodies and to meet our metabolic demands (Ekstrand et al., 2021). However, certain types of foods provide significant health benefits in addition to their nutritional features. These foods are known as functional foods, and they should be defined as any food that improves a person's health, physical performance, or mental condition. These sorts of foods are called functional foods and should be defined as any food that features a positive impact on a person's health, physical performance, or state of mind, additionally to its nutritious value. Other factors, including three additional characteristics in particular, are recommended to define a functional food. It's a food made from natural components (not a pill, tablet, or powder). It can and will be consumed as a part of the daily diet. When consumed, it has a special purpose, serving to manage a specific bodily process, such as: enhancement of biological defense mechanisms, prevention of specific disease, getting well from certain disorders of the physical and mental nature are managed. The aging process is being slowed. Nutraceuticals, medicinal foods, probiotics, designer foods, and pharmaceuticals are all examples of functional foods. Foods, and Vita foods (Bedrosian, et al., 2018; Gandal et al., 2018).

4.2 POTENTIAL HEALTH AREAS OF INTEREST FOR FUNCTIONAL FOOD CONSUMPTION HUMAN PHYSIOLOGY AREAS

Early development and growth, modulation of the immune system are all areas of human physiology that are important to functional foods, basic metabolic processes, defense against oxidative stress, cardiovascular physiology, gastrointestinal physiology, cognitive and mental performance, and physical performance and fitness (Spencer et al., 2017).

4.3 INFLUENCE OF FUNCTIONAL FOOD ON BRAIN HEATH

The notion for the event and functional features of the brain is formed by the individual genotype in combination with a number of environmental circumstances including as nutrition, parental care, social interactions, stress, and illnesses (Sarris et al., 2015; Lanni et al., 2008). The connection between diet and psychological state is becoming increasingly recognized and nutrition has become a crucial ancillary consideration as a neighborhood of the treatment of psychiatric disorders (Donato et al., 2017). This has formed the idea for the concept of "brain foods"—foods that will help optimize mental performance (Paolicelli et al., 2011). We don't present here a scientific summary of particular results from dietary intervention trials in relation to brain development and health but rather a narrative review, in which we talk about the function of nutrition in the brain neurophysiology that are associated with psychological state and performance (Hong and Stevens, 2016; Figure 4.1).

4.4 IMPACT OF THE INFLUENCE OF DIET ON BRAIN DEVELOPMENT

Precursor cells develop into neural and glial cells and move to their ultimate sites throughout the initial stages of brain development. (Eriksson et al., 1998). The brain overproduces neurons and neural connections at this

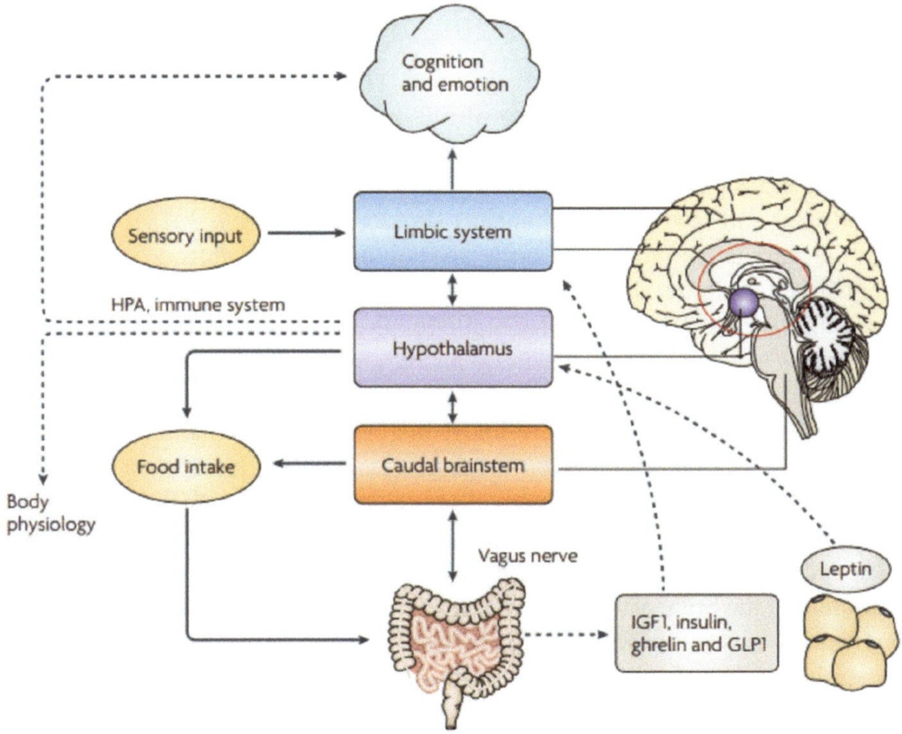

FIGURE 4.1 Effect of functional food in cognition.

time and therefore the peak synapse formation is reached between one and a couple of years aged. Redundant cells and synapses are removed by microglia during a process called synaptic pruning (Gage, 2002). The final stage in fetal brain development is that the formation of myelin sheaths, which creates an insulating layer that allows fast signal transduction. Even in adulthood, stem cells in the central nervous system remain active, developing into neural or glial cells that move into the cerebral white or gray matter (Bourre, 2006; Pilz et al., 2018). It's critical that each of the needed energy and nutrients be often absorbed from the food during the prenatal and perinatal periods of brain development. The mother's diet during pregnancy already is vital for the longer-term brain function of her child (Keunen et al., 2015). Malnutrition during gestation and infancy negatively influences the development of the brain (Nyaradi, et al., 2013; Surana et al., 2021).

4.5 THE FUNCTION OF LIPIDS IN BRAIN STRUCTURE AND PERFORMANCE

Lipids make up about a quarter of the dry weight of the human brain and play a role in vesicle formation, ion flux control, and other functions like unique microenvironments for cell interactions, as well as distinct signaling pathways and feedback mechanisms (Prado, and Dewey, 2014). About 50% of the fatty acids within the brain are polyunsaturated fatty acids (PUFAs), either arachidonic acid (20:4n–6) oromega-3 fatty acid (DHA; 22:6n–3) (Piomelli, et al., 2007). DHA regulates the function of the glutamatergic synapses associated with plasticity and cognitive ability, improves neuronal differentiation (Langelier et al., 2010), and influences organic phenomenon within the brain (Su, 2010). Almost majority of the arachidonic acid and DHA in the brain comes from food, however astrocytes appear to generate some of it from n-6 and n-3 precursors (Robson, 2010). PUFA consumption by the mother has an effect on cell proliferation in the fetus' and neonate's hippocampus (Kitajka, et al., 2004). According to animal research, the lipid content of the brain reflects nutritional consumption. And plasma lipid profile (Urquiza et al., 2000). Phosphatidylserine is a polar lipid found in neuronal membranes that is vital for the transmission and reception of impulses in the brain. Phosphatidylserine supplementation helps to avoid cognitive decline and enhance memory (Williard et al., 2001). Gangliosides are a kind of glycosphingolipid, is particularly abundant within the brain (Tang et al., 2016). Cell-cell recognition, cell adhesion, motility, and proliferation are all mediated by gangliosides. Ganglioside levels are greatest in colostrum and diminish during the transition to mature milk in both human and bovine milk (Giles et al., 2016). Complex milk lipid supplementation improves synaptic plasticity and supports cognitive development in babies (Glade and Smith, 2015). Sialic acid, which is found in gangliosides, is an important structural component in brain growth, learning, and memory (Palmano et al., 2015).

Early infancy supplementation with human milk oligosaccharides as prebiotics may help to build a good gut microbial flora, which may benefit brain development. (Todeschini and Hakomori, 2008).

4.6 BRAIN DEVELOPMENT AND MICRONUTRIENTS

Many micronutrients, such as vitamins and trace minerals, are critical for brain development in the early stages (Martín-Sosa, et al., 2004). B-complex vitamin deficiency has been implicated during a number of mental disorders, and therefore the B vitamins B6, folate (B9), and B12 are the foremost widely studied in reference to brain development (Schnaar, et al., 2014).

Several vitamins and related chemicals, including as folate, vitamin B12, choline, and glycine betaine, are linked to homocysteine levels, suggesting that 1-carbon metabolism is important in brain development, including phospholipid production and DNA methylation (Gurnida et al., 2012). Because of the existence of vitamin D-specific receptors inside the brain and its influence on brain development, vitamin D has been dubbed "the overlooked neurosteroid." It plays a well-known role in calcium absorption in the intestine (Guillermo, et al., 2015), as well as a critical part in synaptic transmission (Wang, 2012). However, because excessive calcium and vitamin D consumption is linked to brain lesions, there is a delicate equilibrium. Iron is an important micronutrient involved in neurodevelopment. It's important for normal brain development, myelination, and neurotransmission (Elison, et al., 2016). Zinc is important for neurogenesis, neuronal migration, synapse genesis, myelination, and regulation of intra- and intercellular signaling throughout development. Severe deficiency illness leads to significant structural abnormalities in the brain (Williams, et al., 2016.). Iodine is essential for the manufacture of thyroid hormones, which are necessary for appropriate neuronal migration and myelination of the brain throughout the fetal and early postnatal periods; iodine shortage ("cretinism") causes permanent brain damage. According to some findings, even mild to severe maternal iodine insufficiency produces cognitive or neurological damage in the kids (Carvajal-Millan et al., 2021; Mattei and Pietrobelli, 2019).

4.7 DIET AND DEPRESSION

Depressive disorders are the foremost common psychological state problem. They incur considerable burden not just for individuals but also for the society. Subclinical signs and symptoms of sadness and anxiety are quite common in the general population, even among those who do not have clinically diagnosed mental illnesses. As a result, innovative supplementary approaches to treat depression and anxiety symptoms in the general population are welcomed. New data reveals that nutrition may have a role in the emergence of mood disorders. Dietary therapies, such as calorie-restricted diets with higher quantities of fruits, vegetables, fatty fish, whole-grain cereals, nuts, and seeds, were found to significantly improve depressed symptoms in a recent meta-analysis of 16 relevant randomized controlled trials (Moretti and Caruso, 2019). It's also been claimed that getting enough n-3 PUFAs helps with mood disorders, such as unipolar and bipolar depression. The content of n-3 PUFA in the blood has an inverse connection with the risk of psychological discomfort (Eyles, et al., 2013). Several substances within the diet has both direct and indirect impacts on depression-related pathways on pathways linked to depression. Several substances within the diet have direct and indirect effects on pathways linked to depression.as an example, b-carboline alkaloids that are synthesized endogenously within the body, functioning because the brain's own benzodiazepine analogs, also are reported to be cosmopolitan in many plant species. Epigallocatechin gallate was discovered by Vignes et al. (EGCG) from tea can induce anxiolytic activity during a dose-dependent manner (Christakos et al., 2011).

4.8 SLEEP AND WAKEFULNESS

As the brain is active in the least times of the day and night, sleep may be avital physiological process with restorative functions. GABA controls the inhibitory/excitatory balance required for proper brain activity in conjunction with the excitatory NT glutamate. Melatonin (N-acetyl-5-methoxytryptamine) may be a compound that has many important functions within the body (Mauceri, et al., 2015) and also plays an important function within the sleep wake cycle. Melatonin levels rise at night, producing tiredness and a drop in blood temperature. Melatonin and its precursor, tryptophan, are found in a variety of foods, including animal sources such as meat, fish, eggs, and milk, as well as plant foods such as cereals, vegetables, and fruit; hence, the diet may have an impact on melatonin levels in the brain. (Payne et al., 2008).

4.9 PROTECTION FROM NEUROINFLAMMATORY PROCESSES BY THE DIET

Acute neuroinflammation may act as a protective mechanism, isolating wounded brain tissue from unaffected parts, destroying harmed cells, and rebuilding the extracellular matrix. Chronic neuroinflammation damages brain tissue over a period of time when inflammatory processes are active (Lozoff et al., 2006). Long-term concern can exacerbate inflammatory processes, raise the likelihood of age-related brain illnesses, and cause cognitive deficiencies. Oxidative stress is a crucial think about age related tissue damage within the brain. The nucleus Niger's dopaminergic neurons have iron-rich melanin granules that inhibit the loci's noradrenergic neurons coeruleus (Lopez et al., 2016). Through oxidative stress, increased iron buildup can impact brain aging. In rodent models, specific types of monosaccharides in the diet alter brain growth or function. In rats, high fructose diet resulted drop in neurogenesis and neuronal loss within the nucleus tractus solitarius, the most region within the brain for vascular regulation (Gower-Winter, and Levenson, 2012).Several naturally occurring chemicals in meals have anti-inflammatory and neuroprotective effects. Sulforaphane, which is found in cruciferous vegetables, is thought to activate anti-inflammatory genes in astrocytes and neurons. The components salidroside, rosavins, and p-tyrosol found in Arctic root (Rhodiolarosea) have been shown to reduce oxidative stress, neuroinflammation, and excitotoxicity. Citrus flavonoids (e.g., nobiletin, tangerine) are neuroprotective in Alzheimer's disease and Parkinson's disease (Delange, 2001).

4.10 FUNCTIONAL FOOD IN BRAIN HEALTH

The following are some of the most popular functional food components for improving mood, attention, and brain health. Many of the benefit claims have been rephrased by manufacturers as structure-function claims, which the FDA has yet to assess.

4.10.1 ASHWAGANDHA

Ashwagandha (*Withania somnifera*), a plant belonging to the nightshade family, is an important component of Ayurvedic medicine. Ashwagandha is known as the "Prince of Herbs." is touted as an apoptogenic herb that aids in the body's adaptive adaptation to a variety of stressful situations, reduces stress and anxiety, and improves memory and cognition. Recent studies have reviewed the neuroprotective compounds in ashwagandha sitoindosides, withaferin A, withanosides IV, withanols, and withanolides. The potential use of anaferine and beta-sitosterol in treating brain diseases is noted, and further research is being done. When compared to a placebo, a concentrated ashwagandha extract was demonstrated to reduce a test score reflecting the intensity of general anxiety symptoms. Ashwagandha is commonly found in teas, coffees, and smoothie powders.(Vermiglio, et al., 2004).

4.10.2 GINKGO

Ginkgo (*Ginkgo biloba*), additionally known as the ginkgo, is one among the oldest tree species in the world. Ginkgo nuts have an extended Chinese medicine has a long history of using it to treat cognitive impairment and other areas of brain health hasn't been scientifically validated. The Ginkgo Memory study, a randomized, double-blinded, placebo-controlled clinical trial including 3,069 older persons monitored for around six years, looked into a variety of ginkgo advantages.

4.10.3 MELATONIN

Melatonin is a hormone that regulates the body's circadian rhythms and sleep cycles. Melatonin has a number of brain-related advantages, including tiredness relief, correction of some sleep-wake problems, and anxiety reduction prior to surgery and hence the sedative zolpidem (Ambien) had similar effects on patients' views of sleep efficacy, sleep disruption, and side effects in a sample of hospitalized patients. Studies on the advantages of melatonin supplementation are still equivocal due to differences in technique and analysis. Melatonin has been added to a few foods and beverages, including chocolates, drinks, and ready-to-drink lattes, although it is more generally accessible as a supplement (Reiter, et al., 2001).

4.10.4 MUSHROOMS

In eastern, western, and indigenous civilizations, mushrooms have a long history of therapeutic usage. Cordyceps, lion's mane, chaga, and reishi mushrooms are being advocated separately or in combination for immune support and brain-related advantages. Memory, cognitive function, and mood are all said to be improved by lion's mane. Functional mushroom research has thus far mostly been done in vitro or on laboratory animals, however it appears that mushrooms may help with cognitive function. More investigation is necessary. Functional mushrooms are commonly used in sports powders with nootropic advantages, protein powders, drinks, and snack bars, in addition to supplements (Freeman, et al., 2006).

4.10.5 OMEGA-3 FATTY ACIDS

The omega-3 fatty acid, because they are of phospholipids components that are found in the brain, EPA and DHA are frequently encouraged to maintain brain health and prevent cognitive decline. According to a summary of studies from the National Institutes of Health Office of Dietary Supplements on consumption of omega-3 as supplements and in fish, omega-3 supplementation hasn't been shown to improve cognitive function in healthy older adults or people with Alzheimer's disease when compared to placebo. However, in persons with moderate cognitive impairment, omega-3 supplementation may enhance various elements of cognitive function, such as attention, processing speed, and recollection, awaiting additional study. Numerous research show that omega-3 fatty acids, especially fish containing omega 3s, can help with depression, however, a 2019 review found insufficient evidence. Omega-3 fatty acids can be found naturally in foods such as fish and walnuts, but they are frequently added to fortify meals and beverages such as eggs, milk, and other animal-based foods via fortified feed (Liperoti, et al., 2009).

4.10.6 PREBIOTICS/PROBIOTICS

According to research on the gut-brain axis, enhancing gut health through the use of prebiotic fibers and probiotics has a good influence on the central nervous system; brain diseases like MS, Parkinson's disease, and Alzheimer's disease; also as mood and anxiety. Signaling between the gut and therefore the brain is assumed to take place straight through the enteric system nervosum, gut bacteria metabolites, the system, and neurochemical signaling (Rabinovitz, 2014). Both inflammation and mood are thought to be linked to gut health. We know that having a bad gut causes inflammation in the body, and inflammation has been linked to anxiety and depression (Vignes et al., 2006). "A gut-healthy diet rich in prebiotic-rich foods, including both prebiotic soluble fibers and resistant starches, can sustain the varied range of bacteria naturally found within the gut, helping to scale down inflammation and so playing a role in mood regulation," Landau explains.

4.10.7 L-THEANINE

L-theanine, also known as gamma ethylamino-L-glutamic acid and found in the leaves of the tea plant (*Camillia sinensis*), is increasingly being recommended for improving sleep, cognition, and mental alertness, as well as reducing anxiety. L-theanine is also touted to induce relaxation without causing drowsiness. Richard emphasizes the need of key dietary components for a healthy gut because the microbiome's wellbeing is so critical for serotonin generation and feeding. Probiotics, fruit and vegetable polyphenols, fermented foods, bacteria with soluble (prebiotic) and insoluble fiber, and fermented meals are all beneficial to the body (Wu and Sun, 2015). L-theanine may aid and increase vigilance, but its other effects haven't yet been proven (Blasko, et al., 2004).

4.11 DIET AND EPIGENETIC

Individual heterogeneity in various protein-coding (Keservani et al., 2015; Jarouliya and Keservani, 2019b) and noncoding areas of the genome is claimed to influence a person's psychological state. Throughout life, epigenetic modulations influence organic phenomenonas a result of endogenous and external regulatory factors. Recently published data on organic phenomenon in reference to brain development and neuropsychiatric disorders provide an in depth publicly available genomic resource. Exons make up about 1.5% of the human genome and 98.5% of noncoding DNA. The noncoding regions of the genome contain the majority of genetic diversity, and mutations in these areas are crucial for brain function and diseases. Early nutrition and stress, for example, have an impact on the serotonergic system, creating epigenetic changes that modify expression and have been related to bipolar disorders and depression later in life (Sandi, and Touyarot, 2006). Methylation of nucleotides, acetylation, phosphorylation, and other posttranslational modifications of histones, microRNA production from noncoding DNA, and changes in the topological layout of three-dimensional DNA structures are all epigenetic processes mediate many of the links between diet and brain health. The mother's eating habits throughout the first trimester of pregnancy, as well as the neonate's early nutrition, can have both permanent and reversible effects on epigenetic programming. Transgenerational consequences might result from a lack of necessary nutrients or a widespread famine in a community. Obesity (Jarouliya and Keservani, 2019a) and a high-fat diet later in life can have an impact on epigenetic programming. Biotin may be a substrate for histone biotinylating, histone adenosine diphosphate ribosylation is aided by niacin. Pantothenic acid may be a component of coenzyme A, which is involved in the transfer of acetyl groups in acetylation. DNA demethylation and histone modification are influenced by genistein from soy and tea catechins, whereas histone acetylation is influenced by resveratrol (red wine), sulforaphane (broccoli), butyrate, diallyl sulfide (garlic), and curcumin (turmeric). The fruit tree (*Olea europaea*) produces chemicals that protect the brain, partly via epigenetic modifications (e.g., oleuropein, tyrosol, and hydroxytyrosol) (Gao, et al., 2006).

4.12 CONCLUSION

Looking at the general connection between nutrition and brain development and performance, the straight forward rule is that "what is sweet for your heart is sweet for your brain." Given the dangers of insulin resistance, an overall healthy diet from a cardiovascular (Keservani et al., 2016a) and weight management perspective will also be helpful for the brain. Early nutrition—the diet of the pregnant mother and therefore the reform the nutritional status of the neonate and the young infant—influences the event of the brain, additional long-term epigenetic processes and other environmental variables. From the perspective of the important balance between the formation of synapses and neurons, and thus the necessary synapses are pruned during childhood and adolescence, the question remains whether one-sided promotion of neuronal growth will be beneficial for brain performance or if it will have unintended negative consequences (Keservani et al., 2016b, 2017). Various NTs and networks between specific centers within the brain control mood and behavior, such as sadness, sleep and wakefulness, pain, and so on, and may be altered by food components that either modify the quantity and effects of active NTs or interfere with relevant receptors. To summarize, the results of several studies support the concept of brain food and therefore the notion that diet has a crucial effect on aspects of brain function, but conclusive evidence remains lacking in many cases. Indeed, alertness and mood are often modulated by the daily diet. Indeed, alertness and mood are often modulated by the daily diet. Although severe malnutrition remains a substantial obstacle to neurocognitive development, it is obvious that people with poor diets may nonetheless operate properly, even at a high level of cognitive demand. Improving vitamin status can help students learn better, especially if they don't have easy access to healthful meals.

ACKNOWLEDGMENT

We thank the MGVs, College of Pharmacy, Panchavati, affiliated under Savitribai Phule Pune University, Nashik.

KEYWORDS

- **functional foods**
- **brain health**
- **cognition**
- **memory**
- **dietary supplement**

REFERENCES

Bedrosian, T. A., Quayle, C., Novaresi, N., & Gage, F. H., (2018). Early life experience drives structural variation of neural genomes in mice. *Science, 359*(6382), 1395–1399.

Benton, D., (2008). The influence of children's diet on their cognition and behavior. *European Journal of Nutrition, 47*(3), 25–37.

Blasko, I., Stampfer-Kountchev, M., Robatscher, P., Veerhuis, R., Eikelenboom, P., & Grubeck-Loebenstein, B., (2004). How chronic inflammation can affect the brain and support the development of Alzheimer's disease in old age: The role of microglia and astrocytes. *Aging Cell, 3*(4), 169–176.

Bourre, J. M., (2006). Effects of nutrients (in food) on the structure and function of the nervous system: Update on dietary requirements for brain. Part 1: Micronutrients. *Journal of Nutrition Health and Aging, 10*(5), 377.

Caballero, B., Trugo, L. C., & Finglas, P. M., (2003). *Encyclopedia of Food Sciences and Nutrition*. Academic.

Carvajal-Millan, E., Yaser, A. Z., & Haghi, A., (2021). *Natural Food Products and Waste Recovery: Healthy Foods, Nutrition Design, and Extraction of Valuable Compounds*. CRC Press.

Christakos, S., Dhawan, P., Porta, A., Mady, L. J., & Seth, T., (2011). Vitamin D and intestinal calcium absorption. *Molecular and Cellular Endocrinology, 347*(1, 2), 25–29.

Delange, F., (2001). Iodine deficiency as a cause of brain damage. *Postgraduate Medical Journal, 77*(906), 217–220.

Donato, F., Jacobsen, R. I., Moser, M. B., & Moser, E. I., (2017). Stellate cells drive maturation of the entorhinal-hippocampal circuit. *Science, 355*(6330), eaai8178.

Ekstrand, B., Scheers, N., Rasmussen, M. K., Young, J. F., Ross, A. B., & Landberg, R., (2021). Brain foods-the role of diet in brain performance and health. *Nutrition Reviews, 79*(6), 693–708.

Elison, E., Vigsnaes, L. K., Krogsgaard, L. R., Rasmussen, J., Sørensen, N., McConnell, B., Hennet, T., et al., (2016). Oral supplementation of healthy adults with 2′-O-fucosyllactose and lacto-N-neotetraose is well tolerated and shifts the intestinal microbiota. *British Journal of Nutrition, 116*(8), 1356–1368.

Eriksson, P. S., Perfilieva, E., Björk-Eriksson, T., Alborn, A. M., Nordborg, C., Peterson, D. A., & Gage, F. H., (1998). Neurogenesis in the adult human hippocampus. *Nature Medicine, 4*(11), 1313–1317.

Eyles, D. W., Burne, T. H., & McGrath, J. J., (2013). Vitamin D, effects on brain development, adult brain function and the links between low levels of vitamin D and neuropsychiatric disease. *Frontiers in Neuroendocrinology, 34*(1), 47–64.

Firth, J., Marx, W., Dash, S., Carney, R., Teasdale, S. B., Solmi, M., Stubbs, B., et al., (2019). The effects of dietary improvement on symptoms of depression and anxiety: A meta-analysis of randomized controlled trials. *Psychosomatic Medicine, 81*(3), 265.

Freeman, M. P., Hibbeln, J. R., Wisner, K. L., Davis, J. M., Mischoulon, D., Peet, M., Keck, Jr. P. E., et al., (2006). Omega-3 fatty acids: Evidence basis for treatment and future research in psychiatry. *Journal of Clinical Psychiatry, 67*(12), 1954.

Gage, F. H., (2002). Neurogenesis in the adult brain. *Journal of Neuroscience, 22*(3), 612, 613.

Gandal, M. J., Haney, J. R., Parikshak, N. N., Leppa, V., Ramaswami, G., Hartl, C., Schork, A. J., et al., (2018). Shared molecular neuropathology across major psychiatric disorders parallels polygenic overlap. *Science, 359*(6376), 693–697.

Gao, Y., Bezchlibnyk, Y. B., Sun, X., Wang, J. F., McEwen, B. S., & Young, L. T., (2006). Effects of restraint stress on the expression of proteins involved in synaptic vesicle exocytosis in the hippocampus. *Neuroscience, 141*(3), 1139–1148.

Giles, C., Takechi, R., Mellett, N. A., Meikle, P. J., Dhaliwal, S., & Mamo, J. C., (2016). The effects of long-term saturated fat enriched diets on the brain lipidome. *PloS One, 11*(12), e0166964.

Glade, M. J., & Smith, K., (2015). Phosphatidylserine and the human brain. *Nutrition, 31*(6), 781–786.

Gower-Winter, S. D., & Levenson, C. W., (2012). Zinc in the central nervous system: From molecules to behavior. *Biofactors, 38*(3), 186–193.

Guillermo, R. B., Yang, P., Vickers, M. H., McJarrow, P., & Guan, J., (2015). Supplementation with complex milk lipids during brain development promotes neuroplasticity without altering myelination or vascular density. *Food & Nutrition Research, 59*(1), 25765.

Gurnida, D. A., Rowan, A. M., Idjradinata, P., Muchtadi, D., & Sekarwana, N., (2012). Association of complex lipids containing gangliosides with cognitive development of 6-month-old infants. *Early Human Development, 88*(8), 595–601.

Hong, S., & Stevens, B., (2016). Microglia: Phagocytosing to clear, sculpt, and eliminate. *Developmental Cell, 38*(2), 126–128.

Jarouliya, U., & Keservani, R. K., (2019a). Pathways leading to child obesity. In: Debasis, B., (ed.), *Global Perspectives on Childhood Obesity: Current Status, Consequences and Prevention* (2nd edn., pp. 137–146). Academic Press, Elsevier, chapter 12. ISBN: 9780128128404.

Jarouliya, U., & Keservani, R. K., (2019b). Protein function as cell surface and nuclear receptor in human diseases. In: Sanjay, B., & Debarshi, K. M., (eds.), *Apple Medicinal Chemistry with Pharmaceutical Product Development* (pp. 1–32). Academic Press, CRC Press, Taylor & Francis Group. Chapter 1. ISBN: 9781771887106.

Keservani, R. K., & Sharma, A. K., (2014). Flavonoids: Emerging trends and potential health benefits. *Journal of Chinese Pharmaceutical Sciences, 23*(12), 815.

Keservani, R. K., Kesharwani, R. K., Sharma, A. K., Vyas, N., & Chadoker, A., (2010b). Nutritional supplements: An overview. International *Journal of Current Pharmaceutical Review and Research, 1*(1), 59–75.

Keservani, R. K., Kesharwani, R. K., Vyas, N., Jain, S., Raghuvanshi, R., & Sharma, A. K., (2010a). Nutraceutical and functional food as future food: A review. *Der Pharmacia Lettre, 2*(1), 106–116.

Keservani, R. K., Sharma, A. K., & Jarouliya, U., (2015). Protein and peptide in drug targeting and its therapeutic approach. *Ars Pharmaceutica (Internet), 56*(3), 165–177.

Keservani, R. K., Sharma, A. K., & Kesharwani, R. K., (2016a). Nutraceutical and functional foods for cardiovascular health. In: *Food Process Engineering* (pp. 257–278, 291–312). Apple Academic Press, CRC Press. ISBN: 9781771884020.

Keservani, R. K., Sharma, A. K., & Kesharwani, R. K., (2016b). Medicinal effect of nutraceutical fruits for the cognition and brain health. *Scientifica*, 10. Article ID 3109254. doi: 10.1155/2016/3109254.

Keservani, R. K., Sharma, A. K., & Kesharwani, R. K., (2017). An overview and therapeutic applications of nutraceutical and functional foods. *Recent Advances in Drug Delivery Technology*, 160–201. ISBN:9781522507543.

Keservani, R. K., Sharma, A. K., & Kesharwani, R. K., (2020). "*Nutraceuticals and Dietary Supplements: Applications in Health Improvement and Disease Management*. CRC Press. ISBN: 9781771888738.

Keunen, K., Van, E R. M., Van, B. F., & Benders, M. J., (2015). Impact of nutrition on brain development and its neuroprotective implications following preterm birth. *Pediatric Research, 77*(1), 148–155.

Kitajka, K., Sinclair, A. J., Weisinger, R. S., Weisinger, H. S., Mathai, M., Jayasooriya, A. P., Halver, J. E., & Puskás, L. G., (2004). Effects of dietary omega-3 polyunsaturated fatty acids on brain gene expression. *Proceedings of the National Academy of Sciences, 101*(30), 10931–10936.

Langelier, B., Linard, A., Bordat, C., Lavialle, M., & Heberden, C., (2010). Long chain-polyunsaturated fatty acids modulate membrane phospholipid composition and protein localization in lipid rafts of neural stem cell cultures. *Journal of Cellular Biochemistry, 110*(6), 1356–1364.

Lanni, C., Lenzken, S. C., Pascale, A., Del, V. I., Racchi, M., Pistoia, F., & Govoni, S., (2008). Cognition enhancers between treating and doping the mind. *Pharmacological Research, 57*(3), 196–213.

Liperoti, R., Landi, F., Fusco, O., Bernabei, R., & Onder, G., (2009). Omega-3 polyunsaturated fatty acids and depression: A review of the evidence. *Current Pharmaceutical Design, 15*(36), 4165–4172.

Lopez, A., Cacoub, P., Macdougall, I. C., & Peyrin-Biroulet, L., (2016). Iron deficiency anaemia. *The Lancet, 387*(10021), 907–916.

Lozoff, B., Beard, J., Connor, J., Felt, B., Georgieff, M., & Schallert, T., (2006). Long-lasting neural and behavioral effects of iron deficiency in infancy. *Nutrition Reviews, 64*(suppl_2), S34–S43.

Martín-Sosa, S., Martín, M. J., Castro, M. D., Cabezas, J. A., & Hueso, P., (2004). Lactational changes in the fatty acid composition of human milk gangliosides. *Lipids, 39*(2), 111–116.

Mattei, D., & Pietrobelli, A., (2019). Micronutrients and brain development. *Curr. Nutr. Rep., 8*(2), 99–107.

Mauceri, D., Hagenston, A. M., Schramm, K., Weiss, U., & Bading, H., (2015). Nuclear calcium buffering capacity shapes neuronal architecture. *Journal of Biological Chemistry, 290*(38), 23039–23049.

Moretti, R., & Caruso, P., (2019). The controversial role of homocysteine in neurology: From labs to clinical practice. *International Journal of Molecular Sciences, 20*(1), 231.

Nyaradi, A., Li, J., Hickling, S., Foster, J., & Oddy, W. H., (2013). The role of nutrition in children's neurocognitive development, from pregnancy through childhood. *Frontiers in Human Neuroscience, 7*, 97.

Ozen, A. E., Del Mar, B. M., Murcia, M. A., Pons, A., & Tur, J. A., (2015). Adherence to the Mediterranean diet and consumption of functional foods among the Balearic Islands' adolescent population. *Public Health Nutrition, 18*(4), 659–668.

Palmano, K., Rowan, A., Guillermo, R., Guan, J., & McJarrow, P., (2015). The role of gangliosides in neurodevelopment. *Nutrients, 7*(5), 3891–3913.

Paolicelli, R. C., Bolasco, G., Pagani, F., Maggi, L., Scianni, M., Panzanelli, P., Giustetto, M., et al., (2011). Synaptic pruning by microglia is necessary for normal brain development. *Science, 333*(6048), 1456–1458.

Payne, M. E., Anderson, J. J., & Steffens, D. C., (2008). Calcium and vitamin D intakes may be positively associated with brain lesions in depressed and nondepressed elders. *Nutrition Research, 28*(5), 285–292.

Pilz, G. A., Bottes, S., Betizeau, M., Jörg, D. J., Carta, S., Simons, B. D., Helmchen, F., & Jessberger, S., (2018). Live imaging of neurogenesis in the adult mouse hippocampus. *Science, 359*(6376), 658–662.

Piomelli, D., Astarita, G., & Rapaka, R., (2007). A neuroscientist's guide to lipidomics. *Nature Reviews Neuroscience, 8*(10), 743–754.

Prado, E. L., & Dewey, K. G., (2014). Nutrition and brain development in early life. *Nutrition Reviews, 72*(4), 267–284.

Rabinovitz, S., (2014). Effects of omega-3 fatty acids on tobacco craving in cigarette smokers: A double-blind, randomized, placebo-controlled pilot study. *Journal of Psychopharmacology, 28*(8), 804–809.

Reiter, R. J., TAN, D. X., & Burkhardt, S., (2001). Free radical-mediated molecular damage: Mechanisms for the protective actions of melatonin in the central nervous system. *Annals of the New York Academy of Sciences, 939*(1), 200–215.

Robson, L. G., Dyall, S., Sidloff, D., & Michael-Titus, A. T., (2010). Omega-3 polyunsaturated fatty acids increase the neurite outgrowth of rat sensory neurones throughout development and in aged animals. *Neurobiology of Aging, 31*(4), 678–687.

Sandi, C., & Touyarot, K., (2006). Mid-life stress and cognitive deficits during early aging in rats: Individual differences and hippocampal correlates. *Neurobiology of Aging, 27*(1), 128–140.

Sarris, J., Logan, A. C., Akbaraly, T. N., Amminger, G. P., Balanzá-Martínez, V., Freeman, M. P., Hibbeln, J., et al., (2015). Nutritional medicine as mainstream in psychiatry. *The Lancet Psychiatry, 2*(3), 271–274.

Schnaar, R. L., Gerardy-Schahn, R., & Hildebrandt, H., (2014). Sialic acids in the brain: Gangliosides and polysialic acid in nervous system development, stability, disease, and regeneration. *Physiological Reviews*.

Spencer, S. J., Korosi, A., Layé, S., Shukitt-Hale, B., & Barrientos, R. M., (2017). Food for thought: How nutrition impacts cognition and emotion. *NPJ Science of Food, 1*(1), 1–8.

Su, H. M., (2010). Mechanisms of n-3 fatty acid-mediated development and maintenance of learning memory performance. *The Journal of Nutritional Biochemistry, 21*(5), 364–373.

Surana, K. R., Ahire, E. D., Sonawane, V. N., Talele, S. G., & Talele, G., (2021). Informatics and methods in nutrition design and development. *Natural Food Products and Waste Recovery: Healthy Foods, Nutrition Design, and Extraction of Valuable Compounds*, 33.

Tang, M., Zhang, M., Cai, H., Li, H., Jiang, P., Dang, R., Liu, Y., et al., (2016). Maternal diet of polyunsaturated fatty acid altered the cell proliferation in the dentate gyrus of hippocampus and influenced glutamatergic and serotoninergic systems of neonatal female rats. *Lipids in Health and Disease, 15*(1), 1–10.

Todeschini, A. R., & Hakomori, S. I., (2008). Functional role of glycosphingolipids and gangliosides in control of cell adhesion, motility, and growth, through glycosynaptic microdomains. *Biochimica et Biophysica Acta (BBA)-General Subjects, 1780*(3), 421–433.

Urquiza, A. M. D., Liu, S., Sjoberg, M., Zetterstrom, R. H., Griffiths, W., Sjovall, J., & Perlmann, T., (2000). Docosahexaenoic acid, a ligand for the retinoid X receptor in mouse brain. *Science, 290*(5499), 2140–2144.

Vermiglio, F., Lo Presti, V. P., Moleti, M., Sidoti, M., Tortorella, G., Scaffidi, G., Castagna, M. G., et al., (2004). Attention deficit and hyperactivity disorders in the offspring of mothers exposed to mild-moderate iodine deficiency: A possible novel iodine deficiency disorder in developed countries. *The Journal of Clinical Endocrinology & Metabolism, 89*(12), 6054–6060.

Vignes, M., Maurice, T., Lanté, F., Nedjar, M., Thethi, K., Guiramand, J., & Récasens, M., (2006). Anxiolytic properties of green tea polyphenol (−)-epigallocatechin gallate (EGCG). *Brain Research, 1110*(1), 102–115.

Wang, B., (2012). Molecular mechanism underlying sialic acid as an essential nutrient for brain development and cognition. *Advances in Nutrition, 3*(3), 465S–472S.

Williams, S., Chen, L., Savignac, H. M., Tzortzis, G., Anthony, D. C., & Burnet, P. W., (2016). Neonatal prebiotic (BGOS) supplementation increases the levels of synaptophysin, Glu N 2 A-subunits and BDNF proteins in the adult rat hippocampus. *Synapse, 70*(3), 121–124.

Williard, D. E., Harmon, S. D., Kaduce, T. L., Preuss, M., Moore, S. A., Robbins, M. E., & Spector, A. A., (2001). Docosahexaenoic acid synthesis from n-3 polyunsaturated fatty acids in differentiated rat brain astrocytes. *Journal of Lipid Research, 42*(9), 1368–1376.

Wu, C., & Sun, D., (2015). GABA receptors in brain development, function, and injury. *Metabolic Brain Disease, 30*(2), 367–379.

CHAPTER 5

FUNCTIONAL FOODS IN THE MANAGEMENT OF WEIGHT LOSS AND OBESITY

KHEMCHAND R. SURANA,[1] ASHWINI SHELKE,[2] TUSHAR LOKHANDE,[3] and SUNIL K. MAHAJAN[3]

[1]*Department of Pharmaceutical Chemistry, Shreeshakti Shaikshanik Sanstha, Divine College of Pharmacy, Satana, Nashik, Maharashtra, India*

[2]*Department of Pharmaceutical Chemistry, Sandip Institute of Pharmaceutical Sciences, Nashik, Maharashtra, India*

[3]*Department of Pharmaceutical Chemistry, MGV's Pharmacy College, Panchvati, Nashik, Maharashtra, India*

ABSTRACT

Obesity is a global issue whose prevalence is rapidly increasing in emerging countries, making it a significant public health problem. In order to reduce the prevalence of various metabolic illnesses, health organizations prioritize weight control and obesity prevention. The main contributive element for obesity is found to be new eating habits, as well as current production and consumption trends. Changes in lifestyle, as well as dietary changes, are essential in the treatment of obesity; simultaneously, it is very important to assess the potential relevance of functional foods in promoting weight loss and preventing obesity. Prevention and treatment of obesity rely heavily on both nutrition and exercise. Because of the potential adverse effects of obesity medicines, the positive effects of natural constituents and functional

Applications of Functional Foods in Disease Prevention. Raj K. Keservani & Eknath D. Ahire, PhD (Eds.)

foods in managing obesity have gotten a lot of attention. Functional foods have a similar appearance to traditional foods, but they provide benefits that go beyond their fundamental nutritional qualities. Functional foods have been studied as a means of improving weight management in those who live a healthy lifestyle. Functional foods have been evaluated for their efficacy as obesity therapies in two ways: first is food that reduces appetite and increases pleasure, and another is foods that increase thermogenesis. Nutraceutical therapies are currently being researched extensively as potential obesity and weight loss treatments. Low-glycemic-index carbohydrates, chromium, green tea extract and 5-hydroxytryptophan are among the substances that have been shown to promote weight loss. In clinical trials, each component has been demonstrated to stimulate weight loss on its own.

5.1 INTRODUCTION

Obesity is a community health issue that causes major social, physical, and psychological consequences, according to WHO (World Health Organization). Obesity (Jarouliya and Keservani, 2019) is specifying as a body mass index (BMI) of 30 kg/m^2, which has been linked to an increased risk of chronic illnesses, morbidity and mortality (Wollner et al., 2017). According to European studies, child obesity is becoming a rising health issue, particularly in Mediterranean countries with high rates of overweight and obesity (Farajian et al., 2011). There are many factors responsible for obesity. Some factors are causative factors, some are consequences of related factors and some are countervail, such as genetic factors, diseases, lifestyle, pharmaceuticals, inflammation, and food diet (Figure 5.1).

Although food portion control is an important aspect of weight management, telling people to eat less of all meals may not have the desired outcomes, because high-energy-dense foods increase energy intake disproportionately more than foods with lower energy density (Konstantinidi and Koutelidakis, 2019). A more appropriate technique could be to encourage people to consume more low-energy-dense foods while also maintaining a balanced diet, such as the Mediterranean diet, and increasing their physical activity (Rolls, 2014). Several organizations and authorities define functional foods in different ways (Bigliardi and Galati, 2013). Functional foods (FFs) were first defined in the 1980s, when Japan's Ministry of Health and Welfare created a regulation system for foods with potential health benefits. Beyond nutritional effects, a food can be deemed 'functional' if it has significant

FIGURE 5.1 Causative, consequence and countervail factors responsible for obesity.

effects on target functions in the human organism, with the goal of improving health and wellbeing and/or reducing chronic diseases. "Functional foods" are defined as foods and food components that give a health benefit beyond basic nutrition, according to The IFT Expert Panel Report, Functional Foods Opportunities and Challenges (for the intended population) (Keservani et al., 2010a, b, 2017, 2020; IFT, 2005). These compounds supply vital nutrients in levels that are often more than those required for normal maintenance, growth, and development, furthermore other biologically active components with health advantages or desirable physiological effects. Functional foods are gaining popularity around the world and are becoming a regular element of people's diets (Brown et al., 2015). Natural health products' or 'healthy foods' are common terms used to describe functional foods. Foods can only be termed functional if, in addition to basic nutritional impact, they display possible beneficial effects on certain functions in the human organism, resulting in improved physical condition and/or decreased probability of chronic disease appearance (Brown et al., 2018; Martirosyan and Singh, 2015). Nutritional and health claims concerning the potential effect of functional foods on illness prevention, in contrast, have only been established in limited circumstances, when sufficient scientific evidence supports their features. Furthermore, scientific evidence suggests that functional foods may benefit human health only when included as part of an effectively balanced diet, such as the Mediterranean diet (Elmaliklis et al., 2019; Koutelidakis and Dimou, 2016; Ntrigiou et al., 2018). Maintaining a stable body weight necessitates an equilibrium between energy expenditure (EE) and energy intake (EI) (Figure 5.2). Even a minor imbalance in this energy equilibrium can cause large changes in body weight over time, leading to obesity.

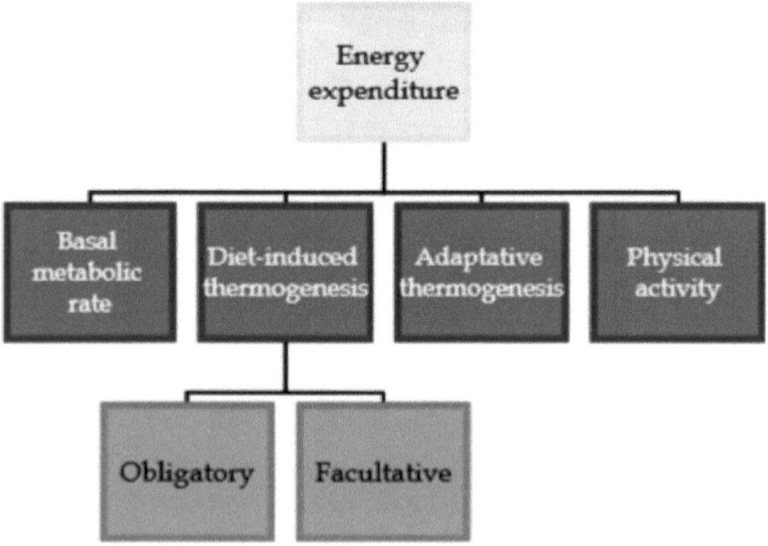

FIGURE 5.2 Principal components of energy expenditure.

Obesity is among the world's biggest health problems, and it is linked to a variety of chronic diseases. Small changes in energy balance can contribute to weight gain, which can be avoided by making minor dietary changes, such as including functional foods for weight management (Koutelidakis and Dimou, 2016). Although it is well established that weight loss can be achieved through by restricting calories and increasing physical exercise, such lifestyle modifications can be challenging to execute and maintain

(Anderson et al., 2001). Inhibiting appetite is also one of the goals of introducing functional meals into weight management. Logically, this should result in a reduction in food consumption and, as a result, weight loss. Hence functional foods that affect metabolic activity and fat partitioning could be beneficial additions to a weight-loss diet. Dietary fats are among the potential functional foods. Dietary fibers of various sorts and sources have been proposed as a technique of achieving these weight-loss objectives. After that capsaicin, conjugated linoleic acid (CLA), *Momordica charantia* (MC), and Psyllium fiber are examples of functional foods with antiobese characteristics (Kasbia, 2005). Also it has been observed that in obese people, a food supplement including chitosan, glucomannan, G sylvestre, fenugreek, and vitamin C lowered body weight and increased fat reduction. Caffeine, gallic acid, catechins, anthocyanins, polyphenols, ascorbic acid, oleuropein, quercetin and capsaicin are some of the bioactive components in those foods that may help with weight control, obesity prevention, and the metabolic implications of obesity (Kasbia, 2005). In this way all possible categories of functional food are explained here systematically (Figure 5.3).

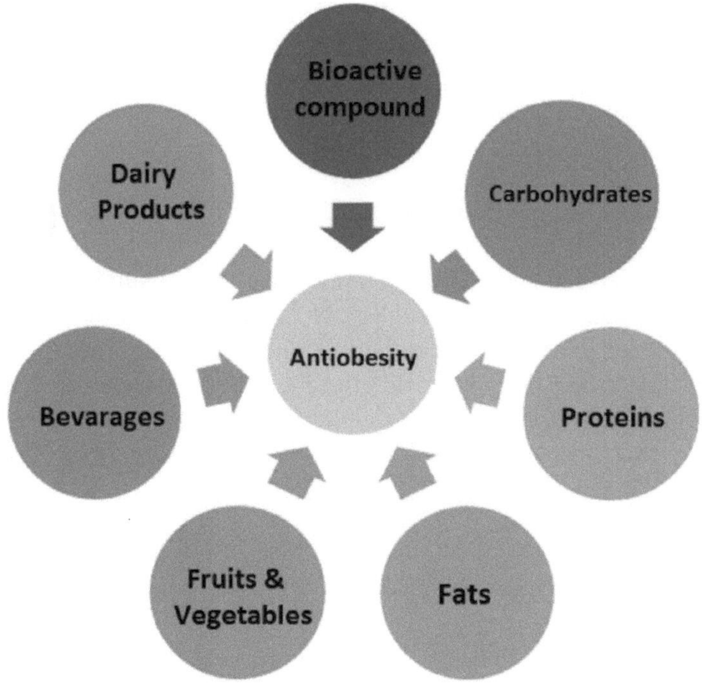

FIGURE 5.3 Various functional foods are useful for weight loss.

5.2 FOODS CONTAINING CARBOHYDRATES

Dietary fiber helps to prevent chronic diseases through various mechanisms. The glycemic index (GI) of carbohydrates and the fiber content of foods has an impact on weight gain. Dietary fiber may influence appetite through the characteristics it imparts to food and also through physiologic effects in the individual. High fiber diets, for example, may offer bulk and cause stomach distension, resulting in a feeling of satiety (Kasbia, 2005). Most studies examining this outcome found increased satiety and/or decreased appetite, which is consistent with this (Dikeman and Fahey Jr, 2006). Dietary macronutrient composition is critical for promoting fullness and suppressing appetite (Larsen et al., 2010). Kozuka et al. investigated the influence of brown rice and γ-oryzanol on dietary fat preference in mice (2012). Brown rice bran contains γ-oryzanol, which is a combination of ferulic acid esters. According to the study, mice fed either brown rice or γ-oryzanol preferred a low-fat control diet to a high-fat diet (Kozuka et al., 2012). Dietary fiber appears to improve satiety and decrease hunger; however this is not always linked to lower energy intake, especially over time. The majority of the research looked at were done on people of average weight. In order to determine the effectiveness of dietary fiber as a weight loss approach, well-designed randomized, controlled clinical trials in overweight and/or obese adults are needed (Surana et al., 2021).

5.3 FOODS CONTAINING PROTEINS

Protein rich diets have been shown to increase satiety, suppress appetite, and reduce food intake (Anderson and Moore, 2004). Because of the more energy expenditure associated with catabolism and protein synthesis, protein has a higher satiety and thermeogenic effect than its macronutrient equivalents carbohydrates and lipids. Protein-rich legumes such as lentils, peas, peanuts, chickpeas, beans, soybeans, and Faba beans boosted energy expenditure by 49% in human subjects (Iwashita et al., 2006).

5.4 FOODS CONTAINING FATS

Dietary fats (e.g., medium-chain triacylglycerols (MCTs) and diacylglycerols), beverages (e.g., tea and milk), and nuts are examples of potential functional foods. MCTs are those triacylglycerols composed of fatty acids that contain 6–12 carbon atoms. LCTs are those triacylglycerols composed of long-chain triacylglycerols. Both differ in their composition as well as way of absorption, transportation. In the gastrointestinal system, MCTs and LCTs are digested into their respective medium and long chain fatty acids (MCFAs and LCFAs, respectively). MCFAs, because of their shorter chain lengths travel directly to the liver through the portal circulation; therefore, they bypass peripheral tissues which make them less susceptible to the actions of hormone-sensitive lipase and to deposition into adipose tissue stores which are not happen in case of LCFAs (Bach and Babayan, 1982). MCTs found in liquid oil extracted from coconut oil could easily be incorporated into the diet as a replacement for other LCT-rich vegetable oils because of a mode of action similar to that of MCTs, diacylglycerols have been appear to promote thermogenesis more than triacylglycerols do and to have the potential to aid in weight loss (Nagao et al., 2000). Diacylglycerols are found in tiny amounts in a variety of edible oils; cottonseed oil has one of the highest concentrations at 9.5%. Cooking oil rich in diacylglycerols has been developed that contains 80% diacylglycerols and looks and tastes like regular oil. To reduce lipid content, weight loss programs prioritize the eating of fatty fish. The addition of salmon to a hypocaloric diet enhanced the plasma lipid profile as well (Mori et al., 2004).

5.6 FRUITS AND VEGETABLES

High eating of fruits and vegetables has been linked to good health because they protect against a variety of chronic diseases. Study reveals that fruits such as Apple, Mango, Bilberries and sea buckthorn reduced fat deposits, body weight gain and glucose (Boqué et al., 2013). Vegetables like bitter gourd (vegetable in Asia), *Piper sarmentosum* (medicinal plant), green tomato, and onion peel tea has been proved to exhibited anti-obesity properties in some studies (Boqué et al., 2013; Chaturvedi, 2012; Choi et al., 2013).

Berries, such as blackberry, blueberry, strawberry, and others, are one of the richest sources of bioactive components in the diet, and they may have a synergistic and cumulative effect on human health promotion and obesity prevention. Berry intake can be thought of as a viable alternative to dietary pattern changes aimed at preventing the onset of metabolic syndrome (Choi et al., 2013; Matsunaga et al., 2014).

Avocado is a berry-like fruit that is high in lipids and contains provitamin A carotenoids. Avocados are a good source of lutein and have a medium calorie density due to the fact that roughly 80% of the edible section of the avocado is made up of 72% of water and 6.8% of dietary fiber. Avocado has been discovered to have similar weight-loss properties to low-fat fruits and vegetables (Kopec et al., 2014).

Pomegranate is a fruit that is high in antioxidants. Pomegranates are high in tannins, ellagitannins, procyanidins as well as flavonoids (Keservani and Sharma, 2014) and all of bioactive substances that may contribute to health benefits. Pomegranate peel makes up 30–40% of the fruit and is regarded as wasted material (Fischer et al., 2011).

5.7 BEVERAGES

Tea (*Camellia sinensis*) is a popular beverage throughout the world. Tea contains caffeine, which has been demonstrated to have thermogenic properties and promote fat oxidation. Tea's caffeine and polyphenols work together to promote fat oxidation. Green tea is a type of tea that has not been fermented and is high in polyphenolic compounds. The majority of theaflavins and polyphenols are found in black tea, which has been fermented. Polyphenols in black tea have been appear to reduce pancreatic lipase, affecting lipid metabolism (Chaloupka et al., 2011; Yuda et al., 2012).

Antioxidant beverage prepared from the coffee and tested for their anti-obesity potential (Zemel et al., 2000). Green coffee bean extract had a higher effect on body weight loss and epididymal and perirenal fat weight loss than roasted coffee and caffeine in multiple studies, probably due to its high chlorogenic acid content (Pan et al., 2016). To conclude, coffee is one of the beverages for which there are some signs of its significance in weight loss.

Juices high in anthocyanins, mulberry juice (MBJ) and blueberry juice (BBJ) were investigated on obesity prevention (Bukowiecki et al., 1983; Wu et al., 2013).

5.8 DAIRY PRODUCTS

Adiposity is inversely connected to a low intake of dairy products and calcium. Dietary calcium is crucial in the regulation of energy metabolism and the risk of obesity (Ackley et al., 1983). Calcium is mostly obtained through dairy products, and various theories for understanding calcium's role have been postulated (Barba and Russo, 2006; Zemel et al., 2000).

5.9 BIOACTIVE COMPOUNDS

In recent years, a lot of work has gone into finding dietary or herbal remedies with bioactive compounds that help with obesity and obesity-related comorbidities. Polyphenols, which can be found in a variety of fruits, vegetables and herbs, are among the most significant chemical groups in bioactives. The Indian gooseberry showed a significant decrease in triglyceride accumulation (Patil et al., 2009). The anti-obesity effect of the natural flavonoid kaempferol was also studied. Furthermore, combining different bioactive compounds may have synergistic effects. Crocin, chlorogenic acid, geniposide, and quercetin were combined to lower triglyceride and cholesterol accumulation in humans. Ginseng seed oil, Ginseng green leaf and Ginseng dry leaf extract was found to be useful in lowering lipid levels. Green tea extract, when combined with diallyl disulfide, a key organosulfur found in garlic, efficiently inhibits adipogenesis and triglyceride formation. One cup of green tea and two garlic cloves every day will provide the recommended dose (Ahire et al., 2021; Bernhoft, 2010; Sandner et al., 2020).

5.10 NUTS

Nuts appear to reduce hunger, which may be due to their high energy content, proteins, and fiber, all of which might create a feeling of satiety (Neale et al., 2017). Several studies suggest that because nuts are high in energy, they should be substituted for other foods in the diet to help control calories and prevent overeating and weight gain. In comparison to the carbohydrate-low-calorie diet group, the almond-low-calorie diet group lost 62% more weight, 50% more waist circumference, and 56% more fat mass, as well as showing better glycemic control (Bitok et al., 2017; Kranz et al., 2014).

5.11 CONCLUSION

Functional foods have been offered as a possible alternate way of weight management and obesity prevention when ingested as part of a balanced diet. Although there is evidence that functional foods have that much potential to be developed, the true difficulty for the food business is to develop goods that are effective through many processes. Obesity develops as a result of the interaction of several complicated mechanisms. Many processes must be targeted rather than a single process. The absence of viable biomarkers to assess the chemicals is another issue in developing functional foods for obesity. Long-term studies should focus on finding the optimal level of ingestion, efficacy, and safety of items. Only a multidisciplinary strategy with severe lifestyle adjustments (diet and physical exercise) may clearly lead to a considerable reduction in obesity.

ACKNOWLEDGMENT

We thank the MGV's, College of Pharmacy, Panchavati, affiliated under Savitribai Phule Pune University, Nashik.

KEYWORDS

- **functional foods**
- **nutrition**
- **obesity**
- **weight loss**

REFERENCES

Ackley, S., Barrett-Connor, E., & Suarez, L., (1983). Dairy products, calcium, and blood pressure. *The American Journal of Clinical Nutrition, 38*, 457–461.

Ahire, E., Sonawane, V., Surana, K., & Talele, G., (2021). Drug discovery, drug-likeness screening, and bioavailability: Development of drug-likeness rule for natural products. *Applied Pharmaceutical Practice and Nutraceuticals.* Apple Academic Press.

Anderson, G. H., & Moore, S. E., (2004). Dietary proteins in the regulation of food intake and body weight in humans. *The Journal of Nutrition, 134*, 974s–979s.

Anderson, J. W., Konz, E. C., Frederich, R. C., & Wood, C. L., (2001). Long-term weight-loss maintenance: A meta-analysis of us studies. *The American Journal of Clinical Nutrition, 74*, 579–584.

Bach, A., & Babayan, V., (1982). Medium-chain triglycerides: An update. *The American Journal of Clinical Nutrition, 36*, 950–962.

Barba, G., & Russo, P., (2006). Dairy foods, dietary calcium and obesity: A short review of the evidence. *Nutrition, Metabolism and Cardiovascular Diseases, 16*, 445–451.

Bernhoft, A., (2010). A brief review on bioactive compounds in plants. *Bioactive Compounds in Plants-Benefits and Risks for Man And Animals, 50*, 11–17.

Bigliardi, B., & Galati, F., (2013). Innovation trends in the food industry: The case of functional foods. *Trends In Food Science & Technology, 31*, 118–129.

Bitok, E., Jaceldo-Siegl, K., Rajaram, S., Serra-Mir, M., Roth, I., Feitas-Simoes, T., Ros, E., & Sabaté, J., (2017). Favorable nutrient intake and displacement with long-term walnut supplementation among elderly: Results of a randomized trial. *British Journal of Nutrition, 118*, 201–209.

Boqué, N., De La Iglesia, R., De La Garza, A. L., Milagro, F. I., Olivares, M., Bañuelos, Ó., Soria, A. C., et al., (2013). Prevention of diet-induced obesity by apple polyphenols in Wistar rats through regulation of adipocyte gene expression and DNA methylation patterns. *Molecular Nutrition & Food Research, 57*, 1473–1478.

Brown, L., Caligiuri, S. P., Brown, D., & Pierce, G. N., (2018). Clinical trials using functional foods provide unique challenges. *Journal Of Functional Foods, 45*, 233–238.

Brown, L., Poudyal, H., & Panchal, S. K., (2015). Functional foods as potential therapeutic options for metabolic syndrome. *Obesity Reviews, 16*, 914–941.

Bukowiecki, L. J., Lupien, J., Follea, N., & Jahjah, L., (1983). Effects of sucrose, caffeine, and cola beverages on obesity, cold resistance, and adipose tissue cellularity. *American Journal of Physiology-Regulatory, Integrative and Comparative Physiology, 244*, R500–R507.

Chaloupka, F. J., Powell, L. M., & Chriqui, J. F., (2011). Sugar-sweetened beverages and obesity prevention: Policy recommendations. *Journal Of Policy Analysis and Management*, 662–664.

Chaturvedi, P., (2012). Antidiabetic potentials of *Momordica charantia*: Multiple mechanisms behind the effects. *Journal Of Medicinal Food, 15*, 101–107.

Choi, K. M., Lee, Y. S., Shin, D. M., Lee, S., Yoo, K. S., Lee, M. K., Lee, J. H., et al., (2013). Green tomato extract attenuates high-fat-diet-induced obesity through activation of the AMPK pathway in C57bl/6 mice. *The Journal of Nutritional Biochemistry, 24*, 335–342.

Dikeman, C. L., & Fahey Jr, G. C., (2006). Viscosity as related to dietary fiber: A review. *Critical Reviews in Food Science and Nutrition, 46*, 649–663.

Elmaliklis, I. N., Liveri, A., Ntelis, B., Paraskeva, K., Goulis, I., & Koutelidakis, A. E., (2019). Increased functional foods' consumption and Mediterranean diet adherence may have a protective effect in the appearance of gastrointestinal diseases: A case-control study. *Medicines, 6,* 50.

Farajian, P., Risvas, G., Karasouli, K., Pounis, G. D., Kastorini, C. M., Panagiotakos, D. B., & Zampelas, A., (2011). Very high childhood obesity prevalence and low adherence rates to the Mediterranean diet in Greek children: The Greco study. *Atherosclerosis, 217*, 525–530.

Fischer, U. A., Carle, R., & Kammerer, D. R., (2011). Identification and quantification of phenolic compounds from pomegranate (*Punica granatum* L.) peel, mesocarp, aril and differently produced juices by HPLC-DAD–ESI/MS^n. *Food Chemistry, 127*, 807–821.

IFT, (2005). *Functional Foods: Opportunities and Challenges*. Institute Of Food Technologists.

Iwashita, S., Mikus, C., Baier, S., & Flakoll, P. J., (2006). Glutamine supplementation increases postprandial energy expenditure and fat oxidation in humans. *Journal Of Parenteral and Enteral Nutrition, 30*, 76–80.

Jarouliya, U., & Keservani, R. K., (2019). Pathways leading to child obesity. In: Debasis B., (ed.), *Global Perspectives on Childhood Obesity: Current Status, Consequences and Prevention* (2nd edn., pp. 137–146) Academic Press, Elsevier, Chapter 12. ISBN: 9780128128404.

Kasbia, G. S., (2005). Functional foods and nutraceuticals in the management of obesity. *Nutrition & Food Science.*

Keservani, R. K., & Sharma, A. K., (2014). Flavonoids: Emerging trends and potential health benefits. *Journal Of Chinese Pharmaceutical Sciences, 23*(12), 815.

Keservani, R. K., Kesharwani, R. K., Sharma, A. K., Vyas, N., & Chadoker, A., (2010b). Nutritional supplements: An Overview. *International Journal of Current Pharmaceutical Review and Research, 1*(1), 59–75.

Keservani, R. K., Kesharwani, R. K., Vyas, N., Jain, S., Raghuvanshi, R., & Sharma, A. K., (2010a). Nutraceutical and functional food as future food: A review. *Der Pharmacia Lettre, 2*(1*), 106–116.*

Keservani, R. K., Sharma, A. K., & Kesharwani, R. K., (2017). An overview and therapeutic applications of nutraceutical and functional foods. *Recent Advances in Drug Delivery Technology*, 160–201. ISBN:9781522507543.

Keservani, R. K., Sharma, A. K., & Kesharwani, R. K., (2020). "*Nutraceuticals And Dietary Supplements: Applications in Health Improvement and Disease Management*. CRC Press. ISBN: 9781771888738.

Konstantinidi, M., & Koutelidakis, A. E., (2019). Functional foods and bioactive compounds: A review of its possible role on weight management and obesity's metabolic consequences. *Medicines, 6*, 94.

Kopec, R. E., Cooperstone, J. L., Schweiggert, R. M., Young, G. S., Harrison, E. H., Francis, D. M., Clinton, S. K., & Schwartz, S. J., (2014). Avocado consumption enhances human postprandial provitamin an absorption and conversion from a novel high-β-carotene tomato sauce and from carrots. *The Journal of Nutrition, 144*, 1158–1166.

Koutelidakis, A., & Dimou, C., (2016). The effects of functional food and bioactive compounds on biomarkers of cardiovascular diseases. In: Martirosyan, D., (ed.), *Functional Foods Text Book* (1st edn., pp. 89–117).

Kozuka, C., Yabiku, K., Sunagawa, S., Ueda, R., Taira, S. I., Ohshiro, H., Ikema, T., Yamakawa, K., Higa, M., & Tanaka, H., (2012). Brown rice and its component, Γ-oryzanol, attenuate the preference for high-fat diet by decreasing hypothalamic endoplasmic reticulum stress in mice. *Diabetes, 61*, 3084–3093.

Kranz, S., Hill, A., Fleming, J., Hartman, T., West, S., & Kris-Etherton, P., (2014). Nutrient displacement associated with walnut supplementation in men. *Journal of Human Nutrition and Dietetics, 27*, 247–254.

Larsen, T. M., Dalskov, S. M., Van, B. M., Jebb, S. A., Papadaki, A., Pfeiffer, A. F., Martinez, J. A., et al., (2010). Diets with high or low protein content and glycemic index for weight-loss maintenance. *New England Journal of Medicine, 363*, 2102–2113.

Martirosyan, D. M., & Singh, J., (2015). A new definition of functional food by FFC: What makes a new definition unique? *Functional Foods in Health and Disease, 5*, 209–223.

Matsunaga, S., Azuma, K., Watanabe, M., Tsuka, T., Imagawa, T., Osaki, T., & Okamoto, Y., (2014). Onion peel tea ameliorates obesity and affects blood parameters in a mouse model of high-fat-diet-induced obesity. *Experimental And Therapeutic Medicine, 7*, 379–382.

Mori, T. A., Burke, V., Puddey, I. B., Shaw, J. E., & Beilin, L. J., (2004). Effect of fish diets and weight loss on serum leptin concentration in overweight, treated-hypertensive subjects. *Journal Of Hypertension, 22*, 1983–1990.

Nagao, T., Watanabe, H., Goto, N., Onizawa, K., Taguchi, H., Matsuo, N., Yasukawa, T., et al., (2000). Dietary diacylglycerol suppresses accumulation of body fat compared to triacylglycerol in men in a double-blind controlled trial. *The Journal of Nutrition, 130*, 792–797.

Neale, E. P., Tapsell, L. C., Martin, A., Batterham, M. J., Wibisono, C., & Probst, Y. C., (2017). Impact of providing walnut samples in a lifestyle intervention for weight loss: A secondary analysis of the health track trial. *Food & Nutrition Research, 61*, 1344522.

Ntrigiou, V., Ntrigios, I., Rigopoulos, N., Dimou, C., & Koutelidakis, A. E., (2018). Functional food consumption correlates with anthropometric characteristics and body composition in healthy adults. *Current Topics in Nutraceutical Research,* 16.

Pan, M. H., Tung, Y. C., Yang, G., Li, S., & Ho, C. T., (2016). Molecular mechanisms of the anti-obesity effect of bioactive compounds in tea and coffee. *Food & Function, 7*, 4481–4491.

Patil, B. S., Jayaprakasha, G. K., Chidambara, M. K., & Vikram, A., (2009). Bioactive compounds: Historical perspectives, opportunities, and challenges. *Journal of Agricultural and Food Chemistry, 57*, 8142–8160.

Rolls, B., (2014). What is the role of portion control in weight management? *International Journal of Obesity, 38*, S1–S8.

Sandner, G., König, A., Wallner, M., & Weghuber, J., (2020). Functional foods-dietary or herbal products on obesity: Application of selected bioactive compounds to target lipid metabolism. *Current Opinion in Food Science, 34*, 9–20.

Surana, K. R., Ahire, E. D., Sonawane, V. N., Talele, S. G., & Talele, G., (2021). Informatics and methods in nutrition design and development. *Natural Food Products and Waste Recovery: Healthy Foods, Nutrition Design, And Extraction of Valuable Compounds*, 33.

Wollner, M., Roberto, B. B. P., Roncally, S. C. A., Jurandir, N., & Edil, L. S., (2017). Accuracy of the who's body mass index cut-off points to measure gender-and age-specific obesity in middle-aged adults living in the city of Rio De Janeiro, Brazil. *Journal of Public Health Research,* 6.

Wu, T., Qi, X., Liu, Y., Guo, J., Zhu, R., Chen, W., Zheng, X., & Yu, T., (2013). Dietary supplementation with purified mulberry (*Morus australis* poir) anthocyanins suppresses body weight gain in high-fat diet fed C57bl/6 mice. *Food Chemistry, 141*, 482–487.

Yuda, N., Tanaka, M., Suzuki, M., Asano, Y., Ochi, H., & Iwatsuki, K., (2012). Polyphenols extracted from black tea (*Camellia sinensis*) Residue by hot-compressed water and their inhibitory effect on pancreatic lipase *in vitro. Journal of Food Science, 77*, H254–H261.

Zemel, M. B., Shi, H., Greer, B., Dirienzo, D., & Zemel, P. C., (2000). regulation of adiposity by dietary calcium. *The FASEB Journal, 14*, 1132–1138.

CHAPTER 6

FUNCTIONAL FOODS IN HEALTH AND DISEASES

AFSAR S. PATHAN,[1] PRADIP P. WAGH,[1] PANKAJ G. JAIN,[1] GANESH B. SONAWANE,[2] and EKNATH D. AHIRE[3]

[1]*Department of Pharmacology, R. C. Patel Institute of Pharmaceutical Education and Research, Shirpur, Maharashtra, India*

[2]*Department of Pharmaceutical Chemistry, Shreeshakti Shaikshanik Sanstha, Divine College of Pharmacy, Satana, Nashik, Maharashtra, India*

[3]*Department of Pharmaceutics, METs Institute of Pharmacy, BKC, Affiliated to Savitribai Phule University, Adgoan, Nashik-3, Maharashtra, India*

ABSTRACT

A functional disorder is a medical condition that impairs the normal functioning of bodily processes rather than affecting the structure of body parts and remains largely undetected under examination, dissection, or even under a microscope. A functional disorder is widely increasing day-by-day in worldwide, and the treatment on that is somewhat helpful, but with some adverse effects and is expensive for poor people. Hence, cost-effective and risk-free management treatments are necessary for functional disease. To resolve this problem, we use functional foods that alter one or more functions in the body with minimum or no risk. Functional food provides us with good nutrition and supplements. It is cost-effective and readily available; we add it to our regular daily diet. In this chapter, we recognized different types of functional food which helpful for the different types of functional diseases.

Applications of Functional Foods in Disease Prevention. Raj K. Keservani & Eknath D. Ahire, PhD (Eds.)

6.1 INTRODUCTION

Functional foods are defined in a variety of ways around the world (Volpe, 2015). According to certain definitions, any food can be marketed as a functional food provided it is marketed correctly. A food can be considered functional if it has been successfully demonstrated to alter one or more specific functions in the body in a way that contributes to improved health and well-being or lower illness risk, in addition to adequate nutritional effects (Doyon and Labrecque, 2008). A functional food should retain to be food and demonstrate its benefits in levels that may be estimated; this is not a pill or a capsule, but rather an element in a daily meal that is consumed. 'Functional food' is not a new concept, but it has recently gained traction as a result of the collapse of most social health systems, as 'functional foods' allow for low-cost illness prevention. 'Functional foods' differ from 'nutraceuticals,' which are drug-based and have a poor taste, but 'functional foods' (Keservani et al., 2010a, b, 2017, 2020) are good food that may be eaten for years and also have a disease-prevention role (Guesry, 2005). Around 26 definitions were chosen from over a 100 for their illustration of the many types of existing definitions for functional foods. After that, three essential concepts were recognized among the selected definitions. (a) The conception of health benefits appears to be essential to most definitions, as evidence by the reality that it appears in 26 definitions chosen (Surana et al., 2021). Only two definitions, however, state that the health benefits must be verified, but do not specify what form of proof is required. When indicated, the health benefits revealed in the criteria could be categorized as improving target function or lowering the risk of specific diseases (Carvajal-Millan et al., 2021). (b) *Food constitution:* The food's nature appears to be an essential element as well. Despite the fact that the majority of definitions include the word "food," nearly a third state that a functional food must be or seem like a traditional food. Some Food must have been fortified, enriched, or contained an additive, according to the definitions. Others refer to allergies or components being removed (Aluko, 2012). Overconsumption is deemed harmful to one's health (e.g., salt and sugar). The nature of the food idea appears in half of the definitions. Level of function. The nutritional value of all foods varies. However, more than half of the definitions (13/26) say that what makes a food functional is the benefits it provides in addition to its basic nutritional functions (Ahire et al., 2021). To emphasize how crucial it is, Roberfroid (2002) recommends seeing food through the lens of its nutritional value. Rather than being a product or having physical characteristics, it serves a function(s). (c)

Pattern of consumption: The last concept uncovered is consumption pattern, which was found in nine of the definitions examined. A functional food must be a part of a typical diet or fit a normal consumption pattern in a certain geographic and/or cultural context, according to this concept. As a result, a food that is considered functional in one country may not be considered functional in another. The differences in consumption patterns across the four definitions are more modest. A functional meal should be a component of a regular or typical diet, according to the four definitions. With its "daily diet and consumption pattern," the definition is more stringent (Johnson and Williamson, 2003).

The major targets for functional food are the following (Roberfroid, 2000):

- Gastrointestinal functions;
- Redox and antioxidant systems;
- Metabolism of the macronutrients;
- Development in fetal and early life;
- Xenobiotic metabolism and its modulation by nonnutritive dietary components, such as some phytochemicals;
- Mood and behavior or cognition and physical performance.

6.2 FUNCTIONAL FOOD DISORDERS AND DISEASES

Incidence and prevalence of functional impairments and diseases affecting this system are significantly higher in older people than in younger people. Symptoms and discomfort are caused by functional problems of the digestive tract when there is no evidence of biological disease. Dysphagia, GERD, irritable bowel syndrome, primary dyspepsia, and chronic primary constipation are among them. Even in the absence of extra digestive diseases that favor their development, these disorders are more common in the elderly (Berstad et al., 2017; Lordan et al., 2011).

6.2.1 DYSPHAGIA

It is a condition that affects many older people, especially those who are physically impaired and require assistance to eat. In the US, dysphagia affects 300,000–600,000 persons yearly (Sura et al., 2012). Functional changes that impact the act of swallowing (*Oropharyngeal dysphagia*) or the transport of swallowed food through the esophagus can cause it. Dysphagia is linked to nutritional deficiencies as well as an increased risk of pneumonia. Dysphagia management in older people necessitates a multidisciplinary approach not just to identifying and treating the illness, but also to ensuring that the patient is well nourished (Humbert and Robbins, 2008).

6.2.2 GASTROESOPHAGEAL REFLUX DISEASE (GERD)

Gastroesophageal reflux disease (GERD) is a prevalent medical condition that affects millions of people throughout the world. Symptoms that are both typical and uncommon help to identify patients. In many people with GERD, acid suppression treatment relieves symptoms and prevents problems. The ability to detect and manage disease consequences has thanks to improved Methods of diagnosis and treatment (Park and Lee, 2019). Gastroesophageal reflux disease is marked by decreased, low-efficiency esophageal peristalsis, which results in delayed food bolus transit, less effective mucosal clearance, lower esophageal sphincter incontinence, and delayed stomach emptying. Shortening of the intra-abdominal section of the lower esophageal sphincter (LES), increased risk of hiatal hernia, decreased defense of the esophageal mucosa, and a higher frequency of duodenogastric reflux, which exposes the mucosa to the bile salts' cytoaggressive effects (Grassi et al., 2011) If left untreated, GERD may lead to esophagitis and Barrett's esophagus, as well as other significant consequences. GI bleeding can also occur as a result of esophagitis. Head of bed (HOB) elevations are the only effective lifestyle adjustment for GERD management (Khan et al., 2012). The most potent class of antacid medicines is proton pump inhibitors. They're taken once or twice a day, and the best time to take them is 30 to 60 minutes earlier than meals. Many individuals will experience a return of symptoms after discontinuing taking PPI, mandating lifetime treatment (Katz, Gerson, and Vela, 2013).

6.2.3 FUNCTIONAL DYSPEPSIA (FD)

The presence of persistent or recurrent upper abdominal symptoms, such as epigastric pain and/or burning, postprandial fullness, and early satiation, is defined as functional dyspepsia (Grassi et al., 2011). Various gastrointestinal illnesses outside the gastroduodenal area, such as other functional disorders, can be confused with functional dyspepsia (Pleyer et al., 2014).

The reason of dyspepsia can usually be determined by an upper gastrointestinal endoscopy, which shows that less than 10% of dyspepsia patients have a peptic ulcer, or less 1% have gastroesophageal adenocarcinoma, and more than 70% of the total have functional dyspepsia (Lichtenstein, 2020). Psychological distress, particularly anxiety, is associated with functional dyspepsia and may precede the onset of the disorder in some persons (Aro et al., 2015). Meal-induced dyspepsia is the most widespread cause of functional dyspepsia. A high-fat meal, for example, might cause changes in gastroduodenal physiology by altering gut-hormone responses, such as increasing cholecystokinin levels. Food intolerance or allergy may play a direct role in functional dyspepsia; however, this theory has received little research (Lichtenstein, 2020).

6.2.4 IRRITABLE BOWEL SYNDROME (IBS)

IBS affects roughly 11% of the overall population, depending on the diagnostic criteria used. Around 30% of persons with IBS symptoms will seek medical help for their problems. There is a female majority in the occurrence of IBS over the world. Over 50-year-olds have 25% fewer IBS diagnoses, and there is no correlation with socioeconomic level (Canavan, West and Card, 2014). Abdominal pain or discomfort that is eased by defecation, changes in stool frequency and/or form, presence of mucus in the feces, and bloating or feelings of abdominal distension are the most typical symptoms (Grassi et al., 2011). Around half of all IBS patients suffer from fibromyalgia, chronic fatigue, back problems, pelvic pain, persistent headache, and temporomandibular joint dysfunction, and they occur nearly twice as frequently as in the common population (Kennedy et al., 1998).

These illnesses have a lot of overlap in terms of symptoms, and their etiologies aren't well understood. Some people believe that these disorders should be lumped together under the phrase "functional somatic syndromes." While these illnesses have certain predisposing risk factors in common, the triggers for each condition differ (Ng et al., 2018; Spiller and Garsed, 2009).

6.2.5 CHRONIC PRIMARY CONSTIPATION

Constipation is a common problem that has a negative blow on people's quality of life. Normal-transit constipation, rectal evacuation problems, and slow-transit constipation are three main chronic constipation that exhibit a lot of overlap. The judgment of primary chronic constipation is a multistep process that begins with the elimination of 'alarm' features (such as accidental weight loss or rectal hemorrhage) that could point out organic diseases (i.e., polyps or tumors) and continues with a therapeutic trial using first-line treatment such as dietary change, way of life modifications, and over-the-counter laxatives (Camilleri et al., 2017). Chronic constipation is thought to be more common as people get older. Chronic constipation is more prevalent in females than in males for the majority of chronic gastrointestinal illnesses, such as IBS and dyspepsia (Kennedy et al., 1998; Lichtenstein, 2020).

6.2.6 DUODENAL ULCERS

Only a small percentage of patients infected with Helicobacter pylori develop duodenal ulcers (DU), implying that other factors are involved. A high-fiber diet lowers the risk of developing DU (Shiotani and Graham, 2002). Vitamin A (Singh et al., 2017a) appears to be protective as well. Fatty acids (Singh et al., 2017b), vitamins C and E, alcohol, and caffeine (Kesharwani et al., 2016) do not appear to be linked to an increased risk of developing DU. Smoking raises the risk of DU, although physical activity appears to protect against it. Dietary fiber, fatty acids, notably linoleic acid, vitamins C, E, and A, alcohol, and caffeine are all factors that have been studied. Smoking, physical activity, and body mass index are examples of lifestyle factors that may take part in a role in the development of DU (Ryan-Harshman and Aldoori, 2004). Dietary fiber may help to avoid the development or recurrence of DU. The types of fiber (soluble or insoluble) and their sources (fruit, vegetables, legumes, or grains) have varied effects on the risk of DU. Soluble fiber-rich foods, such as oranges, carrots, and beans, appear to be more helpful in lowering the risk of DU. There is a lack of data on the impact of nutrition and lifestyle on the risk of DU, particularly in women, and additional research is needed. Vitamin A may also be beneficial in the fight against DU. Other dietary and lifestyle factors have a minor impact on the likelihood of developing DU, however, stopping smoking may aid ulcer repair and avoid recurrence (Kim, Kim, and Lee, 2017).

6.2.7 DIARRHEA

This is the world's second biggest cause of death, and it is a leading cause of death among the elderly in developed countries. Oral or parenteral rehydration is required to prevent general hypotension, organ damage, and failure, regardless of the reason (infection, malabsorption, enzyme shortages, extra-intestinal disorders, etc.) (Liu et al., 2013). Examinations and cultures of the feces are required. Diarrhea can be treated with oral antibiotics, antispasmodics, propulsive medicines, and/or probiotics, depending on the culture results and the clinical course of the disease. Diarrhea is the most common gastrointestinal ailment, accounting for 4 billion episodes and 4% of all fatalities worldwide each year. Antibiotic-associated diarrhea and the use of Probiotics to prevent and cure it, acute infectious diarrhea, traveler's diarrhea, and diarrhea-predominant IBS is backed up by strong evidence from clinical and fundamental research investigations. The most frequently used probiotics include *Bifidobacterium, Lactobacillus,* and *Saccharomyces* (Guarino, Guandalini, and Vecchio, 2015).

6.2.8 DIVERTICULOSIS AND DIVERTICULITIS

Diverticulosis and its complications are referred to by a variety of words. Diverticulosis is defined as the presence of diverticula. Diverticular illness is a symptomatic and clinically severe diverticulosis. Acute diverticulitis is characterized by active inflammation in the diverticula. This can be one-time or recurring, as well as simple or complex. Diverticula are structural alterations in the colonic wall that produce "pockets" known as diverticula in the pas (Thompson, 2016) t. Diverticula occur when the mucosa and submucosa of the colon herniate throughout defects in the rounded muscle layers of the colonic wall. This frequently occurs in the colon at the sites of entering blood vessels (Slack, 1962). Diverticular illness is a prevalent disease that affects around half of all people over the age of 60. Diverticular disease is becoming more common in younger people, who often have a more virulent type and experience more serious consequences. Obesity (Jarouliya and Keservani, 2019a), smoking, and certain medications (such as nonsteroidal anti-inflammatory drugs, corticosteroids, and opiates) are all risk factors for diverticulitis. In cases of mild diverticulitis, antibiotics may not be required. Surgery is no longer regarded as required after two occurrences of diverticulitis since the best timing and need for surgery are unknown. It's unknown what function fiber plays in avoiding diverticulitis and its consequences. Chronic stomach pain and inflammation are linked to diverticular illness (Feuerstein and Falchuk, 2016).

6.2.9 ULCERATIVE COLITIS AND CROHN DISEASE

Ulcerative colitis is a type of inflammatory bowel disease that affects 1–2 million individuals in the United States and many more across the world (Kellock, 1958; Mishina et al., 1996).

Dr. Burrill B. Crohn and colleagues originally described Crohn's disease in 1932. It falls under the category of chronic idiopathic inflammatory bowel illness, along with ulcerative colitis (UC) (IBD). Crohn's disease is an inflammatory bowel disease (IBD) that can affect any region of the gastrointestinal tract. It is usually treated medically, although a large number of people will require surgery. The major goals of treatment are to achieve and maintain a steroid-free remission, reduce the risk of complications and surgery, and enhance overall quality of life (Feuerstein and Cheifetz, 2017).

6.2.10 CANCER OF THE DIGESTIVE TRACT

The elderly have a higher prevalence and incidence of esophageal cancer, which is attributable in part to the higher prevalence of chronic esophagitis and lengthy histories of smoking and/or alcohol misuse in this age group (Kitajima, 2009). Gastric cancer is likewise on the rise in both prevalence and incidence, owing to the higher prevalence of gastric ulcers in these individuals and their prolonged exposure to causative factors (especially H. pylori), which causes atrophic gastritis and mucosal metaplasia. When it comes to organic colon disease, it's vital to remember the high prevalence and incidence of colon cancer, polyps, adenomas (which are typically cancer's early stages), and diverticulosis among the elderly (Grassi et al., 2011).

6.2.11 BILIARY DISEASES

Cholelithiasis is more common in the elderly, with 14.5% of males and 25% of women over 65 years old suffering from it. This trend reflects the cumulative effects of lithogenic variables over time, the spread of the western lifestyle and dietary habits, age-related declines in the bile acid pool, and increased biliary cholesterol concentrations reported in specific ethnic groups. After cholecystectomy, the prevalence of post-cholecystectomy syndromes and recurrent bile stones has also increased. One issue could be the presence of diseases that make cholecystectomy contraindicated or raise the risk of complications (respiratory insufficiency, severe cardiopathy) (Jain and Kaushik, 2020; Thimme, 2020).

6.2.12 LIVER DISEASES

Chronic hepatitis (mostly HCV-related), cirrhosis, and HCC are all more common in the elderly, which is the culmination of a protracted process including the combined impacts of liver-cell degeneration and necrosis, fibrosis, and regenerative mechanisms within the hepatic parenchyma. It is important to note that the use of interferon in these situations is linked to a higher risk of side effects, mostly hematologic and psychological (depression), as well as a higher frequency of contraindications due to the high prevalence of thyroid disease in the elderly (Grassi et al., 2011; Krishnan, 2019).

6.2.13 PANCREATIC DISEASES

Because of the cumulative impact of exogenous toxins like alcohol, Acute and chronic pancreatitis are more common in elderly adults, in addition to the increasing prevalence and incidence of cholelithiasis. Pancreatic cancer is more common and occurs more frequently in the elderly (Kalra et al., 2002; Meier and Beglinger, 2006).

6.3 FUNCTIONAL FOODS AS DIETARY INTERVENTIONS FOR MAINTAINING FUNCTIONAL DISEASE

Instead of taking drugs to manage gastrointestinal illnesses, which in some cases, such as NSAIDs, may cause a lot of problems, dietary intervention with functional foods has created a new window of opportunity for gut health (Brown et al., 2015). Probiotics, prebiotics, and Polyphenols are three well-known examples of functional foods that are legitimate alternatives to pharmaceuticals. Probiotics and prebiotics can improve the intestinal immune system by modulating gut microbiota. Polyphenols, on the other hand, are known for their anti-inflammatory properties (Birch and Bonwick, 2019; Verschuren, 2002).

6.3.1 PROBIOTICS, PREBIOTICS, AND SYNBIOTICS

The human gut is home to a diverse range of bacterial species. Many of which have essential metabolic and immunological functions, all of which have an impact on the host's nutritional and physiological state. As per a consensus definition, probiotics are loosely defined as live microorganisms belonging to natural biota with low or no pathogenicity, but with functions important to the host's health and well-being (Vijayaram and Kannan, 2018). Some probiotic genera and strains are emerging, some with important health benefits, such as *Lactobacillus rhamnosus* GG, *Lactobacillus reuteri*, bifidobacteria, and specific *Lactobacillus casei* strains, the *Lactobacillus acidophilus*-group, *Escherichia coli* strain (Nissle, 1917), certain *Enterococci*, particularly *Enterococcus faecium* SF68. Probiotics were first utilized to influence human health by altering the intestinal flora. Probiotics and their impact on human being health have been studied in a variety of food matrices as well as single or mixed microbial culture preparations (The University of North Dakota, 2011). Furthermore, probiotic strains are now known to have different health-promoting qualities. An international expert group of the International Life Sciences Institute (ILSI) assessed the categorized and published evidence of functionality of different probiotics in four areas of (human) application, namely (i) metabolism, (ii) chronic intestinal inflammatory and functional disorders, (iii) infections, and (iv) allergy (Cencic and Chingwaru, 2010).

A prebiotics is defined as "a selectively fermented substance or fiber that allows for particular alterations in the gastrointestinal microflora's composition and/or activity, resulting in advantages to the host's well-being and health." Other, more specific benefits of prebiotics on health are indirect, such as diarrhea or constipation prevention, modification of gut flora metabolism, cancer prevention, effects on lipid metabolism, mineral adsorption stimulation, and immunomodulatory qualities. Prebiotics are a broad set of carbohydrate (Jarouliya and Keservani, 2019b) components that, despite offering nutraceutical and nutritional benefit, are little understood in terms of their origin, fermentation characteristics, and dosages required for health effects (Douglas and Sanders, 2008).

Only bifidogenic, nondigestible oligosaccharides (especially inulin, its hydrolysis product oligofructose, and (trans) galactooligosaccharides) now meet all of the prebiotic classification criteria. In recent years, successful attempts to make infant formula more breast-milk-like by adding fructo- and (mainly) galactooligosaccharides have been described (Vrese, 2001).

Probiotics and prebiotics play distinct functions in human nutrition, mostly involving population or activity modification of the microbiota that colonizes the human GI tract. Regular probiotic or prebiotic use can boost immune function, increase colonic integrity, reduce the frequency and duration of intestinal infections, reduce allergic reactions, and improve digestion and disposal (Douglas and Sanders, 2008). Human subjects and their intestinal microbiota have developed together to attain a condition of mutual tolerance, which is significant. Inflammatory bowel disease (IBD) is thought to be caused by a breakdown in this mutual connection, according to accumulating evidence from both animal models and human research (Hedin, Whelan and Lindsay, 2007). However, probiotics and prebiotics have been studied in clinical trials as therapies for IBD, with mixed results. While there is widespread agreement on the benefits of probiotics and prebiotics, their impact on Crohn's disease bowl health is less clear. Variations in methodology, discrepancies in the variety of probiotic, prebiotic, and combination (synbiotic) treatments examined, and heterogeneity in test subjects utilized, such as special patient groups, could all contribute to this (Hedin, Whelan and Lindsay, 2007). Synbiotics are probiotic and prebiotic mixtures that work together (Vrese, 2001).

6.3.1.1 MECHANISMS OF PROBIOTIC ACTIONS

While probiotics are gaining popularity as antibiotic or anti-inflammatory medication alternatives, their mechanism of action remains unknown (Oelschlaeger, 2010). Probiotic can affect the immune system of the host, directly affect other microbes, or act on microbial products, host products, or food components. A probiotic's efficiency is determined by its metabolic capabilities, the chemicals present on its surface, the components it secretes, and the microorganism's integral portions, such as DNA or peptidoglycan (Oelschlaeger, 2010).

6.4 CONCLUSION

Functional food is food that alters the physiological role. The consumption of functional food is increasing day by day, and it is necessary for maintaining a healthy lifestyle. Functional foods can be used in a variety of ways and through a variety of processes to combat functional disease. Drug use for the metabolic and functional diseases, which have become more common, produces negative side effects. A nutritious diet that uses functional foods may help to manage and prevent the condition.

ACKNOWLEDGMENT

We thank the MET's, Institute of Pharmacy, BKC, affiliated under Savitribai Phule Pune University, Nashik, and R. C. Patel Institute of Pharmaceutical Education and Research, Shirpur. EDA thanks the Ministry of Tribal Affairs Govt. of India for providing financial assistance in the form of a fellowship (Award No.: 202021-NFST-MAH-01235).

KEYWORDS

- **functional disease**
- **functional food**
- **prebiotic**
- **probiotic**
- **supplements**

REFERENCES

Ahire, E., Sonawane, V., Surana, K., & Talele, G., (2021). Drug discovery, drug-likeness screening, and bioavailability: Development of drug-likeness rule for natural products. *Applied Pharmaceutical Practice and Nutraceuticals.* apple academic press.

Aluko, R. E., (2012). *Functional Foods and Nutraceuticals.* Springer.

Aro, P., et al., (2015). Anxiety is linked to new-onset dyspepsia in the Swedish population: A 10-year follow-up study. *Gastroenterology, 148*(5), 928–937. doi: 10.1053/j.gastro.2015.01.039.

Berstad, A., Raa, J., & Valeur, J., (2017). Functional food for functional disorders. *Microbial Ecology in Health and Disease, 28*, 1281955.

Birch, C. S., & Bonwick, G. A., (2019). Ensuring the future of functional foods. *International Journal of Food Science & Technology, 54*, 1467–1485.

Brown, L., Poudyal, H., & Panchal, S. K., (2015). Functional foods as potential therapeutic options for metabolic syndrome. *Obesity Reviews, 16*, 914–941.

Camilleri, M., et al., (2017). *Chronic Constipation* (Vol. 3, pp. 1–19). Nature Publishing Group. doi: 10.1038/nrdp.2017.95.

Canavan, C., West, J., & Card, T., (2014). The epidemiology of irritable bowel syndrome. *Clinical Epidemiology, 6*(1), 71–80. doi: 10.2147/CLEP.S40245.

Carvajal-Millan, E., Yaser, A. Z., & Haghi, A., (2021). *Natural Food Products and Waste Recovery: Healthy Foods, Nutrition Design, and Extraction of Valuable Compounds*, CRC PRESS.

Cencic, A., & Chingwaru, W., (2010). *The Role of Functional Foods, Nutraceuticals, and Food Supplements in Intestinal Health*, 611–625. doi: 10.3390/nu2060611.

Douglas, L. C., & Sanders, M. E., (2008). Probiotics and prebiotics in dietetics practice. *Journal of the American Dietetic Association, 108*(3), 510–521. doi: 10.1016/j.jada.2007.12.009.

Doyon, M., & Labrecque, J. A., (2008). Functional foods: A conceptual definition. *British Food Journal, 110*(11), 1133–1149. doi: 10.1108/00070700810918036.

Feuerstein, J. D., & Cheifetz, A. S., (2017). Crohn disease: Epidemiology, diagnosis, and management. *Mayo Clinic Proceedings, 92*(7), 1088–1103. doi: 10.1016/j.mayocp.2017.04.010.

Feuerstein, J. D., & Falchuk, K. R., (2016). Diverticulosis and diverticulitis. *Mayo Clinic Proceedings, 91*(8), 1094–1104. doi: 10.1016/j.mayocp.2016.03.012.

Foundation, N. P., (no date). *Common Disorders of the Pancreas National Pancreas Foundation Brochure,* 1–3.

Grassi, M., et al., (2011). Cambios, dolencias funcionales y enfermedades en el sistema gastrointestinal en personas mayores. *Nutricion Hospitalaria, 26*(4), 659–668. doi: 10.3305/nh.2011.26.4.5109.

Guarino, A., Guandalini, S., & Vecchio, A. Lo., (2015). Probiotics for prevention and treatment of diarrhea. *Journal of Clinical Gastroenterology, 49*, S37–S45. doi: 10.1097/MCG.0000000000000349.

Guesry, P. R., (2005). Impact of "functional food." *Forum of Nutrition, 57*, 73–83. doi: 10.1159/000083770.

Hedin, C., Whelan, K., & Lindsay, J. O., (2007). Evidence for the use of probiotics and prebiotics in inflammatory bowel disease: A review of clinical trials. *Proceedings of the Nutrition Society, 66*(3), 307–315. doi: 10.1017/S0029665107005563.

Humbert, I. A., & Robbins, J., (2008). Dysphagia in the elderly. *Physical Medicine and Rehabilitation Clinics of North America, 19*, 853–866.

Jain, S., & Kaushik, R., (2020). *Diseases of Biliary Tract.,* 1–45.

Jarouliya, U., & Keservani, R. K., (2019a). Pathways leading to child obesity. In: Debasis, B., (ed.), *Global Perspectives on Childhood Obesity: Current Status, Consequences and Prevention* (2nd edn., pp. 137–146) Academic Press, Elsevier, chapter 12. ISBN: 9780128128404.

Jarouliya, U., & Keservani, R. K., (2019b). Protein function as cell surface and nuclear receptor in human diseases. In: Sanjay, B., & Debarshi, K. M., (eds.), *Medicinal Chemistry with Pharmaceutical Product Development* (pp. 1–32). Academic Press, CRC Press, Taylor & Francis Group. Chapter 1. ISBN: 9781771887106.

Johnson, I., & Williamson, G., (2003). *Phytochemical Functional Foods.* CRC Press.

Kalra, M. K., Maher, M. M., Sahani, D. V., Digmurthy, S., & Saini, S., (2002). Current status of imaging in pancreatic diseases. *Journal of Computer Assisted Tomography, 26*, 661–675.

Katz, P. O., Gerson, L. B., & Vela, M. F., (2013). Guidelines for the diagnosis and management of gastroesophageal reflux disease. *American Journal of Gastroenterology, 108*(3), 308–328. doi: 10.1038/ajg.2012.444.

Kellock, T. D., (1958). Ulcerative colitis. *International Journal of Food Sciences and Nutrition, 12*(2), 58–62. doi: 10.3109/09637485809142520.

Kennedy, T. M., et al., (1998). Irritable bowel syndrome, gastro-oesophageal reflux, and bronchial hyper-responsiveness in the general population. *Gut, 43*(6), 770–774. doi: 10.1136/gut.43.6.770.

Keservani, R. K., Kesharwani, R. K., Sharma, A. K., Vyas, N., & Chadoker, A., (2010b). Nutritional supplements: An overview. *International Journal of Current Pharmaceutical Review and Research, 1*(1), 59–75.

Keservani, R. K., Kesharwani, R. K., Vyas, N., Jain, S., Raghuvanshi, R., & Sharma, A. K., (2010a). Nutraceutical and functional food as future food: A review. *Der Pharmacia Lettre, 2*(1), 106–116.

Keservani, R. K., Sharma, A. K., & Kesharwani, R. K., (2017). An overview and therapeutic applications of nutraceutical and functional foods. *Recent Advances in Drug Delivery Technology*, 160–201. ISBN:9781522507543.

Keservani, R. K., Sharma, A. K., & Kesharwani, R. K., (2020). *Nutraceuticals and Dietary Supplements: Applications in Health Improvement and Disease Management*. CRC Press. ISBN: 9781771888738.

Kesharwani, R. K., Singh, P., & Keservani, R. K., (2016). Green coffee bean extract and chlorogenic acids: Chemistry and novel antioxidant benefits. In: Bagchi, D., Moriyama, H., & Swaroop, A., (eds.), *Green Coffee Bean Extract in Human Health* (pp. 1–17). Boca Raton, Taylor and Francis, USA. Chapter 1. ISBN: 9781498716376.

Khan, B. A., et al., (2012). Effect of bed head elevation during sleep in symptomatic patients of nocturnal gastroesophageal reflux. *Journal of Gastroenterology and Hepatology (Australia), 27*(6), 1078–1082. doi: 10.1111/j.1440-1746.2011.06968.x.

Kim, J., Kim, K. H., & Lee, B. J., (2017). Association of peptic ulcer disease with obesity, nutritional components, and blood parameters in the Korean population. *PLoS One, 12*(8), 1–14. doi: 10.1371/journal.pone.0183777.

Kitajima, M., (2009). Gastric cancer: Introduction. *Gastric Cancer, 12*(SUPPL. 1), 1, 2. doi: 10.1007/s10120-008-0491-6.

Krishnan, S., (2019). *Liver Diseases – An Overview*. doi: 10.20959/wjpps20191-13036.

Li, L., Hope, L. J., Simon, C., Jamie, P., Susana, S., Joy, E. L., Igor, R., et al., (2013). *Global Diarrhea Burden Diarrhea: Common Illness, Global Killer (*p. 1). U.S. Department of Health and Human Services. Available at: https://www.cdc.gov/healthywater/global/diarrhea-burden.html (accessed on 22 February 2023).

Lichtenstein, G. R., (2020). Functional dyspepsia. *Gastroenterology and Hepatology, 16*(2), 59. doi: 10.5124/jkma.2005.48.1.48.

Lordan, S., Ross, R. P., & Stanton, C., (2011). Marine bioactives as functional food ingredients: Potential to reduce the incidence of chronic diseases. *Marine Drugs, 9*, 1056–1100.

Meier, R. F., & Beglinger, C., (2006). Nutrition in pancreatic diseases. *Best Practice & Research Clinical Gastroenterology, 20*, 507–529.

Mishina, D., et al., (1996). On the etiology of Crohn's disease. *Proceedings of the National Academy of Sciences of the United States of America, 93*(18), 9816–9820. doi: 10.1073/pnas.93.18.9816.

Ng, Q. X., Soh, A. Y. S., Loke, W., Lim, D. Y., & Yeo, W. S., (2018). The role of inflammation in irritable bowel syndrome (IBS). *Journal of Inflammation Research, 11*, 345.

Oelschlaeger, T. A., (2010). Mechanisms of probiotic actions – A review. *International Journal of Medical Microbiology, 300*(1), 57–62. doi: 10.1016/j.ijmm.2009.08.005.

Park, C. H., & Lee, S. K., (2019). Gastroesophageal reflux disease. *The Korean Journal of Gastroenterology = Taehan Sohwagi Hakhoe Chi, 73*(2), 70–76. doi: 10.4166/kjg.2019.73.2.70.

Pleyer, C., et al., (2014). Overdiagnosis of gastro-esophageal reflux disease and underdiagnosis of functional dyspepsia in an uSA community. *Neurogastroenterology and Motility, 26*(8), 1163–1171. doi: 10.1111/nmo.12377.

Roberfroid, M. B., (2000). Concepts and strategy of functional food science: The European perspective. *American Journal of Clinical Nutrition, 71*(6 Suppl.), 1660–1664. doi: 10.1093/ajcn/71.6.1660s.

Ryan-Harshman, M., & Aldoori, W., (2004). How diet and lifestyle affect duodenal ulcers. Review of the evidence. *Canadian Family Physician, 50*, 727–732.

Shiotani, A., & Graham, D. Y., (2002). Pathogenesis and therapy of gastric and duodenal ulcer disease. *Medical Clinics of North America, 86*(6), 1447–1466. doi: 10.1016/S0025-7125(02)00083-4.

Singh, P., Kesharwani, R. K., & Keservani, R. K., (2017a). Antioxidants and vitamins: Roles in cellular function and metabolism. In: *Sustained Energy for Enhanced Human Functions and Activity* (pp. 385–407). Academic Press. ISBN:9780128054130.

Singh, P., Kesharwani, R. K., & Keservani, R. K., (2017b). Protein, carbohydrates, and fats: Energy metabolism. In: *Sustained Energy for Enhanced Human Functions and Activity* (pp. 103–115). Academic Press. ISBN:9780128054130.

Slack, W. W., (1962). The anatomy, pathology, and some clinical features of diverticulitis of the colon. *British Journal of Surgery, 50*(220), 185–190. doi: 10.1002/bjs.18005022012.

Spiller, R., & Garsed, K., (2009). Postinfectious irritable bowel syndrome. *Gastroenterology, 136*, 1979–1988.

Sura, L., et al., (2012). Dysphagia in the elderly: Management and nutritional considerations. *Clinical Interventions in Aging, 7*, 287–298. doi: 10.2147/CIA.S23404.

Surana, K. R., Ahire, E. D., Sonawane, V. N., Talele, S. G., & Talele, G., (2021). Informatics and methods in nutrition design and development. *Natural Food Products and Waste Recovery: Healthy Foods, Nutrition Design, and Extraction of Valuable Compounds*, 33.

The University of North Dakota, (2011). *Prebiotics/Probiotics,* 1–3. Available at: https://und.edu/student-life/dining/_files/docs/fact-sheets/probiotics.pdf (accessed on 22 February 2023).

Thimme, R., (2020). Diseases of the gall-bladder and bile ducts. *Deutsche Medizinische Wochenschrift, 145*(5), 273. doi: 10.1055/a-0952-9479.

Thompson, A. E., (2016). Diverticulosis and diverticulitis. *Journal of the American Medical Association (JAMA), 316*(10), 1124. doi: 10.1001/jama.2016.3592.

Verschuren, P. M., (2002). Functional foods: Scientific and global perspectives. *British Journal of Nutrition, 88*, S126–S130.

Vijayaram, S., & Kannan, S., (2018). Probiotics: The marvelous factor and health benefits. *Biomedical and Biotechnology Research Journal (BBRJ), 2*(1), 1. doi: 10.4103/bbrj.bbrj_87_17.

Volpe, S. L., (2015). Functional food and health. *ACSM's Health and Fitness Journal, 19*(1), 32–34. doi: 10.1249/FIT.0000000000000087.

Vrese, M. De., (2001). *Probiotics, Prebiotics, and Synbiotics—Approaching a Definition, 73*(14).

CHAPTER 7

ROLE OF FUNCTIONAL FOODS IN CARDIOVASCULAR DISEASES

SHUBHAM J. KHAIRNAR,[1] MITHUN RUDRAPAL,[2] EKNATH D. AHIRE,[1] MOHINI R. JAGTAP,[3] and SANJAY J. KSHIRSAGAR[1]

[1]*MET's Institute of Pharmacy, Bhujbal Knowledge City, Adagoan, Nashik, Maharashtra, India*

[2]*Rasiklal M. Dhariwal College of Pharmacy, Chinchwad, Pune, Maharashtra, India*

[3]*JMCTE Institute of Pharmacy, Nashik, Maharashtra, India*

ABSTRACT

Functional foods are foods that include physiologically active dietary components that give health advantages beyond simple nutrition. Many functional foods have been discovered as having the potential to help prevent and treat cardiovascular disease, which is the leading cause of mortality in the United States. These foods include soybeans, oats, psyllium, flaxseed, garlic, tea, salmon, grapes, nuts, and stanol- and sterol-ester enriched margarine. These food products may help to reduce the risk of heart disease by lowering blood lipid levels, improving arterial compliance, lowering low-density lipid oxidation, decreasing plaque formation, free radical scavengers, and inhibiting platelet aggregation when consumed in sufficient amounts on a regular basis.

Applications of Functional Foods in Disease Prevention. Raj K. Keservani & Eknath D. Ahire, PhD (Eds.)

7.1 INTRODUCTION

In the United States, heart disease (cardiovascular diseases (CVD)) has been the leading cause of death since 1918. As we enter the 21st century, we face a similarly bleak outlook for this devastating disease, as CVD continues to claim more lives than the next 14 causes of death combined. Heart disease claims the lives of one people every 33 seconds, accounting for one out of every 2.4 fatalities in 1997(Association, 2004). People are becoming more concerned about the heart disease epidemic, and with the rise of the "self-care" movement, they are beginning to incorporate foods into their diet with the goal of lowering cholesterol and thus lowering their CVD risk. According to a PREVENTION magazine/Food Marketing Institute survey of 1000 shoppers, 72% buy foods to lower cholesterol (Slaughter, 1992). "Foods that have a health advantage are known as "functional foods. Decreasing increased blood total cholesterol (TC) and low-density lipoprotein (LDL) cholesterol, in addition to basic nutrition. According to the International Life Sciences Institute, functional foods give health benefits beyond basic nutrition due to physiologically active dietary components (Clydesdale, 1999). According to the American Dietetic Association's most recent position document, functional foods can be whole, fortified, enriched, or improved meals. They have "a possibly beneficial effect on health when consumed as part of a variety of foods on a regular basis at effective levels," and can include whole, fortified, enriched, or enhanced foods (Hasler et al., 2000). This review summarizes recent literature supporting the efficacy of 10 functional foods in CVD risk reduction, with an emphasis on clinical findings: Soybeans, oats, psyllium, flaxseed, garlic, tea, salmon, grapes, nuts, and margarine fortified with stanol and sterol ester (Ahire et al., 2021; Carvajal-Millan et al., 2021; Surana et al., 2021).

7.2 FOOD USED IN CARDIOVASCULAR DISEASES

7.2.1 SOY PROTEIN

The efficacy of soy protein in lowering the risk of coronary heart disease (CHD) was confirmed in October 1999, when the Food and Drug Administration (FDA) approved a health claim for this diet-disease relationship under the Nutrition Labeling and Education Act of 1990 (NLEA) (FDA, 1999). Protein Technologies International's health claim petition to the FDA summarized more than 40 clinical intervention trials, including a 1995 eta-analysis by Anderson et al. As part of a diet low in saturated fat and cholesterol, the FDA has defined a daily effective dose of soy protein of 25 g/day or 6.25 g/reference amount usually eaten (RACC) (25 divided by four eating occasions per day). Recent research suggests that as little as 25 grams of soy protein per day can effectively lower blood lipids. Washburn et al. found a 7% reduction in LDL and a 6% decrease in TC after 5 weeks in 51 perimenopausal women who consumed 20 g of soy protein per day. Although isoflavones, a class of plant substances functionally similar to estradiol, were proposed in the health claim petition as the physiologically active component in soy responsible for its cholesterol-lowering effect, the FDA rejected this hypothesis. There is very little human clinical data to support that isoflavones can lower TC or LDL cholesterol on their own. Isolated isoflavones have been shown in clinical trials to be ineffective in lowering cholesterol (Hodgson et al., 1998 and Nestel et al., 1999), even at concentrations as high as 160 mg/d. Although a recent clinical trial (Crouse et al., 1999) discovered that 25 grams of isolated soy protein with increasing levels of isoflavones (4, 27, 37, or 62 mg) had a dose-dependent lowering effect on cholesterol lowering in subjects with high cholesterol, the results were modest, with the highest level of isoflavones (62 mg) lowering TC and LDL cholesterol levels by only 4% and 6%, respectively. A higher concentration of isoflavones may be required to have a significant beneficial effect on blood lipids. In a recent study of 13 healthy, normocholesterolemic women who consumed three different levels of isoflavones (10, 64.7, and 128.7 mg), only the highest intake significantly reduced LDL cholesterol by 7.6% to 10% ($P<0.05$), with no effect on TC (Merz-Demlow et al., 2000). Aside from isoflavones, many other components of soy, such as amino acids, saponins, phytic acid, trypsin inhibitors, fiber, and globulins, may help mediate changes in lipid metabolism (Potter et al., 2000). Aside from their ability to lower blood lipids, soy isoflavones may provide a variety of heart health benefits, including improved arterial compliance and reduced LDL oxidation (Tikkanen et al., 2000). Functional foods have a variety of functions and mechanisms that are helpful in the fight against functional disease. Drug use for the functional and metabolic diseases, which are more common and have negative side effects. Use of a functional diet, which includes food, may help to control and avoid the condition. More research is needed to determine the role of soy isoflavones in cardiovascular health (Keservani et al., 2010a, b, 2017; Keservani and Sharma, 2014).

7.2.2 OAT-SOLUBLE FIBER

Viscous soluble fibers, including oat b-glucan, are well known to be hypocholesterolemic (Jenkins et al., 2000). In January 1997, the FDA issued the first food-specific health claim for the association between oat-soluble fiber and reduced risk of CHD in response to a complaint from the Quaker Oats (FDA, 1997). The complaint reviewed 37 human clinical intervention experiments done between 1980 and 1995, 17 of which demonstrated that oat bran or oatmeal reduced total and LDL cholesterol in hypocholesterolemic patients consuming either a conventional American diet or a low-fat diet. According to Quaker Oats, 3 g of b-glucan, oats' principal soluble fiber, is required to achieve a 5% reduction in blood cholesterol, which is around 60 g of oatmeal or 40 g of oat bran (dry weight). As a result, the FDA mandates that a food with the health claim contain 13 g of oat bran or 20 g of oatmeal, as well as at least 1 g of β-glucan-soluble fiber per dose without fortification in the entire oat product. According to a recent meta-analysis of 25 research on oat products (Brown et al., 1999), 3 g of soluble fiber from oats (three servings of oatmeal, 28 g each) lowered TC and LDL by 0.13 mmol/L (2%). Although two recent studies suggest that the presence of various phenolic compounds in oats may reduce LDL oxidation (Emmons et al., 1999), the mechanism by which oat-soluble fiber lowers blood lipids is most likely related to its ability to either reduce cholesterol and bile acid absorption or delay lipid digestion (Wood et al., 1999). Oatmeal has also been demonstrated to prevent arterial constriction, which is an early symptom of heart disease when eaten with a high-fat meal (Anonymous et al., 1999). Around 50 healthy adults ate a meal with 50 g of fat, 1.5 cups of oatmeal with 3.5 g of β-glucan-soluble fiber, 800 IU of vitamin E, or a bowl of hot whole wheat cereal.

Ultrasound measurements revealed that oatmeal reduced the endothelial dysfunction caused by the high-fat meal. Psyllium Seed Husk Another viscous soluble fiber extracted from the husk of the blonde psyllium seed, defined as the dried seed coat (epidermis) of the seed of Plantago ovata, has been shown to be very effective in lowering blood lipids. In response to a petition from the Kellogg Company, the FDA expanded the health claim for oat-soluble fiber and CHD to include fiber from psyllium seed husk in February 1998 (FDA, 1998). The FDA's decision was based on a review of 21 human studies, including a meta-analysis of 404 subjects. They determined that there was strong scientific consensus to support the association between soluble fiber from psyllium husk and a lower risk of CHD based on the totality of publicly accessible scientific information. The FDA determined that 7 g/day of soluble fiber from psyllium was related with a significant reduction in blood lipids (equal to 10.2 g husk) and would cut total cholesterol by 4% to 6% and Cholesterol level by 4% to 8%. According to RACC, a food must contain 1.7 g dietary fibers from psyllium husk to qualify for the psyllium health claim. Because esophageal and gastrointestinal obstructions can occur after consuming psyllium seed husk if it is not consumed with enough liquid, products bearing the health claim must include label statements alerting consumers to the need to consume adequate amounts of liquids. Since the health claim's approval, two additional meta-analyzes have confirmed psyllium's hypocholesterolemic effects, with a percentage reduction in LDL ranging from 6% to 7.25%.

Brown et al. conducted a meta-analysis of 2990 subjects to investigate the hypocholesterolemic effects of various types of soluble fiber (pectin, oat bran, guar gum, and psyllium), of which 17 used psyllium as the fiber source (9.1 g average dose) and included 757 subjects. The results showed that 1 g of psyllium soluble fiber resulted in total cholesterol and LDL cholesterol decreases of –0.028 mmol/L (–1.10 mg/dL) and –0.029 mmol/L (–1.11 mg/dL), respectively. There were no significant effects on triglycerides, but there were significant, albeit minor, reductions in HDL cholesterol. Anderson et al. conducted a second meta-analysis of eight studies involving 384 subjects who consumed 10.2 g of psyllium daily as an adjunct for more than 8 weeks and found that TC and LDL were reduced by 4% ($P<0.0001$) and 7% ($P<0.0001$), respectively, compared to placebo. There were no negative effects on serum HDL or triglycerides. A significant reduction in the ratio of apolipoprotein (apo) B to apo A-I was also observed ($P<0.05$). Psyllium's efficacy in the treatment of hypercholesterolemia, even in people with mild-to-moderate hypercholesterolemia and type 2 diabetes (Anderson et al., 2000), has been demonstrated.

In the meta-analysis by Brown et al. for example, the average length of treatment was less than 2 months (53 days). To evaluate the safety and effectiveness of 10.2 g of psyllium husk fiber (5.1 g twice daily) as an additional therapy for 26 weeks after 8 weeks on an American Heart Association Diet, Anderson et al. recently undertook multicenter research (AHA). The 163 participants who adhered to the trial protocol in full received psyllium therapy, which decreased serum total cholesterol by around 5% and LDL-C by about 7%. Viscous fibers and other functional dietary components clearly present an opportunity to bridge the gap between diet and drug therapy. Flaxseed is an unique plant food with potential cardiovascular health benefits because it contains high levels of two primary dietary constituents: (i) α-linolenic acid, an omega-3 fatty acid, and (ii) lignans, a type of phytoestrogen (Setchell, 1995). Flaxseed has the highest a-linolenic acid content of any seed oil (57%), and its lignan content (milled) is approximately 67,000 mg/100 g—up to 800 times higher than in 66 other plant foods studied (Thompson et al., 1991). Although most research on flaxseed has focused on its cancer chemopreventive activity due to the lignans' weak estrogenic and antiestrogenic activity, data on flaxseed's ability to lower serum cholesterol levels is also emerging. Flaxseed bread is being marketed to women in Australia, the United Kingdom, and New Zealand as an alternative to hormone replacement therapy. Arjmandi et al. conducted a randomized, double-blind crossover study. Consumption of muffins and bread containing 38 grams of either flaxseed or sunflower seed for 6 weeks reduced cholesterol by 6.9% and 5.5%, respectively, in 38 hypercholesterolemic postmenopausal women ($P<0.01$). Both treatments had no effect on serum HDL and triglyceride levels. Only flaxseed reduced serum Lp (a) concentrations, cardiovascular disease is a major predictor, by 7.4% ($P<0.05$).

However, it is unclear whether the cholesterol-lowering effects of flaxseed were caused by the lipid or nonlipid components. Jenkins et al. conducted a controlled clinical intervention trial with partially defatted flaxseed in 29 hyperlipidemic subjects to answer this question (22 men and 7 postmenopausal women). During two 3-week treatment periods in a randomized crossover design, subjects consumed approximately 20 g of fiber per day from either partially defatted flaxseed (approximately 50 g) or wheat bran (control) while consuming self-selected NCEP Step II diets. Defatted flaxseed significantly reduced TC (4.6%, $P = 0.001$), LDL (7.6%, $P = 0.001$), apo B (5.4, $P = 0.001$), and apo A-I (5.8, $P = 0.005$) while having no negative effects on HDL. More clinical trials are needed to confirm flaxseed's cardiovascular benefits and identify the physiologically active constituent or constituents responsible. Garlic (Allium sativum) was the most popular herbal product in the United States in 1999. This functional food's medicinal properties, as well as its distinct flavor and pungency, are due to an abundance of sulfur-containing constituents (Nagourney, 1998). The intact garlic bulb contains alliin, an odorless amino acid that is converted enzymatically by alliinase into allicin when garlic cloves are crushed. This latter compound is responsible for fresh garlic's distinctive odor. Allicin then decomposes spontaneously to form at least 100 sulfur-containing compounds. Although it is unknown which component of garlic is responsible for its cholesterol-lowering effect, many commercial garlic preparations are frequently adjusted or standardized to the "alliin yield" of allicin (Schulz, 1997).

Two meta-analyzes published in the early 1990s that combined the results of 21 clinical trials suggested that a standardized dose of garlic extract ranging from 800 to 900 mg/day could reduce TC by 9 to 12% (Warshafsky et al., 1993 and Silagy, 1994). However, several methodologic flaws in these two reviews, such as the fact that dietary intake, weight, and exogenous garlic ingestion were not always well controlled, cast doubt on the validity of these reports. Recent trials have produced contradictory results. A three-month study of 60 cardiac rehabilitation patients who took two garlic capsules daily (ethyl acetate extract from 1 g peeled and crushed raw garlic) or placebo found a substantial decrease in blood TC and triglycerides, as well as an increase in HDL-cholesterol and fibrinolytic activity (Bordia, 1998). Garlic, on the other hand, has been shown to be ineffective in a number of other randomized, placebo-controlled trials. In a randomized, placebo-controlled, double-blind study conducted by Isaacsohn et al. garlic (900 mg/day) was found to be ineffective as a hypocholesterolemic agent after 12 weeks. Similarly, no significant changes in TC, LDL cholesterol, HDL cholesterol, postprandial triglycerides, apo B, or lp (a) were observed in another three-month double-blind, randomized, placebo-controlled trial involving 50 moderately hypercholesterolemic men who received an equal number of garlic (300 mg garlic tablet three times daily) (Superko et al., 2000). Byrne et al. found negative results in a longer-term (6-month) double blind, randomized, parallel trial of 20 hypercholesterolemic subjects taking 900 mg Kwai garlic tablets. Furthermore, no significant effects of garlic on oxidative markers were observed, which contradicts the increased LDL resistance to oxidation observed by Munday et al. in subjects fed 2.4 g aged garlic for 1 week. Clearly, there is no agreement on the cholesterol-lowering effects of garlic, which can be attributed in large part to differences in study design and the use of different types of preparations in clinical trials.

7.2.3 TEA FLAVONOIDS

Tea, made from the dried leaves of the bush plant Camellia sinensis, is the world's second most popular beverage, trailing only water. In 1996, global tea production totaled 2.61 million metric tons, with black, green, and oolong teas accounting for 76%, 22%, and 2%, respectively (Weisburger, 1999). Green tea is made from fresh tea leaves that have been steamed or dried at high temperatures to prevent the polyphenolic compound from oxidizing. These compounds, which include flavandiols, flavonoids, phenolic acids, and flavonols, account for up to 30% of the total dry weight of fresh tea leaves (commonly known as catechins). Epigallocatechin-3-gallate (EGCG), epigallocatechin (EGC), epicatechin-3-gallate (ECG), and catechin (EC) are the four major green tea catechins (Yang, 1999). Fully dried tea leaves are subjected to a full fermentation process during which the polyphenolic compounds are extensively oxidized during the production of black tea. Thus, catechins are reduced to 3–10% of the remaining solids, while bisflavanols, theaflavins, other oligomers, and thearubigins are formed, accounting for more than 20% of the remaining solids. Oolong tea is a tea that has been partially fermented (Yang, 1999).

Although the majority of tea research has concentrated on its anticarcinogenic potential (Yang et al., 2000), there is emerging evidence that tea consumption may also reduce the risk of heart disease. The Zutphen Elderly Study discovered in 1993 that the intake of several flavonoids (e.g., quercetin, myricetin, kaempferol apigenin and luteolin) was inversely associated in a dose-dependent manner with the incidence of myocardial infarction (MI) in a population of 805 elderly men in the Netherlands (Hertog et al., 1993). After five years of follow-up, the mortality rate incidence (1000 persons/year) for fatal and nonfatal MI was 16.2, 13.8, and 7.6, respectively, for flavonoid intakes of 0 to 19 mg/day, 19.1 to 29.9 mg/day, and more than 29.9 mg/day. Since the publication of the Zutphen Elderly Study, several additional epidemiologic studies have discovered that tea may protect against heart disease. The Boston Area Health Study discovered that drinking one cup of tea per day or more reduced the risk of MI by 44% in 340 cases when compared to age-, gender-, and community-matched non-tea-drinking controls (Sesso et al., 1999). Similarly, the Scottish Heart Health Study found a strong inverse correlation between tea consumption and all-cause mortality, coronary death, or any major coronary event, including nonfatal MI or coronary artery surgery (Woodward et al., 1999). Tea may also protect against aortic atherosclerosis, according to a recent Rotterdam study (Geleijnse et al., 1999). This prospective study included 3454 men and women aged 55 and up. Dietary intakes were assessed at baseline using semi quantitative food frequency questionnaires, and calcified plaques in the abdominal aorta were radiographically detected after two to three years of follow-up. Depending on the length of the calcified area, plaques were classified as "mild," "moderate," or "severe." A substantial inverse connection between tea consumption and severe aortic atherosclerosis was observed after controlling for age, gender, BMI, smoking, education, and intake of alcohol, coffee, vitamins, antioxidants, total fat, and total energy. With 1 to 2 cups of tea per day, the odds of having severe aortic atherosclerosis dropped from 0.54 to 0.31 with 4 or even more cups per day. Women's networks were more powerful than men's. The beneficial effects of tea consumption on heart health could be attributed to a variety of mechanisms (Trevisanato et al., 2000), including inhibition of LDL oxidation. However, evidence from clinical trials on the effects of tea consumption on LDL oxidation remains conflicting. In a recent crossover design study of 18 healthy adults who consumed green tea, normal tea with milk, or water every 2 hours (8 cups/day) for 3 days, LDL resistance to oxidation ex vivo was not increased (van et al., 1999). Similarly, no beneficial effects on LDL oxidation ex vivo, plasma antioxidant levels, or blood lipids were observed in a randomized, placebo-controlled study in which 26 healthy female and male smokers (aged 22 to 46 years) consumed 900 mL black and green tea or received a supplement of 3.6 g of green tea polyphenols per day (equaling 18 cups of green tea/day). These negative findings could be attributed to the difficulty of isolating LDL from polyphenolic compounds in serum. A recent study in 20 healthy men looked at the drinking's immediate consequences 1 cup of either black or green tea on ex vivo Cu^{2+}-induced lipoprotein oxidation without first isolating lipoproteins from serum (Fraser et al., 1992). Drinks were consumed over the course of 30 minutes, and blood samples were collected 60 minutes after the subjects finished the drink. When compared to the water control, black tea had a significantly longer ($P = 0.05$) lag time to LDL diene formation and green tea had a similar but nonsignificant trend.

7.2.4 NUTS

The first published prospective cohort studies highlighting the heart-healthy benefits of nut consumption did not appear until the early 1990s (Hodgson et al., 2000). Although walnuts were the first type of nut studied in this regard (Sabate' et al., 1993), subsequent research has shown that a variety of other nuts, including almonds, macadamia, pistachio, pecans, and hazelnuts, can significantly reduce both TC and LDL cholesterol when consumed as part of a low saturated fat, low cholesterol diet. Five recent clinical studies on nuts (mixed nuts, macadamia nuts, pecans, or walnuts) and cardiovascular disease have all found that nut consumption is protective (Bruce et al., 2000; Morgan et al., 2000; Iwamoto et al., 2000). The first study involving mixed nuts (almonds, hazelnuts, or pecans) included 12 hyperlipidemic women in a crossover design over two four-week periods. Subjects began with a refined-food diet and then transitioned to a phytochemical-rich diet consisting primarily of whole grains, legumes, fruits, vegetables, seeds, and two tablespoons of almonds, hazelnuts, or pecans per day. The phytochemical-rich diet reduced TC by 13% and LDL by 16% when compared to the refined-food diet, with no significant changes in HDL or triglycerides (Iwamoto et al., 2000). A second study on macadamia nuts included 30 subjects who followed three different diets for 30 days each. The first diet was a typical American diet (high in saturated fat), the second the AHA Step 1 diet, and the third a macadamia-based diet high in monounsaturated fatty acids. The macadamia-based diet reduced TC, LDL, and HDL by 5% and triglycerides by 10% when compared to the typical American diet. The AHA diet produced similar results, with the exception that triglycerides increased by 8%. In a third randomized, controlled, parallel study, 19 healthy adults were assigned to either a control diet (no nuts) or a pecan diet containing 68 g of pecans per day (with no additional nuts). Following the pecan diet for 8 weeks resulted in a 6% decrease in LDL cholesterol compared to the baseline value. The effects on TC, HDL, and TG were not altered significantly. The fourth and fifth studies (Bruce et al., 2000) involved walnut consumption. The first was a randomized, crossover feeding study in which 49 adults with hypercholesterolemia were randomly assigned to either a Mediterranean diet emphasizing vegetable products, fish, and olive oil (no nuts) or a walnut diet that replaced 41 to 56 g of walnuts per day in place of olive oil and other fatty foods from the Mediterranean diet. The walnut diet and Mediterranean diet, when compared to baseline values, reduced TC by 9% and 5%, respectively, and LDL by 11% and 6%, but had no effect on HDL. Only the walnut diet reduced triglycerides by 8% and very low-density lipoprotein by 12%. Both diets reduced apo A-1 levels by 5%; apo B levels decreased by 13% on the walnut diet and 8% on the Mediterranean diet, but only the walnut diet reduced Lp (a) levels (by 9%). The second walnut study was a four-week controlled, single-blind, crossover study of 40 Japanese men and women that looked at the effect of adding 43 to 57 g of walnuts per day to the average Japanese diet. The addition of walnuts was offset by eating fewer fatty foods. In comparison to the reference diet (the average Japanese diet), the walnut diet reduced TC by 4% in men and 5% in women. LDL cholesterol levels in women fell by 11%. HDL and triglycerides did not differ significantly between men and women, but apoB decreased by 7% in men and 9% in women (Zambon et al., 2000).

7.2.5 GRAPES AND GRAPE JUICE

Numerous studies have been launched over the last 20 years in an attempt to explain why the French population has a lower incidence of CHD than other Western populations despite eating more fat. The link between wine consumption and lower risk of heart disease, now known as the "French paradox," first became apparent in 1979, when St. Leger et al. discovered a strong negative correlation between wine consumption and death from ischemic heart disease in both men and women from 18 countries. The French paradox is thought to be caused in part by the high concentrations of phenolic antioxidants found in grapes, particularly red grapes, which are incorporated into wine during the manufacturing process. For those who prefer not to drink alcohol, new research on the heart health benefits of grapes is emerging. Grapes are high in antioxidants such as catechin, epicatechin, resveratrol, and proanthocyanidins. Resveratrol (3,4,'5 trihydroxystilbene), a naturally occurring phytoalexin, is found primarily in the skin of grapes, with concentrations ranging from 50 to 100 mg/g. Proanthocyanidins are found primarily in the seeds, whereas resveratrol (3,4,'5 trihydroxystilbene), a naturally occurring phytoalexin, is found primarily in the skin of grapes, with concentrations ranging from 50 to 100 mg/g. Resveratrol may contribute to grapes' heart-health benefits by inhibiting lipid peroxidation, free radical scavenging activity, platelet aggregation inhibition, and vasorelaxation effects (Fremont, 2000). Only a few human studies have looked into the ability of grape products to reduce CVD risk. Commercial grape juice has been shown to significantly inhibit LDL oxidation in vitro and in LDL isolated from human subjects. In one study, 15 adults with angiographically documented CAD were given 7.7 1.2 mL/kg/d of purple grape juice to drink for 14 days. Short-term consumption of purple grape juice decreased LDL oxidation susceptibility and increased flow mediated vasodilation, which increased from 2.2 2.9% to 6.4 4.7% (P = 0.003) (Stein et al., 1999). Platelet aggregation has also been demonstrated to be reduced by grape flavonoids. A recent study looked into whether daily consumption of commercial grape, orange, and grapefruit drinks lowered ex vivo platelet activity. In a randomized crossover design, 10 healthy adult volunteers (aged 26 to 58 years, five of each gender) drank 5 to 7.5 mL/kg/d of purple grape juice, orange juice, or grapefruit juice for 7 to 10 days each. Purple grape juice reduced the aggregation reaction of whole blood platelets to 1 mg/L collagen by 77% (from 17.9 2.3 to 4.0 6.8 ohms, P = 0.0002). In healthy participants, purple grape juice contained nearly three times the total polyphenolic concentration of citrus drinks and was a potent platelet inhibitor, but citrus liquids showed no impact (Keevil et al., 2000).

In one study, 21 healthy volunteers (aged 17 to 27 years) were given either red wine (5 mL/kg body weight) or dried black grapes (1 g/kg body weight) to see how they affected blood antioxidant levels. Both red wine and black grapes had a significant effect on serum antioxidant potential, with the highest values after 2 hours in the wine group and after 4 hours in the black grape group, consistent with the finding that polyphenols from grapes take longer to release and absorb into intestinal cells than polyphenols from red wine (Ozturk et al., 1999). Consistent consumption of red, purple, and black grapes and grape juice may prove to be a practical recommendation as an adjunct therapy for preventing and managing heart disease.

7.2.6 FISH N-3 FATTY ACIDS

The inverse relationship between fatty fish consumption and CVD was first discovered nearly 30 years ago, when it was discovered that Greenland Eskimos had low rates of this disease despite eating a diet high in omega-3 (n-3) fatty acids from marine fish (Ahire et al., 2020). Fatty fish and their oils (salmon, tuna, mackerel, sardines, and herring) are the primary sources of the n-3 fatty acids eicosapentaenoic acid (EPA; 20:5) and docosahexaenoic acid (DHA; 22:6) (Kris-Etherton et al., 2000). One of the primary cardio-protective effects of fish oils has been suggested to be triglyceride reduction (Roche et al., 2000; Keservani et al., 2016). EPA and DHA supplements (1 g/d) have been shown to lower this blood lipid in both diabetic and nondiabetic participants (Yosefy et al., 1999). Even when compared to vegetarians, fish eaters have significantly lower levels of Lp (a) (Marcovina et al., 1999). Platelet aggregation is also reduced by fatty acids found in fish oil. A 6-week intervention with 216 mg EPA, 140 mg DHA, 390 mg gamma linoleic acid, and 3480 mg linoleic acid on in vivo platelet survival and ex vivo markers of platelet activation was undertaken in a placebo-controlled double-blind trial involving 26 hypercholesterolemic individuals. Fish oil increased platelet survival while lowering malondialdehyde production ($P<0.05$). Fish oil improved platelet survival ($P<0.05$) while decreasing malondialdehyde formation. Changes in platelet activation were linked to significant reductions in total cholesterol (–2.9%), LDL cholesterol (–3.5%), and triglycerides (–12.4%) (Pirich et al., 1999). Consumption of moderate amounts of fish (one to two servings per week) has been linked to lower CHD mortality in white men (Albert et al., 1999), as well as lower non-sudden MI mortality (Daviglus et al., 1997). However, fish consumption has not been proven to reduce CVD death in healthy people. In a recent epidemiologic study involving National Health and Nutrition Examination Survey I participants who were followed for an average of 18.8 years, white men who consumed fish once per week had an age-adjusted risk of death that was only about three quarters that of men who never consumed fish (Gillum et al., 2000). However, the risk of cardiovascular death was not reduced significantly in these men. Although more research is needed (there is no data on whether n-3 fatty acids from fish can cause regression of established heart disease), the current consensus is that a moderate intake of fatty fish is beneficial for heart health (Nestel et al., 2000).

7.2.7 PLANT STEROL AND STANOL ESTERS

Plant sterols' ability to lower cholesterol was discovered in the 1950s. Early studies looked into the efficacy of high doses of plant sterols (such as b-sitosterol, campesterol, and stigmasterol), which are found naturally in vegetable fats/oils and pine trees. The average American diet contains about 250 mg of phytosterols, while vegetarians consume twice as much (Nestel et al., 2000). Plant stanols, the saturated derivatives of sterols (i.e., sitostanol), have been shown in studies to lower cholesterol more effectively than plant sterols due to their enhanced ability to reduce intestinal cholesterol absorption while remaining virtually unabsorbed. Because dietary fat appeared to be the most effective vehicle for delivering plant stanols to the small intestine, a method for esterifying and solubilizing plant stanols in fat-based foods without altering the sensory and physical properties of the food was developed (Cater, 1999). Plant sterols and stanols are structurally similar to cholesterol, allowing them to compete with it during absorption in the digestive tract. Although plant stanol or sterol esters have been shown to lower serum carotenoid concentrations, no significant side effects, including gastrointestinal effects, have been observed (Levine et al., 2000). In Finland, stanol esters containing margarine have been on the market since 1995. A landmark clinical trial conducted in Finland in 1995 demonstrated that a plant stanol ester-containing margarine could significantly lower total and LDL cholesterol by 10% and 14%, respectively, in mildly hypercholesterolemic patients (Miettinen et al., 1995). Since then, numerous additional studies have clinically demonstrated the efficacy of plant stanol esters

(Hendriks et al., 1999) and sterols in the treatment of high cholesterol. According to petitions submitted by Lipton, a subsidiary of Unilever (for plant sterol esters), and McNeil Consumer Healthcare, a division of Johnson & Johnson (for plant sterol esters), the FDA announced on September 8, 2000 that they were authorizing the use of health claims for the association between plant sterol/stanol esters and reduced risk of CHD (FDA, 2000). The FDA authorized Lipton's Take Control cholesterol-lowering margarine for sale in the US on April 30, and McNeil's Benecol, a rival product, on May 17, 1999. Lipton submitted 15 scientific studies on the ability of plant sterols to lower cholesterol conducted between 1953 and 2000. The FDA only reviewed the eight studies published since 1982 (considering older literature to be of little relevance because the nature of the substance differed from that included in the petition), plus two additional studies identified through a literature review. The FDA specified the following qualifying criteria for nine of the 10 total studies: (i) provide data as well as adequate descriptions of the study's design and methods; (ii) be accessible in English; (iii) include estimates of, or enough information to estimate intakes of, plant sterols or stanols and their esters; (iv) include direct measurement of blood total cholesterol and other blood lipids related to CHD; and (v) be conducted in people who are representative of the general US population (i.e., adults with blood total cholesterol levels less than 300 mg/dL). McNeil submitted 21 scientific studies on the cholesterol-lowering effects of plant stanol esters conducted between 1993 and 2000; the FDA also identified three additional studies from the literature. Here, 15 of the 24 studies were identified to meet the above-mentioned selection criteria. Based on the entirety of publicly available information, the FDA has found that plant sterol/stanol esters may reduce the risk of CHD, allowing qualified spreads and salad dressings to carry the CHD health claim. According to RACC, foods with a health claim for plant sterols must contain at least 0.65 g of plant sterols (1.3 g divided by two servings per day). According to RACC, foods with a health claim for plant stanol esters must contain at least 1.7 g of stanol esters (3.4 g divided by two servings per day). As a result, the following health claim may now be used on plant sterol-containing food labels: "Foods having at least 0.65 g of plant sterols per serving, consumed twice a day with meals for a daily total of at least 1.3 g, as part of a low-saturated fat and cholesterol diet, may lower the risk of heart disease." contains grams of vegetable oil sterol esters per serving. "Foods containing at least 1.7 grams per serving of plant stanol esters, eaten twice a day with meals for a total daily intake of at least 3.4 grams, as element of a diet low in saturated fat and cholesterol, may reduce the risk of heart disease." A serving of the food contains grams of plant stanol esters (FDA, 2000)."

7.3 CONCLUSIONS

As cardiovascular disease (CVD) remains the leading cause of death in the United States, researchers, practitioners, and consumers are looking for alternatives and supplements to traditional therapies for prevention and treatment. Many consumers and patients may be able to reduce their risk of cardiovascular disease by eating functional foods such soybeans, oats, psyllium, flaxseed, garlic, tea, salmon, grapes, nuts, and stanol- and sterol ester-enhanced margarine. These foods may be beneficial on their own, in combination with, or in addition to cholesterol-lowering medications and other therapies. Although more research is needed, the field of functional foods and CVD prevention is expanding rapidly and will continue to yield practical, food-based recommendations for heart health.

ACKNOWLEDGMENT

We thank the MET's, Institute of Pharmacy, BKC, affiliated under Savitribai Phule Pune University, Nashik. EDA wants to thank the Ministry of Tribal Affairs Govt. of India for providing financial assistance through a fellowship (Award No.: 202021-NFST-MAH-01235).

KEYWORDS

- **cardiovascular disease**
- **food**
- **functional food**
- **low-density lipoprotein**
- **nutrition**
- **total cholesterol**

REFERENCES

Ahire, E. D., Surana, K. R., Patil, C. D., Shah, H. S., Sonwane, G. B., & Talele, S. G., (2020). Role of omega-3 fatty acids in different neurodegenerative disorders. *Applied Pharmaceutical Science and Microbiology: Novel Green Chemistry Methods and Natural Products*, 173.

Ahire, E., Sonawane, V., Surana, K., & Talele, G., (2021). Drug discovery, drug-likeness screening, and bioavailability: Development of drug-likeness rule for natural products. *Applied Pharmaceutical Practice and Nutraceuticals*. Apple Academic Press.

Albert, C. M., Hennekens, C. H., O'Donnell, C. J., et al., (1998). Fish consumption and risk of sudden cardiac death. *JAMA, 279*, 23–28.

American Dietetic Association, (1999). Position of the American Dietetic Association: Phytochemicals and functional foods. *J. Am. Dietetic Assoc., 99*, 1278–1285.

American Heart Association, (2000). *Heart and Stroke Statistical Update.* American Heart Association, Dallas, TX.

Anderson, J. W., Allgood, L. D., Lawrence, A., et al., (2000). Cholesterol lowering effects of psyllium intake adjunctive to diet therapy in men and women with hypercholesterolemia: Meta-analysis of 8 controlled trials. *Am. J. Clin. Nutr., 7*, 472–479.

Anderson, J. W., Allgood, L. D., Turner, J., et al., (2000). Effects of psyllium on glucose and serum lipid responses in men with type diabetes and hypercholesterolemia. *Am. J. Clin. Nutr., 70*(4), 466–473.

Anderson, J. W., Davidson, M. H., Blonde, L., et al., (2000). Long-term cholesterol-lowering effects of psyllium as an adjunct to diet therapy in the treatment of hypercholesterolemia. *Am. J. Clin. Nutr., 71*, 1433–1438.

Anderson, J. W., Johnstone, B. M., & Cook-Newell, M. E., (1995). Meta-analysis of the effects of soy protein intake on serum lipids. *N. Engl. J. Med., 333*, 276–282.

Anonymous, (1999). Oatmeal, vitamin E can prevent reduced blood flow after high-fat meal, study finds. *Food Labeling Nutr. News, 7*, 9.

Arjmandi, B. H., Khan, D. A., Juma, S., et al., (1998). Whole flaxseed consumption lowers serum LDL-cholesterol and lipoprotein(a)concentrations in postmenopausal women. *Nutr. Res., 18*, 1203–1214.

Association, A. H., (2004). *Heart and Stroke Statistical Update—2000.* Dallas, Texas: American Heart Association.

Bordia, A., Verma, S. K., & Srivastava, K. C., (1998). Effect of garlic (*Allium sativum*) on blood lipids, blood sugar, fibrinogen and fibronolytic activity in patients with coronary artery disease. *Prostagland. Leuk. Ess. Fatty Acids., 58*, 257–263.

Brown, L., Rosner, B., Willett, W. W., & Sacks, F. M., (1999). Cholesterol lowering effects of dietary fiber: A meta-analysis. *Am J Clin Nutr., 69*, 30–42.

Bruce, B., Spiller, G. A., Klevay, L. M., & Gallagher, S. K., (2000). A diet high in whole and unrefined foods favorably alters lipids, antioxidant defenses, and colon function. *J. Amer. Coll. Nutr., 19*, 61–67.

Byrne, D. J., Neil, H. A., Vallance, D. T., & Winder, A. F., (1999). A pilot study of garlic consumption shows no significant effect on markers of oxidation or sub-fraction composition of low-density lipoprotein including lipoprotein(a) after allowance for non-compliance and the placebo effect. *Clinical Chim. Acta, 285*, 21–33.

Carvajal-Millan, E., Yaser, A. Z., & Haghi, A., (2021). *Natural Food Products and Waste Recovery: Healthy Foods, Nutrition Design, and Extraction of Valuable Compounds*. CRC Press.

Cater, N. B., (1999). Historical and scientific basis for the development of plant stanol ester foods as cholesterol-lowering agents. *Eur Heart J., 1*, S36–S44.

Clydesdale, F. M., (1999). International life sciences institute North America food component reports. *Crit. Rev. Food Sci. Nutr., 39*, 203–316.

Crouse, J. R., Morgan, T., Terry, J. G., et al., (1999). A randomized trial comparing the effect of casein with that of soy protein containing varying amounts of isoflavones on plasma concentrations of lipids and lipoproteins. *Arch Int Med., 159*, 2070–2076.

Curb, J. D., Wergowske, G., Dobbs, J. C., et al., (2000). Serum lipid effects of a high-monounsaturated fat diet based on *Macadamia* nuts. *Arch Int Med., 160*, 1154–1158.

Daviglus, M. L., Stamler, J., Orencia, A. J., et al., (1997). Fish consumption and the 30-year risk of fatal myocardial infarction. *New Engl. J. Med., 336*, 1046–1053.

Department of Health and Human Services, Food and Drug Administration, (1999). 21 CFR part 101. Food labeling: Health claims; soy protein, and coronary heart disease. *Fed Reg., 64*, 57700–57733.

Department of Health and Human Services, Food and Drug Administration, (1997). 21 CFR part 101: Food labeling: Health claims; oats and coronary heart disease. *Fed. Reg., 62*, 3584–3601.

Department of Health and Human Services, Food and Drug Administration, (1998). 21 CFR part 101 [Docket No.: 96P-0338]. Food labeling: Health claims; soluble fiber from certain foods and coronary heart disease: Final rule. *Fed. Reg., 63*, 8103–8121.

Department of Health and Human Services, Food and Drug Administration (2000). 21 CFR Part 101: Food labeling health claims; plant sterol/stanol esters and coronary heart disease, interim final rule. *Fed. Reg., 65*, 54686–54739.

Emmons, C. L., Peterson, D. M., & Paul, G. L., (1999). Antioxidant capacity of oat (*Avena sativa* L.) extracts: *In vitro* antioxidant activity and contents of phenolic and tocol antioxidants. *J Agric Food Chem., 47*, 4894–4898.

Fraser, G., Sabate, J., Beeson, W., & Strahan, M., (1992). A possible protective effect of nut consumption on risk of coronary heart disease. *Arch Int. Med., 152*, 1416–1424.

Fremont, L., (2000). Biological effects of resveratrol. *Life Sci., 66*, 663–673.

Geleijnse, J. M., Lenore, L. J., Hofman, A., et al., (1999). Tea flavonoids may protect against atherosclerosis: The Rotterdam study. *Arch Int. Med., 159*, 2170–2174.

Gillum, R. F., Mussolina, M., & Madans, J. H., (2000). The relation between fish consumption, death from all causes, and incidence of coronary heart disease: The NHANES I epidemiologic follow-up study. *J. Clin. Epidemiol., 53*, 237–244.

Hasler, C. M., Kundrat, S., & Wool, D., (2000). Functional foods and cardiovascular disease. *Current Atherosclerosis Reports, 2*, 467–475.

Hendriks, H. F. J., Westsrate, J. A., Van, V. T., et al., (1999). Spreads enriched with three different levels of vegetable oil sterols and the degree of cholesterol lowering in normocholesterolaemic and mildly hypercholesterolaemic subjects. *Eur. J. Clin. Nutr., 53*, 319–327.

Hertog, M. G. L., Feskens, E. J. M., Hollman, P. C. H., et al., (1993). Dietaryantioxidant flavonoids and risk of coronary heart disease: The zutphen elderly study. *The Lancet, 342*, 1007–1011.

Hodgson, J. M., Puddey, I. B., Beilin, L. J., et al., (1998). Supplementation with isoflavonoid phytoestrogens does not alter serum lipid concentrations: A randomized controlled trial in humans. *J Nutr., 128*, 728–732.

Hodgson, J. M., Puddey, I. B., Croft, K. D., et al., (2000). Acute effects of ingestion of black and green tea on lipoprotein oxidation. *Am J Clin Nutr., 71*, 1103–1107.

International Life Sciences Institute North America Food Component Reports, (1999). *Crit. Rev. Food Sci. Nutr.*, *39*, 303–316.

Isaacsohn, J. L., Moser, M., Stein, E. A., et al., (1998). Garlic powder and plasma lipids and lipoproteins: A multicenter, randomized, placebo-controlled trial. *Arch Intl. Med., 158*, 1189–1194.

Iwamoto, M., Sato, M., Kono, M., et al., (2000). Walnuts lower serum cholesterol in Japanese men and women. *J. Nutr., 130*(2), 171–176.

Jenkins, D. J. A., Kendall, C. W. C., et al., (1999). Health aspects of partially defatted flaxseed, including effects on serum lipids, oxidative measures, and ex vivo androgen and progestin activity: A controlled crossover trial. *Am. J. Clin. Nutr., 69*, 395–402.

Jenkins, D. J. A., Kendall, C. W. C., Axelsen, J., et al., (2000). Viscous and nonviscous fibers, nonabsorbable and low glycaemic index carbohydrates, blood lipids and coronary heart disease. *Current Opin. Lipid, 11*, 49–56.

Jenkins, D. J. A., Kendall, C. Y. C., Vidgen, E., et al., (2000). The effect on serum lipids and oxidized low-density lipoprotein of supplementing self-selected low-fat diets with soluble-fiber, soy, and vegetable protein foods. *Metabolism, 49*, 67–72.

Keevil, J. G., Osman, H. E., Reed, J. D., & Folts, J. D., (2000). Grape juice, but not orange juice or grapefruit juice, inhibits human platelet aggregation. *J. Nutr., 130*, 53–56.

Keservani, R. K., & Sharma, A. K., (2014). Flavonoids: Emerging trends and potential health benefits. *Journal of Chinese Pharmaceutical Sciences, 23*(12), 815.

Keservani, R. K., Kesharwani, R. K., Sharma, A. K., Vyas, N., & Chadoker, A., (2010b). Nutritional supplements: An overview. *International Journal of Current Pharmaceutical Review and Research, 1*(1), 59–75.

Keservani, R. K., Kesharwani, R. K., Vyas, N., Jain, S., Raghuvanshi, R., & Sharma, A. K., (2010a). Nutraceutical and functional food as future food: A review. *Der Pharmacia Lettre, 2*(1), 106–116.

Keservani, R. K., Sharma, A. K., & Kesharwani, R. K., (2016). Nutraceutical and functional foods for cardiovascular health. In: *Food Process Engineering* (pp. 257–278, 291–312). Apple Academic Press, CRC Press. ISBN: 9781771884020.

Keservani, R. K., Sharma, A. K., & Kesharwani, R. K., (2017). An overview and therapeutic applications of nutraceutical and functional foods. *Recent Advances in Drug Delivery Technology*, 160–201. ISBN:9781522507543.

Kris-Etherton, P. M., Taylor, D. S., Yu-Poth, S., et al., (2000). Polyunsaturated fatty acids in the food chain in the United States. *Am. J. Clin. Nutr., 71*, 179–188.

Levine, B. S., & Cooper, C., (2000). Plant stanol esters: A new tool in the dietary management of cholesterol. *Nutr. Today, 35*, 61–66.

Marcovina, S. M., Kennedy, H., Bon, G. B., et al., (1999). Fish intake, independent of apo(a) size, accounts for lower plasma lipoprotein(a) levels in Bantu fishermen of Tanzania—The lugalawa study. *Arterio Thromb Vasc Biol., 19*, 1250–1256.

Merz-Demlow, B. E., Duncan, A. M., Wangen, K. E., et al., (2000). Soy isoflavones improve plasma lipids in normocholesterolemic, premenopausal women. *Am. J. Clin. Nutr., 71*, 1462–1469.

Messina, M., (1999). Health effects of phytosterols. *Soy Connection, 7*, 3–5.
Miettinen, T. A., Puska, P., Gulling, H., et al., (1995). Reduction of serum cholesterol with sitostanol-ester margarine in a mildlyhy percholesterolemic population. *New Eng. J. Med., 333*, 1308–1312.
Morgan, W. A., & Clayshulte, B. J., (2000). Pecans lower low-density lipoprotein cholesterol in people with normal lipid levels. *J. Amer. Diet Assoc., 100*, 312–318.
Munday, J. S., James, K. A., Fray, L. M., et al., (1999). Daily supplementation with aged garlic extract, but not raw garlic, protects low density lipoprotein against *in vitro* oxidation. *Atherosclerosis, 143*, 399–404.
Nagourney, R. A., (1998). Garlic: Medicinal food or nutritious medicine? *J. Medicinal Food, 1*, 13–28.
Nestel, P. J., (2000). Fish oil and cardiovascular disease: Lipids and arterial function. *Am J Clin Nutr., 71*, 228S–231S.
Nestel, P. J., Pomeroy, S., Kay, S., et al., (1999). Isoflavones from red clover improve systemic arterial compliance but not plasma lipids in menopausal women. *J. Clin. Endo. Metab., 84*, 895–898.
Ozturk, H. S., Kacmaz, M., Cimen, M. Y. B., & Durak, I., (1999). Red wine and black grape strengthen blood antioxidant potential. *Nutrition, 15*, 954–955.
Pirich, C., Gaszo, A., Granegger, S., & Sinzinger, H., (1999). Effects of fish oil supplementation on platelet survival and ex vivo platelet function in hypercholesterolemic patients. *Thromb Res., 96*, 219–227.
Potter, S. M., (2000). Soy-new health benefits associated with an ancient food. *Nutr. Today, 35*, 53–60.
Princen, H. M. G., Van, D. W., Buytenhek, R., et al., (1998). No effect of consumption of green and black tea on plasma lipid and antioxidant levels and on LDL oxidation in smokers. *Arterio Thromb. Vasc. Biol., 18*, 833–841.
Roche, H. M., & Gibney, M. J., (2000). Effect of long chain n-3 polyunsaturated fatty acids on fasting and postprandial triacylglycerol metabolism. *Am. J. Clin. Nutr., 71*, 232–237.
Sabate, J., Fraser, G. E., Burke, K., et al., (1993). Effects of walnuts on serum lipid levels and blood pressure in normal men. *New Eng. J. Med., 328*, 603–607.
Schulz, V., Hänsel, R., & Tyler, V. E., (1997). Cardiovascular system. In: *Rational Phytotherapy: A Physicians' Guide to Herbal Medicine* (pp. 107–123). New York: Springer-Verlag.
Sesso, H. D., Gaziano, J. M., Buring, J. E., & Hennekens, C. H., (1999). Coffee and tea intake and the risk of myocardial infarction. *Amer J. Epidemiol., 149*, 162–167.
Setchell, K. D. R., (1999). Discovery and potential clinical importance of mammalian lignans. In: Cunnane, S., & Thompson, L. U., (eds.), *Flaxseed in Human Nutrition* (pp. 82–98) Edited by Champaign, IL, AOCS Press.
Shopping for Health, (1999). *The Growing Self-Care Movement.* Prevention Magazine and the Food Marketing Institute. Data collected by Roper Starch Worldwide Inc., New York, NY.
Silagy, C., & Neil, A., (1994). Garlic as a lipid-lowering agent a meta-analysis. *J. Royal Coll. Physicians Lond., 28*, 39–45.
Slaughter, E., (1992). *Prevention Magazine and the Food Marketing Institute Survey of Public Concern Regarding Good Nutrition.* Princeton, NJ.
St. Leger, A. S., Cochrane, A. L., & Moore, F., (1979). Factors associated with cardiac mortality in developed countries with particular reference to the consumption of wine. *The Lancet, 313*(8124), 1017–1020.

Stein, J. H., Keevil, J. G., Wiebe, D. A., et al., (1999). Purple grape juice improves endothelial function and reduces the susceptibility of LDL cholesterol to oxidation in patients with coronary artery disease. *Circulation, 100*, 1050–1055. One of the few existing human clinical intervention trials demonstrating that commercial purple grape juice can significantly improve flow-mediated vasodilation and reduce LDL oxidation.

Superko, H. R., & Krauss, R. M., (2000). Garlic powder, effect on plasma lipids, postprandial lipemia, low-density lipoprotein particle size, high density lipoprotein subclass distribution and lipoprotein(a). *J Amer Coll Cardiol., 35*, 321–326.

Surana, K. R., Ahire, E. D., Sonawane, V. N., Talele, S. G., & Talele, G., (2021). Informatics and methods in nutrition design and development. *Natural Food Products and Waste Recovery: Healthy Foods, Nutrition Design, And Extraction of Valuable Compounds*, 33.

Thompson, L. U., Robb, P., Serraino, M., & Cheung, F., (1991). Mammalian lignan production from various foods. *Nutr. Cancer, 16*, 43–52.

Tikkanen, M. J., & Adlercreutz, H., (2000). Dietary soy-derived isoflavone phytoestrogens: Could they have a role in coronary heart disease prevention? *Biochem. Pharmacol., 60*, 1–5.

Trevisanato, S. I., & Kim, Y. I., (2000). Tea and health. *Nutr. Rev., 58*, 1–10.

Van, H. H. K. H., Wiseman, S. A., Yang, C. S., & Tijburg, L. B. M., (1999). Plasma and lipoprotein levels of tea catechins following repeated tea consumption. *Proc Soc Exp Biol Med., 220*, 203–209.

Warshafsky, S., Kamer, R. S., & Sivak, S. L., (1993). Effect of garlic on total serum cholesterol: A meta-analysis. *Ann. Int. Med., 119*, 599–605.

Washburn, S., Burke, G. L., Morgan, T., & Anthony, M., (1999). Effect of soy protein supplementation on serum lipoproteins, blood pressure, and menopausal symptoms in perimenopausal women. *Menopause, 6*, 7–13.

Weisburger, J. H., (1999). Second international scientific symposium on tea and health: An introduction. *Proc. Soc. Exp. Bio. Med., 220*, 193–194.

Wood, P. J., & Beer, M. U., (1998). Functional oat products. In: Mazza, G. L., (ed.), *Functional Foods: Biochemical & Processing Aspects* (pp. 1–38). Technomic Publishing Co., Inc.

Woodward, M., & Tunstall-Pedoe, H., (1999). Coffee and tea consumption in the Scottish heart health study follow up: Conflicting relations with coronary risk factors, coronary disease, and all-cause mortality. *J. Epidemiol. Commun. Health, 53*, 481–487.

Yang, C. S., (1999). Tea and health. *Nutrition, 15*, 946–949.

Yang, C. S., Chung, J. Y., Yang, G., et al., (2000). Tea and tea polyphenolsin cancer prevention. *J Nutr., 130*, 472S–478S.

Yosefy, C., Viskoper, J. R., Laszt, A., et al., (1999). The effect of fish oil on hypertension, plasma lipids and hemostasis in hypertensive, obese, dyslipidemic patients with and without diabetes. *Prostaglan Leuk Essen Fatty Acids, 61*, 83–87.

Zambon, D., Sabate, J., Munoz, S., et al., (2000). Substituting walnuts for monounsaturated fat improves the serum lipid profile of hypercholesterolemic men and women—A randomized crossover trial. *Ann. Int. Med., 132*, 538–546.

CHAPTER 8

FUNCTIONAL FOODS IN PREVENTION OF DIABETES MELLITUS

AFSAR S. PATHAN,[1] MANJUSHA R. AHIRE,[1] SHIWANI A. DIWANE,[1] PANKAJ G. JAIN,[1] PRATIKSHA M. PANDAGALE,[2] and EKNATH D. AHIRE[3]

[1]*Department of Pharmacology, R. C. Patel Institute of Pharmaceutical Education and Research, Shirpur, Maharashtra, India*

[2]*Department of Pharmacology, Oriental College of Pharmacy, Navi Mumbai, Maharashtra, India*

[3]*Department of Pharmaceutics, METs Institute of Pharmacy, BKC, Affiliated to Savitribai Phule University, Adgoan, Nashik-3, Maharashtra, India*

ABSTRACT

Diabetes mellitus (DM) is a group of metabolic diseases characterized by persistent hyperglycemia. DM is a major global health problem; 451 million people were projected to have diabetes in 2017, and type 2 diabetes mellitus (T2DM) affects more than 90% of them. The basic pathophysiology behind DM is the destruction of beta cell, insulin resistance and glucose metabolism, there are several treatment and management therapy, but they have some disadvantages. DM is widely connected with other life-threatening diseases, and its a common co-morbid condition now a day. Functional food is food that may provide the nutritional effect to the body with minimum risk. Functional food in daily diet improve mental health and also for managing disease. We use functional food in DM to manage DM in different ways. Functional food has different activities which are necessary for the treatment

Applications of Functional Foods in Disease Prevention. Raj K. Keservani & Eknath D. Ahire, PhD (Eds.)

of DM. In this chapter, we focus on the different types of DM and functional food and their role in DM.

8.1 INTRODUCTION

Diabetes mellitus is a group of metabolic diseases characterized by persistent hyperglycemia brought on by insulin release, insulin action, or a combination of the two (Kharroubi, 2015). Diabetes is caused by a number of different pathogenic mechanisms. They can also include autoimmune destruction of pancreatic cells, resulting in insulin insufficiency, to anomalies that result in insulin resistance. Diabetes causes anomalies in glucose, lipid, and protein metabolism due to insulin's ineffective action on target tissues (Of et al., 2005). Diabetes mellitus is a major global health problem; 451 million people were projected to have diabetes in 2017, type 2 diabetes mellitus (T2DM) affects more than 90% of them, and the number is anticipated to climb to 693 million by 2045 (Li et al., 2019). Diabetes mellitus type 2 is a severe metabolic illness characterized by microvascular and macrovascular dysfunction. Hyperglycemia is the most important cause of microvascular problems, and lowering HbA1c lowers the risk of retinopathy, nephropathy, and neuropathy. Microvascular problems are lowered by 25% to 35% for each 1% decrease in HbA1c. Despite the fact that microvascular complications are a major cause of morbidity, macrovascular problems are the most cause of death, accounting for over 80% of all deaths (DeFronzo, 2010); the function of insulin inside glucose transportation in cells and regulation of glucose metabolism (Hanif Mughal, 2019).

8.2 TYPES OF DIABETES MELLITUS

8.2.1 TYPE 1 DIABETES MELLITUS

It is recognized as autoimmune diabetes, and this diabetes is a chronic condition which caused by the insulin insufficiency due to pancreatic β-cell dysfunctioning and leads to hyperglycemia (DiMeglio, Evans-Molina and Oram, 2019). Type 1 diabetes can be classified into three stages depending on the existence or absence of hyperglycemia and hyperglycemia-related symptoms (such as polyuria and thirst) (Katsarou et al., 2017).

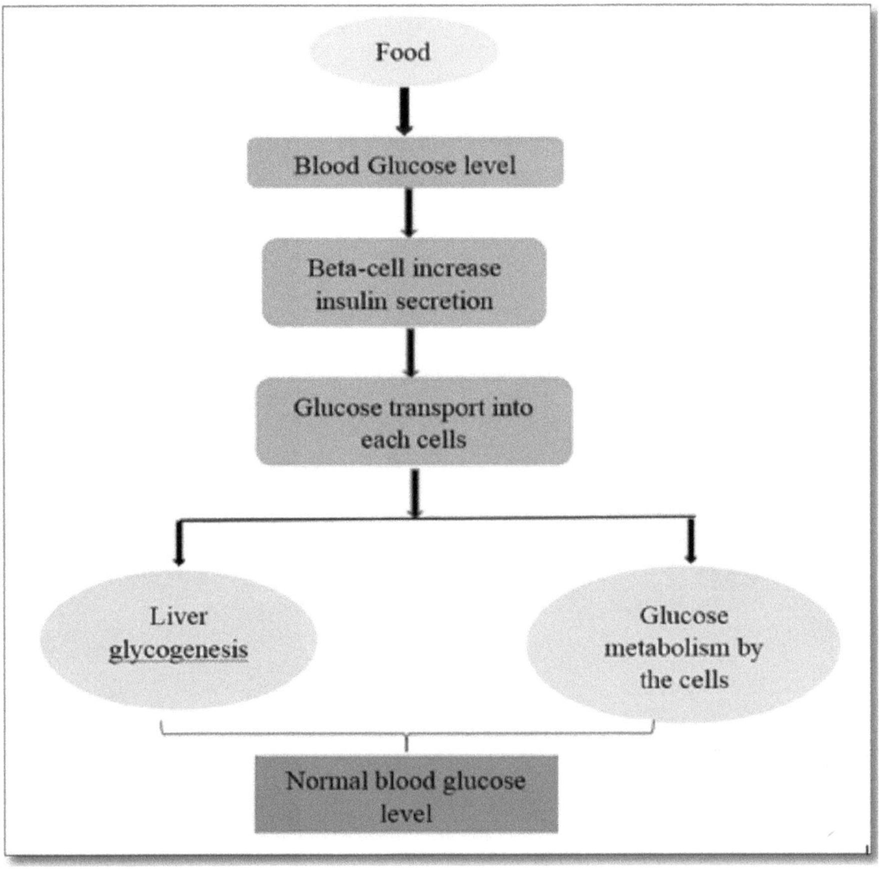

FIGURE 8.1 Blood glucose level maintaining cycle.

8.2.2 TYPE 2 DIABETES MELLITUS

Adult-onset diabetes is another name for type 2 diabetes mellitus. Insulin insensitivity is a symptom of Type 2 diabetes, which is caused by insulin resistance, decreased insulin production, and eventually pancreatic beta-cell failure (Padhi, Nayak and Behera, 2020). Type 2 diabetes is caused by both lifestyle and genetic factors. In terms of lifestyle, lack of physical activity, poor diet, smoking habits, and excessive alcohol consumption are key causes. Obesity has been found to contribute to approximately 55% of cases of type 2 DM (Olokoba, Obateru and Olokoba, 2012).

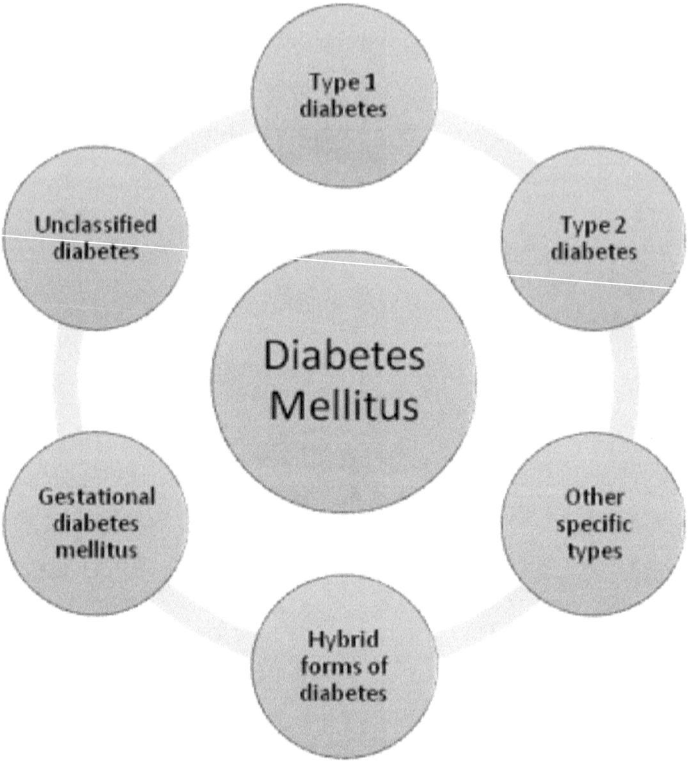

FIGURE 8.2 Types of diabetes mellitus.

8.2.3 HYBRID FORM OF DIABETES

This is a new type of diabetes, and as the name suggest, it is a hybrid form. Hyperglycemia in overweight children with positive autoimmunity against the beta cells is referred to as "hybrid diabetes (HD)." There is a lack of information about the clinical presentation and disease progression in children with this kind of diabetes.

8.2.3.1 GRADUALLY EVOLVING, IMMUNE-MEDIATED DIABETES OF ADULTS

Slowly developing immune-mediated diabetes in adults, also known as latent autoimmune diabetes, is defined by the existence of autoantibodies (ATs) against glutamic acid decarboxylase (GADA) and a patient's age at onset of more than 35 years (Ragimov et al., 2021).

8.2.3.2 KETOSIS-PRONE TYPE 2 DIABETES

Ketosis-prone type 2 diabetes (also known as "Flatbush diabetes") has characteristics of both type 1 and type 2 diabetes and is quite common in African communities. The condition, which is ketosis-prone but not persistently insulin phobic or autoimmune, is characterized by a high proclivity for ketosis (Sjöholm, 2019).

8.2.3.3 GESTATIONAL DIABETES MELLITUS

Moderate to severe maternal hyperglycemia during pregnancy poses a distinct set of diabetes-related complications since it affects two people rather than simply one: the expectant mother and her unborn child (Mirghani Dirar and Doupis, 2017). To distinguish it from pre-diagnosed type 1 or type 2 diabetes or maturity-onset diabetes of the young (MODY) in women who get pregnant, gestational diabetes (GDM) has been described as any degree of glucose intolerance with onset or first identification during pregnancy (Table 8.1).

TABLE 8.1 Diagnostic Criteria of Diabetes Mellitus

Diagnosis	Range
Random blood glucose level	≥ 11.1 mmol/l (≥ 200 mg/dl)
Fasting blood glucose level	≥ 7.0 mmol/l (≥ 126 mg/dl), (no caloric intake 8 h)
120 min blood glucose in OGTT	≥ 11.1 mmol/l (≥ 200 mg/dl)
HbA1c	≥ 6.5%

Source: Adapted from Mirghani & Doupis (2017).

8.2.4 OTHER SPECIFIC TYPES

8.2.4.1 MONOGENIC DIABETES

Monogenic diabetes is a rare form of diabetes produced by a single gene alteration or mutation. Insulin production is low in monogenic diabetes. Glucose (blood sugar) levels rise as a result of this. However, not all cases of monogenic diabetes need insulin therapy. Monogenic diabetes is divided into two types.

i. **Neonatal Diabetes Mellitus (NDM):** This occur in newborns and infants.

ii. **Maturity-Onset Diabetes of the Young (MODY):** It is more common than NDM, and occurs in adolescents and young adults.

a. **Diseases of the Exocrine Pancreas:** Hyperglycemia can be caused by a number of conditions that affect the pancreas (Kazi and Blonde, 2001).
b. **Endocrine Disorders:** Occurs in illnesses characterized by excessive secretion of insulin antagonistic hormones (Kazi and Blonde, 2001).
c. **Drug- or Chemical-Induced:** Hyperglycemia can be caused by a variety of factors, including changes in insulin secretion and sensitivity, direct cytotoxic effects on pancreatic cells, and enhance glucose production (Fathallah et al., 2015). Some class of drugs and chemicals affect insulin secretion or action; some can destroy the β-cells action (Ferner, 1992).
d. **Infection-Related Diabetes:** Some viruses play a role in the progression of diabetes mellitus. Many different viruses from various genera have the ability to harm beta cell (Szopa et al., 1993). Due to viruses immune system also affect and associated various diseases (Toniolo et al., 2019).
e. **Uncommon Specific Forms of Immune-Mediated Diabetes:** Diabetes has been linked to immunity in some situations. They are the cause of rare immune-mediated illnesses (Eizirik and Mandrup-Poulsen, 2001).
f. **Unclassified Diabetes:** Diabetes that does not fall within any of the other categories (Kazi and Blonde, 2001).

8.3 FUNCTIONAL FOODS IN DIABETES MELLITUS

Foods or food components that may provide health advantages by lowering the risk of certain diseases or other health issues. Any foods and food element that is supposed to give health advantages in conjunction to its nutritional value are considered functional food. The phrase "functional food" and its concept were first presented in 1984, when the Ministry of Education, Science, and Culture sponsored a systematic and large-scale investigation into the new food function. The phrase "functional food" was coined in Japan in the 1980s to describe manufactured foods that, in adding up to being nutritious, contain substances that enhance various physiological functions. Japan is currently a country no more than that has developed a legal approval for functional foods (Kaur and Das, 2011). The desire to create functional foods is based on the eastern belief that "medicine and food have the same genesis." It is widely accepted that the quality of any meal should be measured in terms of several of dynamic functions that the food performs after being consumed. Because any food is essentially a source of nutrients for the body, its dietary function is automatically seen as crucial (Arai, 1996). These chemicals with tertiary functionalities can be used to develop and construct functional foods. They occupy a space between traditional foods and pharmaceuticals. 'These advantages could include disease prevention or management, in addition to improved physical or mental performance (Huggett, 1996). Functional foods are foods that improve health in addition to providing basic nutrients. These foods have the potential to improve health in ways that orthodox nutrition research has not predicted (Sanders, 1998). Functional foods are products that resembled traditional recipes but have been demonstrated to have physiological benefits, according to Health Canada (Shahidi, 2009). Functional foods are those that have been sufficiently demonstrated to support one or more bodily functions in addition to the proper nutritional impacts in a manner that either improves health and welfare or lowers illness risk (Blades, 2000). Functional foods that have been modified to make them more functional are known as functional foods. A functional food can be a real food, a food to which a health-promoting element has been incorporated or where a component has been omitted for particular health purposes, a nutrition in which an aspect has been converted by technological or chemical means to provide a specific clinical benefit, a food wherein the biocompatibility of a component has been altered, or any mixture of these (Norsuzila Ya'acob et al., 1989). Functional foods might be "designed to have health benefit and/or decrease the hazard of stroke ailment beyond basic nutritional features, and may be similar in design to processed foods and consumed as part of a regular diet," as well as "intended to have beneficial properties and/or slow the progression disease beyond basic nutritional functions" (Doyon and Labrecque, 2008) A functional food must stay food and demonstrate its benefits in levels that may be expected to be ingested in the diet: it is not a tablet or a capsule, but rather an ingredient in a regular diet (ILSI, 1999). Nutraceuticals, a pharma food, or a food supplements are clearly not the same thing as a functional food. It is a food, not a medicine, because it has health benefits that are usually disease preventative rather than therapeutic preventive and not therapeutic properties (Keservani et al., 2010a, b, 2017, 2020; Norsuzila Ya'acob et al., 1989) (Table 8.2).

TABLE 8.2 Definitions of Functional Food According to Different Government

Sl. No.	Definitions	References
1.	Foods that may be eaten regularly as part of a normal diet, that have been designed specifically to provide a physiological or medical benefit by regulating body functions to protect against or retard the progression of diseases such as coronary heart disease, cancer, hypertension, diabetes, and osteoporosis.	Walker, Dunshea, & Doyle (2004)
2.	A functional food is a conventional food or a food similar in appearance to a conventional food, it is part of a regular diet and has proven health-related benefits and (or) reduces the risk of specific chronic diseases above its basic nutritional functions.	Tracy et al. (2006)
3.	The Japanese government defines FOSHU as foods which are expected to have certain health benefits, and have been licensed to bear a label claiming that a person using them for a specified health use may expect to obtain the health use through the consumption thereof.	Sanders (1998)
4.	International Food Information Council (IFIC), that is functional foods are, 'foods or dietary components that may provide a health benefit beyond basic nutrition.'	Bagchi (2008)
5.	The European Commission Concerted Action on Functional Food Science in Europe regards a food as functional, 'if it is satisfactorily demonstrated to affect beneficially one or more target functions in the body, beyond adequate nutritional effects, in a way that is relevant to either an improved state of health and well-being and/or reduction of risk of disease.	Europe et al. (1999)
6.	The Food and Nutrition Board (FNB) of the National Academy of Sciences defines a functional food as one that encompasses potentially healthful products, including any modified food or food ingredient that may provide a health benefit beyond that of the traditional nutrients it contains.	Milner (2000)

Source: Keservani et al. (2010a, b, 2017, 2020); and Norsuzila et al. (1989).

8.3.1 FUNCTIONAL FOODS OF PLANT OR ANIMAL ORIGIN

It can be classified into the following categories based on their actions:

- Supplementation with vitamins and minerals (e.g., vit C, folic acid, calcium, iron);
- Lowering cholesterol levels (e.g., omega-3 fatty acids, phytosterols);
- Fiber in the diet (e.g., nondigestible carbohydrates and lignin).

Phytochemicals are a type of phytochemicals that be able to be found in plants (e.g., phenolic compounds, carotenoids, lycopene (Norsuzila Ya'acob et al., 1989).

8.3.1.1 SNACK BAR

Food bars are snacking foods with good sensory attributes due to the high protein, fat, and crude fiber of their ingredients. Although chocolate cereal bars are preferred by the majority of customers, they are no safer than whole fiber plain snack foods. According to chemical research, cereal bars are slightly better than favorite traditional snacks in terms of sugar, fat, salt, and fiber content considerable affinity exists among children and teenagers (Boustani and Mitchell, 1990).

8.3.1.2 ALGAE

Algae can be used as a natural source of beneficial substances in the diet. The low toxicity of the selected kinds was a significant consideration, as any possible new active ingredient obtained from them could be utilized in the future implementation of novel functional foods (Plaza, Cifuentes and Ibáñez, 2008) (Table 8.3).

TABLE 8.3 Different Algae and Their Functional Food Applications

Sl. No.	Algae	Applications as Functional Food
1.	*S. vulgare*	Major component is carbohydrate, and its synthesis is favored by intensity of light and temperature. Also contain a high percentage of fiber and proteins with essential amino acids in significant amounts.
2.	*Porphyra* sp. and *C. crispus* (red algae)	Poryphera contains proteins and PUFA like eicasopentaenoic acid, sterols palmitic, palmitoleic, oleic, and arachidonic mainly present are demosterols fibers agar or carrageenans.
3.	*H. elongata* and *U. pinnatifida* (brown algae)	Valuable protein content, polyunsaturated omega-3-fatty acids, as for example, the eicosapentaenoic acid (EPA). *H. elongata* also has high levels of a-tocopherol, sterolsmainly present is fucosterol, fibers alginates, fucans and laminarin, sulfated polysaccharides.
4.	*Cystoseira* spp.	High concentration of aryl terpenes with valuable curative value and no side effects. The marine algae *Cystoseira josteroides* and *Sargassum macrocarpum* are good sources of zosterdiol A, zosterdiol B, zosteronol and zosteronediol.
5.	Ulva species	Main sterol is cholesterol and isofucosterol. *Ulva oxysperma* and *Ulva* spp showed high mineral levels.

Source: Adapted from Plaza, Cifuentes, & Ibáñez (2008).

Due to composition and nutritional value of algae values that make them good candidates as source of proteins, carbohydrates, fiber, minerals, vitamins, and, besides, they present a low content in lipids. The high lipid content of fiber in above algae has anti-cancer, anti-constipation, anti-hyper-cholesterolemia, anti-obesity, anti-diabetes properties. Algae have antiviral activity due to the presence of significant concentrations of frucans, and a sulfated polysaccharide called porphyran has shown some potential apoptotic activity. There are a variety of antioxidant (Singh et al., 2017a) compounds found in algae, including vitamin E (or atocopherol) and carotenoids in the fat-soluble fraction, and polyphenols, phycobiliproteins, and vitamins in the water-soluble fraction (vitamin C).

8.3.1.3 PROBIOTICS

Probiotic foods are described as foods that contain an unique or mixed culture of bacteria that improve the consumer's gut microbial balance and hence benefit their health (Fuller and Gibson, 1997). Probiotic bacteria strains change the gut's natural bacterial population in a positive way. LC1 is a yogurt-like foodstuff (Blades, 2000).

According to scientific research, prebiotics may help with lactose metabolism, cancer prevention, immunological modulation, and serum cholesterol lowering (Gilliland, 1990; Fooks et al., 1999; Salminen et al., 1998). The dairy industry has pioneered the use of probiotic strains in traditional food products, resulting in the creation of probiotic yogurt, discovered new forms of fermented milk and cheese (Gomes and Malcata, 1999).

Prebiotic is a food element that is not hydrolyzed by the human digestive system and promotes the growth of microorganisms that are beneficial to the host's health (Gibson and Roberfroid, 1995). A bacterium must possess a number of characteristics in order to be considered a probiotic. The bacterium must thrive in an acidic upper GIT environment that is hostile to pathogenic and cancerous cells (Havenaar, 1992).

Cereals are non-dairy probiotic products that produce high-quality functional nutrients from natural resources. They are high in protein (Singh et al., 2017b), dietary fiber (Keservani et al., 2020), energy, minerals, and vitamins, all of which are essential for human health (Saarela et al., 2000).

8.3.1.4 PREBIOTICS

A Prebiotics is a non-digestible food element that has a favorable effect on the host by encouraging the development of activity in one or a minute

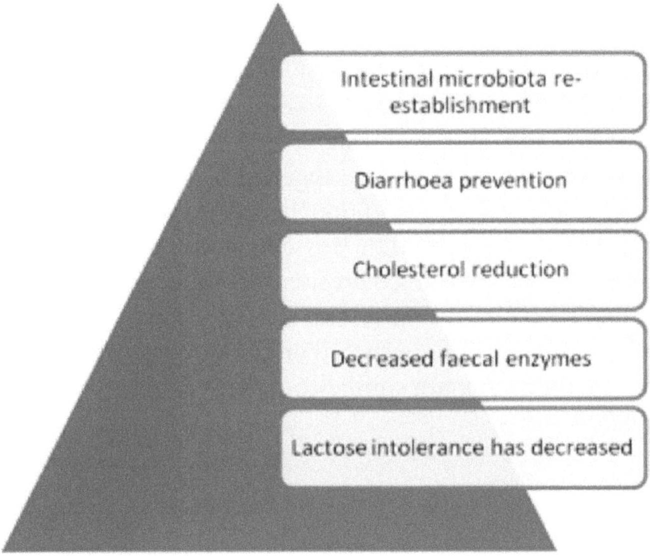

FIGURE 8.3 Role of probiotics in diabetes mellitus.

number of colonic bacteria (Gibson and Roberfroid, 1995). Only two Prebiotics, inulin and trans-galactooligosaccharide and, fit Roberfroid's criteria in 2007. Prebiotics take advantage of the development of specific enzymes by gut microbes that may provide health benefits to the host (Ziemer and Gibson, 1998). Nondigestible oligosaccharides aid in the establishment of probiotic bacterial populations, as well as the substrates that probiotic bacteria live on (Blades, 2000). Delay in gastric emptying, regulation of GIT transit time, and enhanced glucose tolerance are all validatory health benefits of nondigestible carbohydrate on the GIT (Joshi, Roy and Banerjee, 2018). Fructooligosaccharides (FOS) are the carbohydrate that has been studied the most. Enzymatic manufacture of fructo-oligosaccharides is also possible, which is advantageous for large-scale production (Williams and Stevensen, 1990). Health payback of prebiotics are discussed below:

- Acute gastroenteritis;
- Cancer reduce the genotoxicity and inhibit genotoxic enzymes (Ito et al., 1993);
- Mineral absorption increased calcium and magnesium intake;
- Lipid metabolism reduces blood cholesterol and triglycerides (Chonan, Matsumoto, and Watanuki, 1995).

Oligosaccharide prebiotics' greatest power is the growth of bifidobacteria as a consequence of a functional diet. FOS feeds bifidobacteria and lactobacilli, as well as increasing SCFA levels (Fuller and Gibson, 1997). Non-digestible fermentable fructans like inulin and oligofructose are among the most researched and well-established prebiotics (Panesar and Bali, 2015). These chemicals, in adding together to being Prebiotics, have been proven to promote calcium absorption, hence improving both increased bone content and bone density (BMD) (Bosscher, Van Loo, and Franck, 2006). In the manufacturing of low-fat ice creams and yogurts, beta-glucans are used. Has been the subject of recent research. Low-fat dairy products with beta-glucans and added soluble dietary fiber can have mouthfeel, scoop ability, and sensory qualities that are similar to full-fat dairy products (Brennan and Cleary, 2005).

8.4 EPIDEMIOLOGY OF FUNCTIONAL FOOD IN DIABETES MELLITUS

8.4.1 PREVALENCE OF DIABETES MELLITUS

Type 2 diabetes is widely recognized as a critical public health issue that has a significant health-care expenses and influence on human life. Rapid economic development is occurring in many parts of the world. Diabetes has become more common as a result of globalization and urbanization (Abdul et al., 2020). Diabetes affected 8.5% of persons aged 18 and above in 2014. Diabetes was indeed the direct cause of 1.5 million mortality in 2019, with 48% of all diabetes-related deaths occurring before the age of 70 (https://www.who.int/news-room/fact-sheets/detail/diabetes). Diabetes, termed "the disease of the 21st century," has become a concern for modern generations as its prevalence has increased. According to WHO research, 108 million persons had diabetes in 1980 and 422 million had diabetes in 2014 ('Ada Krawęcka.pdf,' no date). In 2009, 285 million people were predicted to have diabetes (including T1D and T2D), rising to 366 million in 2011, 382 million in 2013, 415 million in 2015, and 425 million in 2017 (Saeedi et al., 2019). According to the International Diabetes Federation, 415 million people had diabetes in 2015, with that population predicted to climb to 642 million by 2040 (Tripathy et al., 2017). The most major risk factor for patients with type 2 diabetes, as per the World Health Organization's Global Report on Diabetes, is being overweight or obese (T2DM), and T2DM and prediabetes are becoming more common in children, adolescents, and young adults (Liu et al., 2020). In 2017, the global incidence, prevalence, death, and disorder life-years (DALYs) of diabetes were 22.9 million, 476.0 million, 1.37 million, and 67.9 million, respectively, with predictions of 26.6 million, 570.9 million, 1.59 million, and 79.3 million in 2025(Lin et al., 2020). Diabetes is becoming more prevalent at an alarming rate all across the world. Rapid urbanization, aging populations, socioeconomic change, and diet and nutrition factors are all major contributors to the increase in disease burden. Diabetes develops as a result of a complicated combination of genetic, behavioral, and environmental variables (Moeti et al., 2020).

8.4.2 PREVALENCE OF FUNCTIONAL FOOD

The invention of functional foods is frequently credited to Japan in the late 1980s. Functional foods, on the contrary, have a significant history. Many foods have traditionally been associated with distinct health benefits in China, Japan, and other Asian countries (Weststrate, van Poppel, and Verschuren, 2002). Functional foods were described in 1994 by the Food and Nutrition Board of the National Academy of Sciences as "any tailored food or food ingredient that may deliver a health benefit in addition to the traditional nutrients it supplies" (Hasler, 2002). The older adult population segment is quickly growing, with forecasts of individuals aged 65 and up in Canada increasing from 4.2 million to 9.8 million sandwiched between 2005 and 2036, accounting for up to 25% of the population by 2041. The most often mentioned elements that would boost functional food consumption among participants were increased awareness and knowledge (85.5% of participants) and the influence of a health professional (71.0%). With no significant differences detected in gender, education, or economic classes, the existence of a nutrient content claim on a functional food label would increase the likelihood that people would use that functional food (Vella et al., 2014). In 2018, the retail value of India's functional or fortified food marketplace was predictable to be over 3.9 billion US dollars. In 2023, it was expected to reach about $6 billion. Functional foods, after naturally healthy foods, had the biggest retail market worth in the Indian health and wellness sector throughout the time period studied (https://www.statista.com/statistics/1061728/india-functional-food-retail-market-value/).

8.5 ROLE OF FUNCTIONAL FOOD IN DIABETES MELLITUS

Diabetes mellitus (DM) and its consequences are a major global public health concern around the world. This is the first assessment to focus on significant functional foods and herb with anti-diabetic action, with a special emphasis on type 2 diabetes mellitus (T2DM) and clinical studies, as well as potential anti-diabetic processes. Functional foods are foods that include biologically active components that have been allied to physiological health advantages in the prevention and management of chronic diseases like T2DM (Alkhatib et al., 2017). Eating of functional foods on a regular basis may be linked to improved anti-inflammatory, anti-oxidant, insulin sensitivity, and anti-cholesterol activities, all of which are important in the prevention and management of T2DM (Jing et al., 2021). According to the World Health Organization (WHO), 50% of T2DM patients do not follow professional recommendations for life-style and eating habits. As an effect of this situation, nutrition science is faced with a challenge: finding novel foods and/or dietary elements that improve health and lower the risk of certain diseases (Ahire et al., 2021). A food is deemed "functional" if it has been proved that, in addition to its nutritional impacts, it positively influences one or more functions of the organism in a way that enhances health or well-being or decreases illness risk (Roberfroid, 2000).

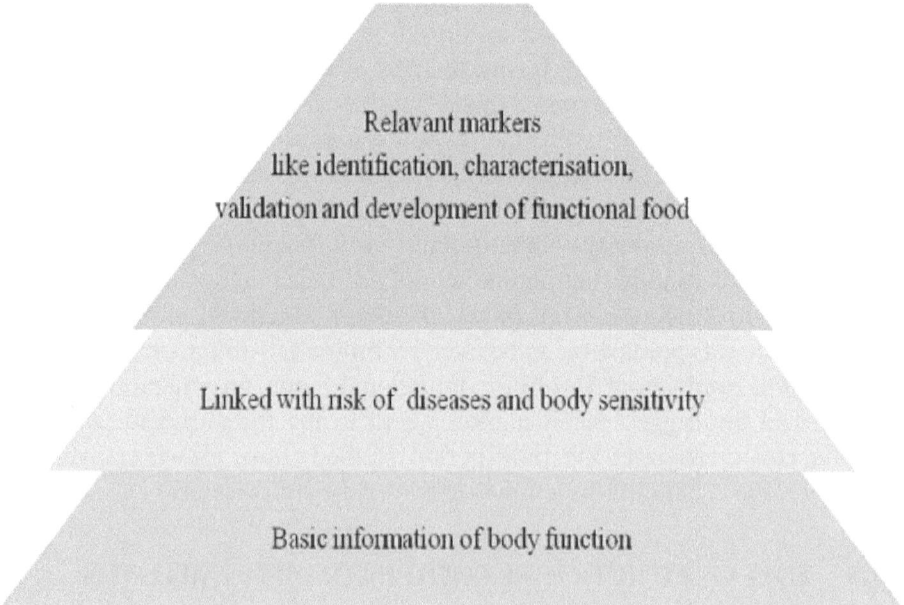

FIGURE 8.4 Steps in the development of functional food.

Prebiotics, probiotics, and synbiotics offer a variety of health benefits to their users, including improved glycemic management, lower blood triglycerides (TG), cancer prevention, and improved mineral availability (Pandey, Naik and Vakil, 2015). Prebiotics are fibers and natural sugars that stimulate the good bacteria in the gut. These food includes legumes, oats, garlic, etc.

8.5.1 LEGUMES

Legumes (peas, beans, lentils, peanuts) are high in dietary protein, non-digestible carbohydrates such as dietary fiber, resistance starches, oligosaccharides, and bioactive components such as functional fatty acids (linoleic acid, -linolenic acid), isoflavones (daidzein, genistein, and glycitein), phenolic acids, saponins, and phytic acid. Legumes are a part of a balanced diet, and there is a lot of evidence that moderate intake of legumes reduces the risk of overweight, type 2 diabetes, and cardiovascular disease (Polak, Phillips and Campbell, 2015). Other bioactive peptides of grain legumes, such as the 7S globulin chain and conglutin, have special features to control lipid metabolism and normalize lipid and lipoprotein levels; -amylase inhibitory peptides are one of the bioactive components in legumes and beans that decrease absorption of nutrients of dietary carbohydrates and modulatory postprandial glycemic response (Mirmiran, 2014).

8.5.2 OATS

Oat is one of the majority farmed and consumed food crops and ancient grains on the world. It's appropriate more popular as a result of its nutritional profile and the multipurpose effects of some bioactive components. Oat grains contain beta-glucan, which is an essential component of dietary fiber. It's the main active ingredient in oats, and it's been shown to have cholesterol-lowering and anti-diabetic properties (Paudel et al., 2021).

8.5.3 FATS

Obesity, dyslipidemia, and atherosclerotic heart disease High-fat diets can exacerbate all of these problems. Reducing saturated fat consumption can reverse or improve certain metabolic disorders (Ashwlayan and Nimesh, 2018).

8.5.4 WHOLE GRAINS AND FIBER

Past research has shown that dietary carbohydrate dysregulation in people with metabolic syndrome leads to beneficial metabolic effects, such as decreased adipocyte size, insulin sensitivity, and changes in adipocyte gene expression involved in insulin signaling (such as insulin receptors, hormone-sensitive lipase, and insulin-like tissue binding protein-5 (Kallio et al., 2007; Laaksonen et al., 2005). Cereals, whole grain, and bakery products fortify the (Martínez et al., 2013; Shao et al., 2011) found that beta glucan and inulin function as prebiotics that modulate gut microbiota by promoting the proliferation and impact of lactic acid bacteria and bifidobacteria. Increased use of oat products has been demonstrated to improve insulinemic, lipidemic, and diabetic response in diabetic patients, and act as an active constituent in lowering post-prandial blood glucose levels and inhibiting pancreatic-cell death (Tapola et al., 2005). Whole grains contain a high level of fiber, which helps to prevent glucose intolerance and diabetes.

8.5.5 POLYPHENOLS

Phenolic acids, flavonoids (Keservani and Sharma, 2014), stilbenes, and lignans are the different types of phenolic acids. T2DM (Lasa et al., 2014; Zamora-Ros et al., 2013) by stimulating the nuclear factor erythroid 2-related factor 2 (Nrf2) pathway and upregulating antioxidant response element (ARE) gene expression, catechin and epigallocatechin gallate (ECGC) decrease oxidative stress, inflammation, and cell death. A potential antioxidant defense role in preventing several chronic diseases that are often characterized by an increased production of reactive oxygen species (ROS). Insulin sensitivity, vascular endothelial function, lipid and carbohydrate metabolism, and inflammatory mediators are all influenced by procyanidins and chlorogenic acid (Alkhatib et al., 2017).

8.5.6 SOYABEAN

Soybeans have been demonstrated to improve lipid homeostasis and insulin sensitivity via activating the peroxisome proliferator-activated receptor (PPAR) and liver X receptors, as well as inhibiting the sterol regulatory element binding protein-1c (Signaling et al., 2009) Beta conglycinin, found in soy beans, has an anti-atherogenic action via regulating lipogenesis, lowering lipogenic enzyme activity, inhibiting fatty acid production in the liver, and facilitating fatty acid beta oxidation. Soybeans offer anti-inflammatory, antioxidant, and hypotensive properties (Feng and Ming, 2011) Soy selectively binds to both alpha and beta estrogen receptors; ER is subsequently linked to insulin production and secretion, beta cells survival, as well as lipid and glucose metabolism (Fu, Gilbert, and Liu, 2012).

8.5.7 POLYUNSATURATED FATTY ACIDS

Some experts believe that PUFA consumption can help with glucose management (Coelho et al., 2017). Improvement in insulin sensitivity Omega-3 fatty acids can help with diabetes treatment by lowering TG levels, boosting HDL cholesterol (HDL-C), raising glucose-induced insulin release, and perhaps lowering insulin resistance (Ahire et al., 2020). Consumption of high levels of omega-3 fatty acids improved insulin sensitivity and decreased inflammatory markers (Rudkowska, 2009).

8.5.8 NUTS AND LEGUMES

Bioactive compounds found in beans and legumes, such as – amylase, 7S globulin chain, and conglutin, reduce the breakdown and absorption of dietary carbohydrates, alter the postprandial glycemic response, maintain lipid and lipoprotein levels, and regulate lipid metabolism (López-Amorós, Hernández and Estrella, 2006). Beans have a hypoglycemic effect by inhibiting alpha-amylase and beta-glucosidase (Olagunju et al., 2020). In T2DM patients, pinto, dark red kidney, and black beans were linked to better dyslipidemia, postprandial glycemic responses, and weight control. Nut consumption improves post-prandial glucose after high-carbohydrate meals, lowers post-prandial oxidative stress, improves insulin resistance, reduces inflammatory processes, reduces lipid atherogenicity, and modulates blood lipids levels. Control the glycemic index as well. Nuts improve postprandial glycemic response in T2DM patients after consuming high carbohydrate meals, reduce postprandial oxidative stress and inflammation, and improve cholesterol, decrease lipid atherogenicity, and sensitivity to insulin.

By activating FFAR2 and FFAR3, which contribute to immune system modulation and the pathophysiology of autoimmune illnesses like T1D, particular probiotic strains could boost the generation of SCFAs (e.g., butyrate) and therefore regulate intestinal cellular homeostasis [20].

The activation of FFAR2/3 by SCFAs may also increase the synthesis of glucagon-like peptide-1 (GLP-1) by intestinal L-cells. GLP-1 is a hormone that promotes insulin secretion by pancreatic β-cells, lowering blood sugar levels (known as the 'incretin effect' (Mishra et al., 2019). In diabetic patients, it also lowers HbA1 and improves cholesterol levels (Wang et al., 2021).

8.5.9 PROBIOTIC AND PROBIOTIC

They were observed to improve glycosylated hemoglobin (HbA1c) and insulin resistance, as well as lower fasting blood glucose and insulin levels (as measured by lowering of HOMA-IR Homeostatic model assessment for insulin resistance values. Increases in high density polymer and antioxidant enzyme levels (Shah and Swami, 2017). The probiotic preparation includes different concentrations of *Lactobacillus acidophilus, Lactobacillus casei, Lactobacillus rhamnosus, Lactobacillus bulgaricus, Bifidobacterium breve, Bifidobacterium longum*, and *Streptococcus thermophilus* strains, as well as fructo-oligosaccharide. Probiotic treatment increased glutathione levels and improved antioxidant status in T2D patients. HbA1c, TC, and blood pressure levels were all shown to be significantly lower. Probiotics from (Kesika, Sivamaruthi and Chaiyasut, 2019) enhanced glucose tolerance, however the live probiotic treated hypoglycemia symptoms in the host more efficiently by reducing insulin resistance (Zepeda-Hernández et al., 2021). Inulin-type fructans and GOS are the two most regularly used prebiotics. Prebiotics have the most positive and helpful influence on the proliferation and activity of specific beneficial gut microorganisms, as well as maintaining (or restoring) proper immunomodulation in the intestine. Increased Bifidobacterium spp. levels in the gut have been linked to increased glucose-induced insulin secretion, glucose tolerance, decreased adipose tissue, pro-inflammatory cytokines, and endotoxin release (Mishra et al., 2019). Obese and diabetic mice with prebiotic therapy had lower intestinal permeability and greater GLP-2 secretion, as well as lower liver expression of caused by oxidative stress markers, as well as lower LPS levels. Prebiotics can also help with blood sugar control and insulin resistance (Wang et al., 2021). Enhance blood glucose control. Hyperinsulinemia and glycosylated hemoglobin are reduced. Increase SCFA levels and the health of your gut microbiota. Improve the damage to the liver and kidneys (Zepeda-Hernández et al., 2021). Enhanced insulin selectivity in skeletal muscles, improved glycemic management, and improved insulin sensitivity are some of the other mechanisms (Table 8.4).

TABLE 8.4 Functional Foods and Their Effect in Diabetic Patient

Sl. No.	Functional Foods	Effect in Diabetic Patients	References
1.	Cinnamon	Due to presence of water-soluble polyphenols decreased the insulin receptors phosphorylation was observed in patients with type 2 DM improved insulin resistance and a reduction in the glycation of proteins.	Olagunju et al. (2020); Sikand, Kris-Etherton, & Boulos (2015); Hamasaki (2016); Rudkowska (2009)
2.	Fruits and vegetables	Improvements were noted in Hb A1C and triglyceride (TG) levels, antioxidants, oxidative stress and inflammatory markers, diabetic retinopathy, and risk of carotid atherosclerosis with fruits and vegetables intake.	Sikand, Kris-Etherton, & Boulos (2015); Mirmiran (2014); Keservani et al. (2016b)
3.	Green tea	Tea catechins are flavonoids that contain EGCG (50–80%), epicatechin 3-gallate (ECG), epigallocatechin (EGC), epicatechin (EC), and catechin acute intake of catechin-rich green tea provides benefits for postprandial glucose and redox homeostasis among postmenopausal women.	Chen et al. (2014); Sikand, Kris-Etherton, & Boulos (2015); Kesharwani et al. (2016)
4.	Dark chocolate	Dietary flavonols in cocoa may also lower insulin resistance and reduce the risk for T2D due to antioxidant effects.	Sikand, Kris-Etherton, & Boulos (2015)
5.	Garlic	↓ Hyperglycemia, induce insulin secretion from β-cell, increase insulin sensitivity.	Ballali & Lanciai (2012)
6.	Vitamins	Vitamin C acts as an antioxidant, causes inhibition of aldose reductase and hence regulates high blood sugar. Vitamin C is the scavenger of cellular free oxygen radicals and being structurally similar to glucose, replaces glucose in many biochemical reactions; hence it is effective in the prevention of glycosylation of proteins.	Pathak (2014)
7.	Lycopene	↓ LDL oxidation, improve diabetes-induced lipid disorders, ↓ insulin resistance and adipocyte size, ↓ Triglyceride levels	Pathak (2014)

8.6 FUTURE PROSPECT AND CHALLENGE

The development of innovative foods with health advantages is aided by scientific advances in understanding the relationship between nutrition and disease, as well as rising health-care expenses and consumers' wants to change their lifestyles (functional foods) (Singletary and Morganosky,

2004). The global functional food and nutraceutical market is being driven by consumers' growing awareness of the health advantages of foods and their nutritional benefits for disease prevention and enhancement (Daliri and Lee, 2015). Another area of concern for the development of functional beverages is food products. For example, one difficult issue in the beverage industry is to convert sucrose with naturally taking place sweeteners, which might approach involving sugar intake concerns in diabetic patients while also providing a healthier option for people (Haghi, 2020).

8.7 CONCLUSION

Worldwide, an increase in the cases of DM and side effects of oral hypoglycemic agents are the most attractive concepts now a day, and by use of functional food, we can manage and prevent DM. The various types of functional food are cost-effective in our normal diet. The effects like antioxidants, anti-hypoglycemic, and others contribute to managing DM. We can use as much nutrition as required for our bodies and have fewer side effects than synthetic agents.

ACKNOWLEDGMENT

We thank the MET's, Institute of Pharmacy, BKC, affiliated under Savitribai Phule Pune University, Nashik, and R. C. Patel Institute of Pharmaceutical Education and Research, Shirpur. EDA thanks the Ministry of Tribal Affairs Govt. of India for providing financial assistance in the form of a fellowship (Award No.: 202021-NFST-MAH-01235).

KEYWORDS

- **diabetes mellitus**
- **functional food**
- **glucose metabolism**
- **nutrition**
- **persistent hyperglycemia**

REFERENCES

Ahire, E. D., Surana, K. R., Patil, C. D., Shah, H. S., Sonwane, G. B., & Talele, S. G., (2020). Role of omega-3 fatty acids in different neurodegenerative disorders. *Applied Pharmaceutical Science and Microbiology: Novel Green Chemistry Methods and Natural Products*, 173.

Ahire, E., Sonawane, V., Surana, K., & Talele, G., (2021). Drug discovery, drug-likeness screening, and bioavailability: Development of drug-likeness rule for natural products. *Applied Pharmaceutical Practice and Nutraceuticals.* Apple Academic Press.

Alkhatib, A., et al., (2017). Functional foods and lifestyle approaches for diabetes prevention and management. *Nutrients, 9*(12), 1–18. doi: 10.3390/nu9121310.

Arai, S., (1996). Studies on functional foods in Japan-state of the art. *Bioscience, Biotechnology and Biochemistry, 60*(1), 9–15. doi: 10.1271/bbb.60.9.

Ashwlayan, V. D., & Nimesh, S., (2018). Nutraceuticals in the management of diabetes mellitus. *Pharmacy & Pharmacology International Journal, 6*(2), 114–120. doi: 10.15406/ppij.2018.06.00166.

Ballali, S., & Lanciai, F., (2012). Functional food and diabetes: A natural way in diabetes prevention? *International Journal of Food Sciences and Nutrition, 63*(SUPPL. 1), 51–61. doi: 10.3109/09637486.2011.637487.

Blades, M., (2000). Functional foods or nutraceuticals. *Nutrition & Food Science, 30*(2), 73–76. doi: 10.1108/00346650010314313.

Bosscher, D., Van, L. J., & Franck, A., (2006). Inulin and oligofructose as prebiotics in the prevention of intestinal infections and diseases. *Nutrition Research Reviews, 19*(2), 216–226. doi: 10.1017/S0954422407249686.

Brennan, C. S., & Cleary, L. J., (2005). The potential use of cereal (1→3,1→4)-β-d-glucans as functional food ingredients. *Journal of Cereal Science, 42*(1), 1–13. doi: 10.1016/j.jcs.2005.01.002.

Carballo, J., Cofrades, S., & Jime, F., (2001). Healthier meat and meat products: Their role as functional foods. *Meat Science, 59*, 5–13.

Chen, G., et al., (2014). Nutraceuticals and functional foods in the management of hyperlipidemia. *Critical Reviews in Food Science and Nutrition, 54*(9), 1180–1201. doi: 10.1080/10408398.2011.629354.

Chonan, O., Matsumoto, K., & Watanuki, M., (1995). Effect of galactooligosaccharides on calcium absorption and preventing bone loss in ovariectomized rats. *Bioscience, Biotechnology, and Biochemistry, 59*(2), 236–239. doi: 10.1271/bbb.59.236.

Coelho, O. G. L., et al., (2017). Polyunsaturated fatty acids and type 2 diabetes: Impact on the glycemic control mechanism. *Critical Reviews in Food Science and Nutrition, 57*(17), 3614–3619. doi: 10.1080/10408398.2015.1130016.

Das, D., Vimala, R., & Das, N., (2010). Functional foods of natural origin-An overview. *Indian Journal of Natural Products and Resources, 1*(2), 136–142.

DeFronzo, R. A., (2010). Insulin resistance, lipotoxicity, type 2 diabetes and atherosclerosis: The missing links. The Claude Bernard Lecture 2009. *Diabetologia, 53*(7), 1270–1287. doi: 10.1007/s00125-010-1684-1.

DiMeglio, L. A., Evans-Molina, C., & Oram, R. A., (2019). Type 1 Diabetes. HHS Public Access. *Lancet (London, England), 176*(3), 139–148. doi: 10.1016/S0140-6736(18)31320-5. Type.

Diplock, A. T., & Action, E. C., (1999). Scientific concepts of functional foods in Europe consensus document. *Br. J. Nutr., 81*, S1–S27.

Doublier, J. L., & Wood, P. J., (1995). Rheological properties of aqueous solutions of (1->3) (1->4)-ß-D-glucan from oat (*Avena sativa* L.). *Cereal Chemistry, 72*(4), 335–340.

Doyon, M., & Labrecque, J. A., (2008). Functional foods: A conceptual definition. *British Food Journal, 110*(11), 1133–1149. doi: 10.1108/00070700810918036.

Eizirik, D. L., & Mandrup-Poulsen, T., (2001). A choice of death – the signal-transduction of immune-mediated beta-cell apoptosis. *Diabetologia, 44*(12), 2115–2133. doi: 10.1007/s001250100021.

Europe, P. I., et al., (1999). Scientific concepts of functional foods in Europe consensus document. *British Journal of Nutrition, 81*(4), S1–S27. doi: 10.1017/S0007114599000471.

Fathallah, N., et al., (2015). Drug-induced hyperglycaemia and diabetes. *Drug Safety, 38*(12), 1153–1168. doi: 10.1007/s40264-015-0339-z.

Feng, C., & Ming, T., (2011). Beneficial effects of bioactive peptides derived from soybean on human health and their production by genetic engineering. *Soybean and Health*. doi: 10.5772/18678.

Ferner, R. E., (1992). 8 Drug-induced diabetes. *Bailliere's Clinical Endocrinology and Metabolism, 6*(4), 849–866. doi: 10.1016/S0950–351X(05)80170–3.

Fu, Z., Gilbert, E. R., & Liu, D., (2012). Regulation of insulin synthesis and secretion and pancreatic beta-cell dysfunction in diabetes. *Current Diabetes Reviews, 9*(1), 25–53. doi: 10.2174/157339913804143225.

Fuller, R., & Gibson, G. R., (1997). Modification of the intestinal microflora using probiotics and prebiotics. *Scandinavian Journal of Gastroenterology, Supplement, 32*(222), 28–31. doi: 10.1080/00365521.1997.11720714.

Gibson, G. R., & Roberfroid, M. B., (1995). Dietary modulation of the human colonic microbiota: Introducing the concept of prebiotics. *Journal of Nutrition, 125*(6), 1401–1412. doi: 10.1093/jn/125.6.1401.

Gilliland, S. E., (1990). Health and nutritional benefits from lactic acid bacteria. *FEMS Microbiology Letters, 87*(1, 2), 175–188. doi: 10.1016/0378-1097(90)90705-U.

Gomes, A. M. P., & Malcata, F. X., (1999). Bifidobacterium spp. and *Lactobacillus acidophilus*: Biological, biochemical, technological and therapeutical properties relevant for use as probiotics. *Trends in Food Science and Technology, 10*(4, 5), 139–157. doi: 10.1016/S0924-2244(99)00033-3.

Haghi, A., (2020). *Biologically Active Natural Products: Microbial Technologies and Phytopharmaceuticals in Drug Development*, Apple Academic Press.

Hamasaki, H., (2016). Functional foods for type 2 diabetes. *AIMS Medical Science, 3*(3), 278–297. doi: 10.3934/medsci.2016.3.278.

Hanif, M. M., (2019). Spices: A review on diabetes mellitus. *Biomedical Journal of Scientific & Technical Research, 15*(5). doi: 10.26717/bjstr.2019.15.002770.

Hasler, C. M., (2002). Functional foods: Benefits, concerns and challenges – a position paper from the American council on science and health. *Journal of Nutrition, 132*(12), 3772–3781. doi: 10.1093/jn/132.12.3772.

Havenaar, R., (1992). *Probiotics: A General View, 1*, 151–170.

Haverkorn, M. J., (1994). Reduction of bacteriuria and pyuria using cranberry juice. *JAMA: The Journal of the American Medical Association, 272*(8), 590a–590. doi: 10.1001/jama.272.8.590a.

Ito, M., et al., (1993). Effects of transgalactosylated intestinal microflora and their on the human metabolism. *Journal of Nutritional Science and Vitaminology, 39*, 279–288.

Jing, F. Y., et al., (2021). Enhancing antioxidant and anti-hyperglycaemic functions of gingko biloba L. seeds using thermal detoxification. *Journal of Functional Foods, 87*, 104819. doi: 10.1016/j.jff.2021.104819.

Joshi, D., Roy, S., & Banerjee, S., (2018). *Prebiotics: A Functional Food in Health and Disease, Natural Products and Drug Discovery: An Integrated Approach.* Elsevier. doi: 10.1016/B978-0-08-102081-4.00019-8.

Kallio, P., et al., (2007). Dietary carbohydrate modification induces alterations in gene expression in abdominal subcutaneous adipose tissue in persons with the metabolic syndrome: The FUNGENUT study. *American Journal of Clinical Nutrition, 85*(5), 1417–1427. doi: 10.1093/ajcn/85.5.1417.

Katsarou, A., et al., (2017). Type 1 diabetes mellitus. *Nature Reviews Disease Primers, 3*(1), 17016. doi: 10.1038/nrdp.2017.16.

Kaur, S., & Das, M., (2011). Functional foods: An overview. *Food Science and Biotechnology, 20*(4), 861–875. doi: 10.1007/s10068-011-0121-7.

Kazi, A. A., & Blonde, L., (2001). Classification of diabetes mellitus. *Clinics in Laboratory Medicine*. doi: 10.5005/jp/books/12855_84.

Keservani, R. K., & Sharma, A. K., (2014). Flavonoids: Emerging trends and potential health benefits. *Journal of Chinese Pharmaceutical Sciences, 23*(12), 815.

Keservani, R. K., Kesharwani, R. K., Sharma, A. K., Vyas, N., & Chadoker, A., (2010b). Nutritional supplements: An overview. *International Journal of Current Pharmaceutical Review and Research, 1*(1), 59–75.

Keservani, R. K., Kesharwani, R. K., Vyas, N., Jain, S., Raghuvanshi, R., & Sharma, A. K., (2010a). Nutraceutical and functional food as future food: A review. *Der Pharmacia Lettre, 2*(1), 106–116.

Keservani, R. K., Sharma, A. K., & Kesharwani, R. K., (2016a). Nutraceutical and functional foods for cardiovascular health. In: *Food Process Engineering* (pp. 257–278, 291–312). Apple Academic Press, CRC Press. ISBN: 9781771884020.

Keservani, R. K., Sharma, A. K., & Kesharwani, R. K., (2016b). Medicinal effect of nutraceutical fruits for the cognition and brain health. *Scientifica*, 10. Article ID 3109254. doi: 10.1155/2016/3109254.

Keservani, R. K., Sharma, A. K., & Kesharwani, R. K., (2017). An overview and therapeutic applications of nutraceutical and functional foods. *Recent Advances in Drug Delivery Technology*, 160–201. ISBN:9781522507543.

Keservani, R. K., Sharma, A. K., & Kesharwani, R. K., (eds.). (2020). *Nutraceuticals and Dietary Supplements: Applications in Health Improvement and Disease Management*. CRC Press. ISBN: 9781771888738.

Kesharwani, R. K., Singh, P., & Keservani, R. K., (2016). Green coffee bean extract and chlorogenic acids: Chemistry and novel antioxidant benefits. In: Bagchi, D., Moriyama, H., & Swaroop, A., (eds.), *Green Coffee Bean Extract in Human Health* (pp. 1–17). Boca Raton, Taylor and Francis, USA. Chapter 1. ISBN: 9781498716376.

Kesika, P., Sivamaruthi, B. S., & Chaiyasut, C., (2019). Do probiotics improve the health status of individuals with diabetes mellitus? A review on outcomes of clinical trials. *BioMed Research International*. doi: 10.1155/2019/1531567.

Kharroubi, A. T., (2015). Diabetes mellitus: The epidemic of the century. *World Journal of Diabetes, 6*(6), 850. doi: 10.4239/wjd.v6.i6.850.

Laaksonen, D. E., et al., (2005). Dietary carbohydrate modification enhances insulin secretion in persons with the metabolic syndrome. *American Journal of Clinical Nutrition, 82*(6), 1218–1227. doi: 10.1093/ajcn/82.6.1218.

Lasa, A., et al., (2014). Comparative effect of two Mediterranean diets versus a low-fat diet on glycaemic control in individuals with type 2 diabetes. *European Journal of Clinical Nutrition, 68*(7), 767–772. doi: 10.1038/ejcn.2014.1.

Lau, T. C., et al., (2012). Functional food: A growing trend among the health conscious. *Asian Social Science, 9*(1), 198–208. doi: 10.5539/ass.v9n1p198.

Leger, A. S. S., Cochrane, A. L., & Moore, F., (1979). Factors associated with cardiac mortality in developed countries with particular reference to the consumption of wine. *The Lancet, 313*(8124), 1017–1020. doi: 10.1016/S0140-6736(79)92765-X.

Li, S., et al., (2019). Effect of CAPE-pNO_2 against type 2 diabetes mellitus via the AMPK/GLUT4/GSK3β/PPARα pathway in HFD/STZ-induced diabetic mice. *European Journal of Pharmacology, 853*(2), 1–10. doi: 10.1016/j.ejphar.2019.03.027.

López-Amorós, M. L., Hernández, T., & Estrella, I., (2006). Effect of germination on legume phenolic compounds and their antioxidant activity. *Journal of Food Composition and Analysis, 19*(4), 277–283. doi: 10.1016/j.jfca.2004.06.012.

Martínez, I., et al., (2013). Gut microbiome composition is linked to whole grain-induced immunological improvements. *ISME Journal, 7*(2), 269–280. doi: 10.1038/ismej.2012.104.

Michnovicz, J. J., & Bradlow, H. L., (1991). Altered estrogen metabolism and excretion in humans following consumption of indole-3-carbinol. *Nutrition and Cancer, 16*(1), 59–66. doi: 10.1080/01635589109514141.

Milner, J. A., (2000). Functional foods: The US perspective. *American Journal of Clinical Nutrition, 71*(6 suppl.), 1654–1659. doi: 10.1093/ajcn/71.6.1654s.

Mirghani, D. A., & Doupis, J., (2017). Gestational diabetes from A to Z. *World Journal of Diabetes, 8*(12), 489–511. doi: 10.4239/wjd.v8.i12.489.

Mirmiran, P., (2014). Functional foods-based diet as a novel dietary approach for management of type 2 diabetes and its complications: A review. *World Journal of Diabetes, 5*(3), 267. doi: 10.4239/wjd.v5.i3.267.

Mishra, S., et al., (2019). Probiotics and prebiotics for the amelioration of type 1 diabetes: Present and future perspectives. *Microorganisms, 7*(3), 5–9. doi: 10.3390/microorganisms7030067.

Muhas, C., & Naseef, P. P., (2016). A review article-gestational diabetes mellitus. *International Journal of Current Pharmaceutical Research, 9*(1), 1. doi: 10.22159/ijcpr.2017v9i1.16615.

Of, C., et al., (2005). Diabetes mellitus and other categories of description of diabetes. *World Health, 28*(Suppl 1), 224102. doi: 10.2337/diacare.27.2007.S5.

Olagunju, A. S., et al., (2020). Functional-foods use: A novel dietary regimen for type 2 diabetes treatment and management. *International Journal of Innovative Science and Research Technology, 5*(6), 340–345. doi: 10.38124/ijisrt20jun078.

Olokoba, A. B., Obateru, O. A., & Olokoba, L. B., (2012). Type 2 diabetes mellitus: A review of current trends. *Oman Medical Journal, 27*(4), 269–273. doi: 10.5001/omj.2012.68.

Padhi, S., Nayak, A. K., & Behera, A., (2020). Type II diabetes mellitus: A review on recent drug based therapeutics. *Biomedicine and Pharmacotherapy, 131*, 110708. doi: 10.1016/j.biopha.2020.110708.

Pandey, K. R., Naik, S. R., & Vakil, B. V., (2015). Probiotics, prebiotics and synbiotics: A review. *Journal of Food Science and Technology, 52*(12), 7577–7587. doi: 10.1007/s13197-015-1921-1.

Panesar, P. S., & Bali, V., (2015). Prebiotics. *Encyclopedia of Food and Health, 18*(2), 464–471. doi: 10.1016/B978-0-12-384947-2.00560-2.

Pathak, M., (2014). Diabetes mellitus type 2 and functional foods of plant origin. *Recent Patents on Biotechnology, 8*(2), 160–164. doi: 10.2174/1872208309666140904120633.

Paudel, D., et al., (2021). A review of health-beneficial properties of oats. *Foods, 10*(11), 1–23. doi: 10.3390/foods10112591.

Plaza, M., Cifuentes, A., & Ibáñez, E., (2008). In the search of new functional food ingredients from algae. *Trends in Food Science and Technology, 19*(1), 31–39. doi: 10.1016/j.tifs.2007.07.012.

Polak, R., Phillips, E. M., & Campbell, A., (2015). Legumes: Health benefits and culinary approaches to increase intake. *Clinical Diabetes, 33*(4), 198–205. doi: 10.2337/diaclin.33.4.198.

Raghuveer, C., & Tandon, R., (2009). Consumption of functional foods and our health concerns. *Pak J. Physiol., 5*(1), 76–83.

Ragimov, M., et al., (2021). Slowly evolving, immune-mediated diabetes in 14-year-old patient: A case report. *Diabetes Mellitus, 24*, 70–73. doi: 10.14341/DM12397.

Roberfroid, M. B., (2000). Concept and strategy of food science. *The European Perspective. Am. J. Cli. Nutr., 71*(6), 1660–1664.

Rodríguez, E., Arqués, J. L., Rodríguez, R., Nuñez, M., Medina, M., Talarico, T. L., Casas, I. A., Chung, T. C., Dobrogosz, W. J., Axelsson, L., & Lindgren, S. E., (1989). We are IntechOpen, the world's leading publisher of Open Access books Built by scientists, for scientists TOP 1%. *Intech., 32(Tourism)*, 137–144.

Ronis, M. J., Chen, Y., Badeaux, J., & Badger, T. M., (2009). Dietary soy protein isolate attenuates metabolic syndrome in rats via effects on PPAR, LXR, and SREBP signaling. *The Journal of Nutrition, 139*(8), 1431–1438.

Rudkowska, I., (2009). Functional foods for health: Focus on diabetes. *Maturitas, 62*(3), 263–269. doi: 10.1016/j.maturitas.2009.01.011.

Saarela, M., et al., (2000). Probiotic bacteria: Safety, functional and technological properties. *Journal of Biotechnology, 84*(3), 197–215. doi: 10.1016/S0168-1656(00)00375-8.

Salminen, S., et al., (1998). Demonstration of safety of probiotics – A review. *International Journal of Food Microbiology, 44*(1, 2), 93–106. doi: 10.1016/S0168-1605(98)00128-7.

Sanders, M. E., (1998). Overview of functional foods: Emphasis on probiotic bacteria. *International Dairy Journal, 8*(5, 6), 341–347. doi: 10.1016/S0958-6946(98)00056-9.

Sangwan, S., & Seth, R., (2020). Functional food of plant origin. *Research Journal of Animal Husbandry and Dairy Science, 11*(1), 9–16. doi: 10.15740/has/rjahds/11.1/9-16.

Satokari, R., et al., (2009). Bifidobacterium and *Lactobacillus* DNA in the human placenta. *Letters in Applied Microbiology, 48*(1), 8–12. doi: 10.1111/j.1472-765X.2008.02475.x.

Shah, N. J., & Swami, O. C., (2017). Role of probiotics in diabetes: A review of their rationale and efficacy. *European Medical Journal*, 104–110.

Shahidi, F., (2009). Nutraceuticals and functional foods: Whole versus processed foods. *Trends in Food Science and Technology, 20*(9), 376–387. doi: 10.1016/j.tifs.2008.08.004.

Shao, Y., et al., (2011). Association mapping of grain color, phenolic content, flavonoid content and antioxidant capacity in dehulled rice. *Theoretical and Applied Genetics, 122*(5), 1005–1016. doi: 10.1007/s00122-010-1505-4.

Sikand, G., Kris-Etherton, P., & Boulos, N. M., (2015). Impact of functional foods on prevention of cardiovascular disease and diabetes. *Current Cardiology Reports, 17*(6). doi: 10.1007/s11886-015-0593-9.

Singh, P., Kesharwani, R. K., & Keservani, R. K., (2017a). Antioxidants and vitamins: Roles in cellular function and metabolism. In: *Sustained Energy for Enhanced Human Functions and Activity* (pp. 385–407). Academic Press. ISBN:9780128054130.

Singh, P., Kesharwani, R. K., & Keservani, R. K., (2017b). Protein, carbohydrates, and fats: Energy metabolism. In: *Sustained Energy for Enhanced Human Functions and Activity* (pp. 103–115). Academic Press. ISBN:9780128054130.

Siró, I., et al., (2008). Functional food. Product development, marketing and consumer acceptance-A review. *Appetite, 51*(3), 456–467. doi: 10.1016/j.appet.2008.05.060.

Sjöholm, Å., (2019). Ketosis-prone type 2 diabetes: A case series. *Frontiers in Endocrinology, 10*, 1–7. doi: 10.3389/fendo.2019.00684.

Tapola, N., et al., (2005). Glycemic responses of oat bran products in type 2 diabetic patients. *Nutrition, Metabolism and Cardiovascular Diseases, 15*(4), 255–261. doi: 10.1016/j.numecd.2004.09.003.

Toniolo, A., et al., (2019). The diabetes pandemic and associated infections: Suggestions for clinical microbiology. *Reviews in Medical Microbiology, 30*(1), 1–17. doi: 10.1097/MRM.0000000000000155.

Tracy, B. L., et al., (2006). Assessment and management of residential radon health risks: A report from the health Canada radon workshop. *Journal of Toxicology and Environmental Health – Part A, 69*(7, 8), 735–758. doi: 10.1080/15287390500261281.

Walker, G. P., Dunshea, F. R., & Doyle, P. T., (2004). Effects of nutrition and management on the production and composition of milk fat and protein: A review. *Australian Journal of Agricultural Research, 55*(10), 1009–1028. doi: 10.1071/AR03173.

Wang, X., et al., (2021). Probiotics, pre-biotics and synbiotics in the treatment of pre-diabetes: A systematic review of randomized controlled trials. *Frontiers in Public Health*, 9. doi: 10.3389/fpubh.2021.645035.

Williams, P. C., & Stevensen, S. G., (1990). Near-infrared reflectance analysis: Food industry applications. *Trends in Food Science and Technology, 1*(C), 44–48. doi: 10.1016/0924-2244(90)90030-3.

Zamora-Ros, R., et al., (2013). The association between dietary flavonoid and lignan intakes and incident type 2 diabetes in European populations: The EPIC-inter act study. *Diabetes Care, 36*(12), 3961–3970. doi: 10.2337/dc13-0877.

Zepeda-Hernández, A., et al., (2021). Probiotics, prebiotics, and synbiotics added to dairy products: Uses and applications to manage type 2 diabetes. *Food Research International, 142*. doi: 10.1016/j.foodres.2021.110208.

Ziemer, C. J., & Gibson, G. R., (1998). An overview of probiotics, prebiotics and synbiotics in the functional food concept: Perspectives and future strategies. *International Dairy Journal, 8*(5, 6), 473–479. doi: 10.1016/S0958-6946(98)00071-5.

CHAPTER 9

FUNCTIONAL FOODS FOR MICROBIAL INFECTIONS

ADITI M. KOKANDE,[1] KHEMCHAND R. SURANA,[2] SUNIL K. MAHAJAN,[3] EKNATH D. AHIRE,[4] and SHITAL J. PATIL[3]

[1]*Department of Pharmaceutical Chemistry, Oriental Education Society College of Pharmacy, Navi Mumbai, Maharashtra, India*

[2]*Department of Pharmaceutical Chemistry, Shreeshakti Shaikshanik Sanstha, Divine College of Pharmacy, Satana, Nashik, Maharashtra, India*

[3]*Department of Pharmaceutical Chemistry, MGV's Pharmacy College, Panchvati, Nashik, Maharashtra, India*

[4]*Department of Pharmaceutics, METs Institute of Pharmacy, BKC, Affiliated to Savitribai Phule Pune University, Adgaon, Nashik, Maharashtra, India*

ABSTRACT

Foods that have a potentially positive impact on health beyond basic nutrition are known as functional foods. Phytochemicals, dietary fibers, structural lipids, bioactive peptides, and polyunsaturated fatty acids are among them. Proponents of functional foods believe that they can help people achieve maximum health and lower their risk of illness. Food may also be found in honey, peanuts, mushrooms, quinoa, sheep milk, probiotics, folate, and lactobacillus. Infection happens when a bacterium enters a person's body and causes harm while also affecting the human body's immune system. Honey's immune-stimulating, anti-inflammatory, and prebiotic characteristics can

Applications of Functional Foods in Disease Prevention. Raj K. Keservani & Eknath D. Ahire, PhD (Eds.)

help combat microbial illness. In ill and postoperative patients, prebiotics and probiotics have been proven to lower the risk of infection. Prebiotics are nondigestible fiber molecules that pass through the upper gastrointestinal tract undigested, stimulating bacteria development and colonization in the large intestine. Sheep milk has a higher concentration of essential proteins, vitamins, minerals, and lipids, as well as a caloric value that reflects energy. Lentinan, Lentinusedodes mycelium, KS-2, and eritadenine are antibacterial characteristics found in mushrooms. When compared to the control, both pasteurized and unpasteurized yogurt suppressed tumor cell proliferation by 28%. Green tea possesses anticarcinogenic, anti-inflammatories, and antioxidant effects, according to studies. Some of these may have antiviral properties, or they may play a role in immune system modulation and cell defense against oxidative stress. The only antibacterial action is found in hesperidin, which is found in citrus fruits.

9.1 INTRODUCTION

The world population is becoming more conscious of the relationship between nutrition and good health. This has increased research on the identification of food and food components that have special benefits to the consumer. With these efforts, functional food comes into society, which is more beneficial against microbial infection. These include foods containing phytochemicals, dietary fibers, structural lipids, bioactive peptides, polyunsaturated fatty acids (Hasler, 2002). Functional foods are ingredients which have a probably high-quality impact on fitness beyond primary nutrition. Proponents of functional foods promote the finest fitness and assist in lessening the risk of disorder. Functional food enhances bioregulation consisting of stresses, appetite, absorption, and biodefense, such as immunity and suppression of allergic reactions, which prevents disorder that includes diarrhea, constipation, cancer, and diabetes and suppresses mutagenesis, oxidation process. These sicknesses arise due to adjustments in nutritional habits (Salminen et al., 1998; Pang et al., 2012; Nataraj et al., 2020).

After Second World War, Japanese eating habits were westernized. This resulted withinside the reason of loss of life being more like the ones in Europe and America. It had end up properly installed that food additives make contributions to physiological and biological properly being moreover the cancer pattern in Japanese modified stomach cancer reduced and colorectal cancer, breast cancer unexpectedly improved. So, those advances in clinical technology withinside the beyond five decades have improved the human lifestyles span that is the best example from Japan. Recent advances in intestinal flowers studies also are withinside the background for the advent of functional food. Optimal intestinal flowers is achieved through a nutritionally well-balanced weight loss plan of foods consisting of oligo, nutritional fiber, fermented milk, which promote useful microorganism or suppress dangerous microorganism (Smilowitz et al., 2014; Surana et al., 2021). From the year 1991, the ministry of fitness and welfare commenced food for unique fitness use device. This device needs to lifestyle-relay style-related diseases and decrease their consequences and foods included in them are equal to functional foods. The microbial contamination happens whilst the microorganisms enter right into a person's body and make damage to it. It uses the person's body to preserve itself, reproduce, and colonize. These infectious microscopic organisms are called pathogens, and that they multiply quickly. The primary element from this functional food in microbial contamination happens with the ability foods like honey, peanut, mushroom, quinoa, sheep milk, probiotics, folate, and lactobacillus containing food also (Gobbetti et al., 2010).

9.2 MICROBIAL INFECTION

An infection takes place while a microorganism enters a person's body and reasons damage and influences the immunity system of the human body. The microorganisms use that person's body to preserve themselves and they could multiply quickly. The microorganisms lie bacteria, virus and which might be the primary reason of microbial infection. They can spread in numerous exclusive approaches it inclusive of skin contact, contact with feces, the switch of the frame fluids, touching an item that someone wearing the pathogen carrying that pathogen has additionally touched. This microbial infection generally influences the immune system. However, pathogens might also additionally from time to time overwhelm the immune system's capacity to combat them off. Among those viral infection happens because of infection with a virus. Viruses invade a number and fasten themselves to a cell. As they input into the cell, they release their genetic material. This material forces the cell to replicate the virus, and the virus multiplies. When the cell dies, it releases new viruses, which infect new cells. Also, the opposite microbial infection takes place like amebic dysentery that is an example of the protozoan. The results of an infection, which includes swelling or a runny nose, arise because of the immune systems' try to take away the invading organism. A wound fills the pus that is an example of when the white blood cells rush to the site of an injury to fight the foreign microorganism. Apart from this there are a form of infection that is as a single celled organism containing the nucleus that can reasons the protozoan infection. Protozoa which usually show numerous characteristic just like the animals, inclusive of they could live on outside the human body. In the most typically transfer to the humans through the feces. The causes of these types of infection in the entrance of any pathogen into the body and the symptoms of the infection which is depend on the organism responsible at the site of action. A person who is suffering from the bacterial infection will often experience heat, swelling, redness, fever and the pain. This is how the microbial infection spread into the communicable diseases (Bogdanov, 2012).

9.3 TYPES OF FUNCTIONAL IN PREVENTION OF MICROBIAL INFECTION

9.3.1 HONEY

Honey is the natural sweetener obtained from honey bees *Apismellifera*. Homo sapiens worked from the early Stone Age. According to the bible, the wise Solomon has said, Eat honey my son because it is good. It was a belief in the people, that honey is a nutrient, drug and also can be ointment which is been carried out for several years. From the history of humans, honey has been not only a medicine but also a natural sweetener and the medicine branch of honey is called apitherapy. Which has been developed in recent years, offering the treatments of many diseases by honey and other bee products. Nowadays the production rate of honey is 1.2 million annually and which is less than 1% of the total sugar production. But now in recent studies, honey is the only untreated natural food. The major production of honey occurred in China and Argentina (Arya et al., 2016).

9.3.1.1 COMPOSITION AND NUTRITIONAL REQUIREMENT OF HONEY

9.3.1.1.1 Carbohydrates

The main sugar or the carbohydrates found in the honey is fructose and glucose and beyond this, almost 25 oligosaccharides have been found.

Among this 25 oligosaccharide, the main blossoms are the disaccharide that is sucrose, maltose, trehalose, and turanose. Honeydew honey compared to blossom honey contains higher amounts of oligosaccharides and trisaccharides, such as melezitose and raffinose (Ziemer and Gibson, 1998).

9.3.1.1.2 Enzyme and Amino Acid

Honey includes approximately 0.5% proteins, particularly enzymes, and amino acids. There are three major honey enzymes are observed this is diastase, decomposing starch or glycogen into smaller sugar units, invertase that is once more decomposing sucrose into fructose and glucose in addition to glucose oxidase that is generating hydrogen peroxide.

9.3.1.1.3 Vitamins, Minerals, and Trace Compounds

The little or no quantity of nutrients and minerals present in the honey. Honey consists of different trace elements from a dietary point of view which can be the minerals which might be manganese and selenium, additionally Sulphur, boron, cobalt, fluorine, iodine, and molybdenum (Keservani et al., 2010a, 2017; Singh et al., 2017a).

9.3.1.1.4 Aroma Compounds

Polyphenols are the group of compounds foin the honey concerning functional and taste properties. Because beyond sugars, honey aroma depends on the quantity and quality of honey acids and amino acids.

9.3.1.1.5 Glycemic Index

It is very necessary to discuss glycethemic index of honey for diabetes patientsglycemic index of honey is 19 and for sucrose is 68.

9.3.1.1.6 Functional Properties

The functional properties of the honey are tested in animal and cell columns. It shows various antimicrobial act action as antimicrobial action. In honey, there are primarily two forms of antibacterial activity. By activating the immune system, reducing inflammation, and promoting good bacteria, honey helps combat microbial infection indirectly. Moreover, honey directly stops fungus and bacteria from growing. Both the *in vitro* Rubella virus and the anti-HIV-1 activity as assessed by PCR have been demonstrated to be inhibited by the antiviral action of honey. Honey inhibits quorum sensing and virulence. Honey reduces the capacity of the pathogen microorganisms to obtain iron from their host is detected to both colonization and increase in the form of infection (Arya et al., 2016).

9.3.2 PEANUTS

Peanuts are also called groundnuts which are the edible seed of legumes. India is the second-largest country in theproduces peanuts, with a total production of approximately 7.131 million metric tons per hectare. Peanuts are consumed all over the world in a wide variety of forms, most of which are traditional cuisine (Chang, 1996).

9.3.2.1 NUTRITIONAL PROFILE

9.3.2.1.1 Fat

According to the Peanut Council America, the fat content of peanuts nut is about 50% monosaturated fatty acids, 33% paraformaldehyde, and 14% saturated heart-friendly fatty acids (Singh et al., 2017b). The high monosaturated fat peanut diets are lowered their total body cholesterol to up to 11% and bad low-density lipoprotein upto 14% (Balthazar et al., 2017).

9.3.2.1.2 Protein

Legumes of the protein contains protein in n a higher amount than any other nuts. From the extraction process of peanut oil, the protein content observed is 50%. Because protein contains all the 20 amino acids beneficially. Amino acid Arginine is found in high amounts. Since, the proteins in peanuts in plant-based, it carries with is additional benefits that have positive health benefits like fiber and unique bioactive compound, unlike animal protein (Graf et al., 2015; Jarouliya and Keservani, 2019b).

9.3.2.1.3 *Peanut Digestibility*

Peanuts contain the ng sustain substances digestible. The very true protein digestibility of peanuts is comparable with that of animal protein. Fat digestibility is based on the structure of different fatty acids. Peanuts contain over 50% monosaturated fats so, they are highly digestible.

9.3.2.1.4 *Fiber*

Peanuts also contain fibers. According to FDA, sucrose and starch constitute major nutrients such as calcium, potassium, manganese, and vitamin A and E.

9.3.2.1.5 *Vitamins*

Vitamins detected in peanuts are 75% niacin, 60% folate, 53% thiamine, 10% riboflavin, 35% pantothenic acid, 27% pyridoxine, 55% vitamin A. As per the study, peanuts are a good source of niacin and it prevents Alzheimer's disease.

9.3.2.1.6 *Antimicrobial Action*

As a multidisciplinary that simultaneously addresses infection and nutrition. Much progress has been made in that the treatment in childhood infection in recent decades that the formulation of peanut with the potent drug shows antibiotic activity in the pediatrics patient.

9.3.3 *PROBIOTICS*

Probiotics are live microorganisms promoted to offer fitness advantages whilst consumed, usually via way of means of enhancing or restoring the intestine flora. They are considered that they are commonly stable to eat but may moreover purpose bacteria host interactions and unwanted side consequences which might be decided withinside the uncommon cases. The first discovered probiotic become a certain strain of bacillus in Bulgarian yogurt, referred to as Lactobacillus bulgaricus. The present day idea is usually attributed to the Russian Nobel, who discovered around 1907 that yogurt-consuming Bulgarian humans stay longer. A developing probiotics marketplace needs stricter requirements for medical substantiation of putative benefits conferred through microorganisms claimed to be probiotics. The benefits are advertised in the direction of the use of consumer probiotic products, which includes decreasing gastrointestinal discomfort, enhancing immune health, relieving constipation, or averting the common cold (Agrawal, 2005).

9.3.3.1 PROBIOTICS PRODUCTS

Probiotics require a vehicle to reach the site of action in an active form, which is GI the reaction of the human body. The vehicle is generally a food product, which contains live bacteria. From this, the product should have a good shelf life and should have a cell count. Several new probiotic products based on cereals, fruits, vegetables, and meat are in the development stages. These products undergo medical human trials before they are medically accepted (Elmer, 2001).

9.3.3.2 NONMILK PROBIOTIC PRODUCTS

Apart from the milk products, attempts are being made to develop non-milk probiotic products for the treatment of lactose intolerance and control of cholesterol, which is a drawback in milk-based products. Arihara & Itoh (2000) have isolated UV-induced *L. gasseri* mutants that are resistant to sodium chloride and sodium nitrate. These can be used as starter cultures for the preparation of meat fermentation. They found traditionally prepared sausages with a long ripening time have a better sensory quality than when an UV-induced mutant strain was used (Vanga et al., 2018).

9.3.3.3 FERMENTED MILK PRODUCTS

In the year 1996, Nighswogner et al. stated that carcasses *L. acidophilus* survive better in cultured buttermilk than in yogurt. And studied the cell count in milk which has above 8.7–9.18 log CFU/ml after 6–16 hrs of incubation (Syngai et al., 2016).

9.3.3.4 CLINICAL STUDIES

This is an area of probiotic remedy that requires distinctive attention. If the research is done systematically, they'll show beneficial in the treatment of diverse scientific problems along with exclusive forms of di-diarrhea gastro-intestinal issues, and levels of cholesterol in the blood. This component is different from the only wherein probiotics are used for well-being; the latter does now no longer require a scientific prescription. Probiotics also are used for their immunological and anti-microbial properties. There are many reports in the literature concerning clinical trials however most of the research did now no longer have proper controls like blindfold research and consequently couldn't be taken ahead (Peter, 2004).

9.3.3.5 USES OF PROBIOTICS IN MICROBIAL INFECTION

Probiotics and prebiotics have been shown to reduce the rate of infection in sick and postoperative patients. Compounds such as antibiotics and pharmacological agents have detrimental effects on mucosal flora and immune systems. Recent studies in patients with severe and acute pancreatitis abdominal surgery and liver transplant patients have shown some beneficial effects with the use of probiotic *Lactobacillus plantarum* 299V along with oral fiber as a prebiotic (Fooks and Gibson, 2002; Salminen et al., 1999). In another study of abdominal surgery patients, when the scientist L. Plantarum 299V was fed with oat fiber, no significant differences were found as compared with the control possibly due to the low level of lactic acid bacteria and oat fiber employed in the study. The scientist Bengmarkin the year 2003 and Edgar in the year 2005 has reviewed the use of probiotics in many other surgical cases (Reygaert, 2018; Walker, 2008; Rautava, 2007).

9.3.3.6 ANTIMICROBIAL ACTION

Concerning the probiotics the emergence of antibiotic-resistant microorganisms, the concept of probiotic therapy is gaining ground. According to the medical experts, correct identification of the strain with medical evidence along with clinical trials will allow the authorities to recommend them for therapeutic purposes.

The lowering of pH by lactic acid and acetic acid has bactericidal and bacteriostatic effects. With these they also produce hydrogen peroxide and bacteriocins. In 1998 Kasper determined antimutagenic properties by acids produced against several mutagenic compounds. Again study shows that LAB ingestion reduced the activity of enzymes like beta-glucuronidase, nitroreductase, and azoreductase which of these transform pre-carcinogen into the active form. Lactic acid bacteria are useful in the promotion of human health (Kubanek et al., 2003).

9.3.4 MUSHROOM

Mushrooms are a big source of protein. Commonly mushrooms are known for edible and functional properties also (Bellavite and Donzelli, 2020; Mizuno, 1995). There are some species of mushrooms are Articularia, Flammulina, Pleurotus, and Tremella. But the most edible mushrooms don't have medicinal value and some mushrooms have a medicinal value that cannot be eaten. Along this many cultures have used mushrooms both for food and medicine, the use of mushrooms as a functional food is most notable in the East, where the application of mushrooms is to maintain health was formally recorded in China. Mushrooms have been incorporated into healthy tonics, tinctures, teas, and soups. Also in the daily healthy as a functional food (Keservani et al., 2010a, b, 2017, 2020) and herbal formulae (Sharma et al., 2020; Rane et al., 2020a–c). From the framework of the traditional medicine system, mushrooms have been used to lubricate the lungs, tonify the kidney, invigorate the spleen and reduce excessive dampness (Lu et al., 2020). Recent studies about mushrooms say that functional mushrooms have been shown to modulate the immune system, lower blood pressure, and also lipid concentration. This is a model of functional mushroom with demonstrated bioactivity because it is widely consumed as a food in the East and increasingly in the West and the extensive research on its bioactivity has led to the isolation of pure compounds that have reached pharmaceutical status (Alissa and Ferns, 2012; Manzi et al., 2001).

9.3.4.1 NUTRITIONAL VALUE

Mushrooms contains protein, lipidcontain carbohydrate, fiber, minerals, Vitamin B1, B2, and C also the ergosterol, Vit D provitamin (Knorr, 1998; Chang and Miles, 2004).

9.3.4.2 ANTIMICROBIAL ACTION

Mushrooms show anti-microbial properties including lentinan, *Lentinuse-dodes mycelium*, KS-2, and eritadenine. The antimicrobial effects of lentinan are also believed to be enacted via immunologic potentiation of host defenses, which has been demonstrated against bacteria, viral and parasitic infections. Mushrooms have other biologically active polysaccharide fractions that have also been isolated from shiitake. LEM is derived from shiitake mycelium and contains a heteroglycan protein and another monosaccharide. Other activities of mushrooms are about anti-tumors. Scientist Chihara et al. isolated an antitumor polysaccharide from the shiitake and named it lentinan. The lentinan is a beta-D glucan as shown by electrophoresis, ultracentrifugation, and other instrumental analysis. Lentinan has been found to activate macrophages, T-lymphocytes, and other immune effector cells that modulate the release of cytokines, which may in turn to for its indirect antitumor and antimicrobial properties (Knorr, 1998; Surana et al., 2021).

9.3.5 SHEEP MILK

The history of sheep milk dates back to the Neolithic period when hominids transitioned from nomadic hunting to collecting cultures. It was able to gather and domesticate animals for employment and food after the development of a fixed living location, thereby creating agricultural methods. Goats and sheep were the first domesticated animals in the Middle East because of their size, toughness, adaptability, temperament, and sociable nature, which made human control easier. Goats and sheep have long been used as a source of food (meat and milk) and clothing (wool). Milk has become a desirable and valued source of nutrition throughout the years, resulting in the formation of herds and dairy breeds. Cow milk is the most widely consumed milk on the planet, accounting for 782 million tons of milk production in 2013. Cattle milk accounts for 85% of global milk output, with buffalo milk accounting for 11%, goat milk for 2.3%, sheep milk for 1.4%, and camel milk accounting for 1.4%. Sheep milking farms, on the other hand, cover a large portion of the country. Sheep milk output in Europe was at 9.1 million tons in 2009, however liquid consumption is uncommon. Sheep milk is now regarded as a delicacy in a number of nations, including the United States. As a result, sheep dairy products have grown in popularity owing to their excellent quality, high production, and nutritious value; the nutritional value comes from the greater quantities of proteins, lipids, vitamins, and minerals in sheep milk compared to other domestic animals' milk. Scientists mostly utilize sheep milk to make exquisite cheeses, yogurt, and whey cheeses. Casein is an ideal matrix for cheese manufacturing because of its high quantities of protein, fat, and calcium. Despite the fact that sheep milk production is on the rise, due to seasonality, sheep milk does not produce as much as cow and goat milk in terms of livestock output (Iyer and Tomar, 2009; Mohapatra et al., 2019).

9.3.5.1 CHARACTERISTICS OF SHEEP MILK

Milk was the first dietary source after birth for humans who evolved in close touch with nature. Breast milk has traditionally been a child's only source of nutrition. After the domestication of animals, which began with goats and sheep around 13,000 years ago and was followed by the domestication of cows about 4000 years ago, milk was only introduced after nursing (Moatsou and Sakkas, 2019). Fresh sheep milk chemical composition changes over time and across animals based on a variety of parameters including lactation stage, parity, season, ambient temperature, lactation efficiency, animal age and nutrition, genetic characteristics (species and breed), and illnesses. Due to changes in the forage content of range-fed animals, seasonal fluctuations have a significant impact on fatty acid composition (Yan and Polk, 2006; De Renobales et al., 2012).

9.3.5.2 NUTRITIONAL CHARACTERISTICS OF SHEEP MILK

Sheep milk has a better nutritional value than goat and sheep milk, with higher quantities of vital proteins, lipids, minerals, and vitamins, as well as a caloric value that corresponds to energy. Because they are hydrolyzed preferentially from TAGs and transported straight from the colon to the portal circulation without resynthesis of TAG, they are highly digestible. As a result, they provide a quick source of energy, especially for those suffering from malnutrition or fat malabsorption syndrome, and they have a low potential to build adipose tissue. In 2016, Balthazar and colleagues discovered that oleic acid is the most prevalent fatty acid in sheep milk yogurt, followed by palmitic acid and myristic acid. Human meals rich in oleic acid have been shown to lower LDL cholesterol levels while having no effect on HDL cholesterol levels (Raynal-Ljutovac et al., 2008). Linoleic and -linolenic acids, as well as their positional and geometric isomers, make up the majority of the polyunsaturated fatty acids (PUFAs) in sheep milk fat. According to Balthazar and other experts, mono and polyunsaturated fatty acids in sheep milk may help to avoid cardiovascular illnesses (Keservani et al., 2016a) by lowering atherogenic and thrombogenic indices (Walls and Scott, 1997; Balthazar et al., 2017).

9.4 POTENTIALLY FUNCTIONAL FOOD AND PROBIOTICS

Functional foods have the ability to promote health via mechanisms not discovered in traditional nutrition, and the results are limited to promoting health and well-being, for this reason maximizing the physiological features of the person in preference to curing diseases. Prebiotics and probiotics are a success examples of functional food groups that promote health and health in an person by Mohammadi and any other scientist in the year 2011.

The stages in sheep milk may be changed by the prebiotic and probiotic inulin fiber in the product formulation, thereby providing a functional and technological nature even as keeping the softer texture of the product due to the fact this fiber is extensively used as a fats alternative. Moreover, as a fats alternative in dairies, inulin seems to make a contribution to higher mouthfeel that's observed by the scientist Meyer and others in 2011, Karimi and others in the year 2015. When inulin is brought to the milk product at low concentrations, the rheological properties and sensory characteristics of the product aren't affected because of its impartial or barely candy taste, and confined impact at the viscosity by the scientist Franck in 2002 and Kalyani Nair and others in 2010. However, in keeping with the European rules by the scientist Saad in 2013, to attain the functional impact of nutritional fiber, inulin quantities starting from 3 to 6 g per 100 gm make sure their bifidogenic impact (Fuller, 1989).

9.4.1 ANTIMICROBIAL ACTION

The various types or for various bacteria the sheep milk shows antimicrobial activity. As seep milk is a dairy product, it has been for probiotic and prebiotic or fiber for a long time. From concerning using probiotic microorganism in the sheep milk dairy industry, the excessive protein and fats levels in sheep milk permit the manufacturing of cheese and ice cream with denser matrices, which may also bring about greater powerful protection for probiotic microorganism towards passage through the gastrointestinal tract and throughout commercial storage of the product because of the decreased oxygen permeability. The excessive protein degree is likewise a positive function for the manufacturing of fermented milk due to the fact excessive protein levels confer vitamins which are to be had for microbial growth, main to a reduction in fermentation time with no direct effect on the industrial production process (Ivey et al., 2013; Carvajal-Millan et al., 2021; Haghi, 2020).

9.5 QUINOA

As mentioned in advance human health and food safety have become an increasing number of critical with the advents of weather change, elevated human populace growth, both of metabolic disease, and growth in the median age of the populace. Quinoa (Chenopodium quinoa Willd). It is a grain-like food crop that has supplied nutrients and sustenance to Andean indigenous cultures for hundreds of years and now performs an growing role in human diets worldwide. Quinoa has been promoted as an opportunity agricultural crop because of its stress-tolerant characteristics and advertised as a "superfood" for its nutritious qualities. This brilliant food has a lot of biological and antimicrobial activity also (Ivey et al., 2013).

9.5.1 NUTRITIONAL VALUE

9.5.1.1 PROTEIN

The protein has an amount and quality of quinoa are usually advanced to the ones of cereal grains even as providing gluten-free property and excessive digestibility. Quinoa seeds are wealthy in protein, nutritional fiber, vitamin B and the nutritional minerals and the overall protein content material same to that of wheat and it additionally carries essential amino acids (Fitzpatrick et al., 2019).

9.5.1.2 CARBOHYDRATE AND FIBER

Quinoa starch comprises 58.1% to 64.2% of dry seed weight, but has a low glycemic index. The starch is constituted mainly by D-xylose and maltose with low glucose and fructose content. The starch is highly branched and consists of small granules (particle size less than 2 μm in diameter), which are smaller than the particle sizes of common cereal grains this has been discovered by scientist named as Vega-Galvez and others in 2010 (Fitzpatrick et al., 2019).

9.5.1.3 LIPID

The oil content in quinoa ranges from 2% to 10% (average 5% to 7%), which is higher than that of maize (3% to 4%). A recent study found that quinoa seed oil contains 89.4% unsaturated fatty acids and 54.2% to 58.3% of polyunsaturated fatty acids (Salminen et al., 1998).

9.5.1.4 VITAMINS

Quinoa seeds are a wealthy supply of nutrients, which can be required in the human food plan to behave as enzymatic cofactors in metabolism, adjust cell growth and development, shield against oxidative damage, enhance vision, and play beneficial roles in diverse different physiological processes. Documented nutrients in quinoa encompass vitamin A precursor β-carotene, thiamine or vitamin B1, riboflavin or vitamin B2, niacin or vitamin B3, pantothenic acid or vitamin B5, pyridoxine or vitamin B6, folic acid or vitamin B9, ascorbic acid or vitamin C, and tocopherols or vitamin E (Nwodo et al., 2012).

9.5.1.5 SAPONINS

Saponins own numerous common properties, consisting of excessive foaming potential in aqueous solutions, hemolytic activity while in direct contact with blood cells, and the formation of complexes with cholesterol and different steroidal additives of cell membranes. Therefore, they're beneficial for crop safety from microbial infection and insect or bird herbivory, which allows different natural manufacturing of this crop (Nwodo et al., 2012).

9.5.1.6 MINERALS

Quinoa has a higher total mineral or ash content material than rice, wheat, and different cereals. The micronutrients calcium, copper, iron, magnesium, phosphorus, potassium and zinc are found in enough quantities in quinoa to hold a balanced human diet (Florowska et al., 2016).

9.5.1.7 ANTIMICROBIAL ACTION

As the main antioxidant compounds in quinoa extracts might be flavonoid but nonphenolic. However, quinoa seed extracts. It has antimicrobial properties but is not for foodborne bacteria (Florowska et al., 2016; Keservani and Sharma 204; Ahire et al., 2021).

9.5.1.8 LACTIC ACID

The lactic acid is always produced from the muscle cells and which are red blood cells. The lactic acid form when the body breaks down the several macronutrients like carbohydrates to use for the energy consumption when the oxygen levels are low (Salminen et al., 1998).

Lactic acid bacteria fermentation products (lactic fermentation products) are products fermented with lactic acid bacteria or lactic acid bacteria and yeasts. A representative fermented milk product is yogurt, which is manufactured by fermenting milk with lactic acid bacteria. Fermented milk such as kefir containing alcohol is made by adding yeast. The component of fermented milk is more than, the same as in milk. The health consequences of fermented misreported primarily based totally on good enough scientific records or animal records consist of prevention and remedy of diarrheal disease, prevention of systemic infections, control of inflammatory bowel disease, immunomodulation, prevention and remedy of allergies, anticancer outcomes, remedy of cholesterolemia, and comfort of lactose intolerance. In our research in healthful volunteers, it become discovered that not unusual place outcomes of fermented milk and lactic milk liquids at the intestinal surroundings had been an growth withinside the counts of bifidobacteria and a lower withinside the counts of dangerous microorganism which include C. perfringens and E Coli and that as a result, the degrees of, fecal putrefactive product decreases. The consequences of eating pasteurized lactic milk drink Changes withinside the intestinal microbiota in the course of the experimental period. The numbers of Enterobacteriaceae, Streptococcus a, and Lactobacillus did now no longer drastically fluctuate many of the three groups. It was then combined with water and fed to the mice. The improvement of tumor cells withinside the mice given pasteurized bitter milk become slower than that during the ones given the manage diet. Both the pasteurized and unpasteurized yogurt inhibited the boom of tumor cells with the aid of using 28% as compared with the manage, while the bitter milk inhibited the boom with the aid of using 42% as compared with the manage. It has been acknowledged that the peptidoglycan of the bacterial cell partitions has adjuvant activity (Salminen et al., 1998).

9.5.1.8.1 Mechanism of Action of Lactic Acid Fermented Products

Functional foods promote fitness in phrases of each enhancing wellbeing and decreasing the risk of diseases. Lactic fermentation products are categorized into three parts (Earnshaw, 1992). Probiotics are described as live microbial food components that beneficially have an effect on the host via way of means of enhancing its intestinal microbial balance; candidate's probiotics consist of lactic acid microorganism, bacilli, and yeasts (Cross et al., 2001; Leroy and De Vuyst, 2004). Prebiotics are non-digestible food elements that beneficially have an effect on the host via way of means of selectively stimulating the increase of useful microorganism and/or via way of means of suppressing that of dangerous microorganism withinside the colon, that have the capability to enhance host fitness; the candidate prebiotics consists of oligosaccharides, nutritional fiber, and resistance starch. Biogenics are meals elements that beneficially have an effect on the host via way of means of direct immunostimulation, suppression of mutagenesis, tumorigenesis, peroxidation, hypercholesterolem, in, or intestinal putrefaction; applicants biogenic consist of organic reaction modifiers (BRM), carotenoids, flavonoids, EPA, DHA, etc. Probiotics and prebiotics act at the intestinal flora (Salminen et al., 1998; Licandro et al., 2020).

9.6 OREGANO

Oregano is the common name for a general aroma flavor primarily derived from a plethora of plant genera and species used all over the world as a spice but usually refers to the genus Origanum, the European oregano, the name of which is derived from the Greek words oros, mountain and hill, and ganos, ornament. Oregano contains the various types of polyphenol, including the flavones. It also contains the essential oil and which is composed of monoterpenoids and the monoterpens with the relative concentration of each compound which is varying from the geographical origin. Drying of the plant material affects both the quantity and distribution of the volatile content with the methods of using higher heat and longer drying times having the greater reduction impacts (Hasler, 2002; Leroy and De Vuyst, 2004).

9.6.1 NUTRITIONAL VALUE

The dietary value of oregano is quite high: it contains significant amounts of vitamins E, B6, riboflavin, niacin, folate, pantothenate, and biotin which is discovered by the scientist Holland et al. in the year 1991. Relatively high values (expressed as mg/100 g fresh leaves) have also been reported for vitamin C, thiamin, and carotene (0.81) (O'Bryan et al., 2015).

9.6.2 ANTIMICROBIAL ACTION

In conjunction with the antioxidant properties of the herb, there are abundant reports of the microbial inhibitory effects of oregano essential oil or its components. These effects are generally classified either as antifungal or antibacterial. There is a link between the chemical structure of the most abundant essential oil components and their antifungal and anti-aflatoxigenic efficacy, which is also significantly related to the concentration of the essential oil or active ingredient and the pH of the medicine being tested. Phenolic compounds are probably responsible for the high inhibitory activity of carvacrol or thymol chemotypes of oregano against fungal growth, conidial germination, and production of Penicillium species (Ziemer and Gibson, 1998; Ahire et al., 2020a, b).

9.7 SEAWEED

The seaweed is the huge kind of species of marine vegetation and algae that is grown in the sea and in the rivers, lakes. Some seaweeds are the microscopic, which ones are phytoplankton which have been stay the suspended into the water our bodies and that they offer the bottom for the most of the marine food chains. The pathogenic microbes can devastate populations of marine vegetation and animals. Yet, many sessile organisms which include seaweeds and sponges go through remarkably low degrees of microbial infection, no matter missing cell-primarily based totally immune systems. Antimicrobial protection of marine organisms is basically uncharacterized, despite the fact that from a small wide variety of research it seems that chemical defenses may also enhance host resistance (Chang, 1996).

9.8 GREEN TEA

Green tea has been proven to have anti-carcinogenic, anti-inflammatory, and antioxidant properties and is useful in cardiovascular disease (CVD)

(Keservani et al., 2016a), diabetes and obesity (Jarouliya and Keservani 2019a), and neurologic and oral health. The anticarcinogenic properties consist of controlling cell proliferation, apoptosis, and ang, biogenesis in tumor cells. Inflammation is a part of many conditions and illnesses such as aging, arthritis, cancer and obesity. HIV or AIDS and malaria are most of the top 10 infectious illnesses in the world, and the most common kinds of infections are respiration tract and diarrheal illnesses. With the arrival of antimicrobial agents in the mid-1900s got here the desire that eradication of infectious illnesses changed into close (Roberfroid, 2004).

9.8.1 COMPOSITION OF GREEN TEA

The components in green tea which are the most medically applicable are polyps. The most pertinent polyphenols are flavonoids, and the most pertinent flavonoids are the catechins. The catechins include of the flavonoids and the water-soluble solids in green tea. Green tea includes more catechins than the opposite teas, especially due to the manner it is processed after harvesting. The four primary catechins discovered in green tea are epicatechin (EC), epigallocatechin (EGC), epicatechin-3-gallate (ECG), and epigallocatechin-3-gallate (EGCG). The most important quantity of catechin is EGCG, and the following maximum important is EGC, then ECG, and EC (Graf et al., 2015).

9.8.2 ANTIMICROBIAL ACTION

The antimicrobial effects of green tea catechins (GTCs) on microorganisms have been studied for plenty of years. Green tea has been proven to fight those with organisms in numerous ways, directly and indirectly, and has been proven to paintings synergistically with a few antibiotic agents. Other known health advantages of green tea are the anti-inflammatory and antioxidant effects can also make contributions to the antimicrobial effects. There is a huge quantity of research that has assessed the antimicrobial outcomes of inexperienced tea catechins on a huge form of microorganisms, inclusive of many gram-poor and gram-positive bacteria, a few viruses, fungi, and prions.

9.9 CITRUS FRUIT

The functional food which is very rich in substances with a potential beneficial effect on health, and some of these could have an antiviral action or this is to be important in modulating the immune system and in defending cells from oxidative stress (Keservani et al., 2016b). Coronavirus disease 2019 (COVID-19), being a new and largely unknown disease, has provided doctors with the need to investigate and try new approaches and interventions. In the early stages of the pandemic, there is a high demand for this citrus containing food or medicine which are containing vitamin C chewable tablet. On the prevention front, various measures have been put in place, and recommendations have been issued, although not all based on rigorous evidence. Many of the people had been hope for the efficient vaccines which may save the lives of many people. Although clinical trials are underway to test several antivirals and other agents, an important question for the population is whether any nutrients and food/nutrition patterns can prevent viral infection or mitigation the severity of the disease. From this, the citrus fruits or the vitamin C played a vital role in preventing microbial infection (Sanders and Veld, 1999).

9.10 CONCLUSION

From this chapter, we concluded that functional food which is trending nowadays and people used to intake that functional food is preventing the microbial infection. The functional food prevents microbial infection by antimicrobial activity like increasing the immune system because most of the functional foods also show antioxidant action. So, these antioxidants stop the free radical formation and help to boost the immunity system. Among these functional foods, most of them are plant-based functional food and also the herbal drug, so from this, they do not show any other infectious action on the human body. As the world is facing new viruses many times so everyone should prefer this functional food to boost their immunity and prevent the self from any microbial infection.

ACKNOWLEDGMENT

We thank the MET's, Institute of Pharmacy, BKC, affiliated under Savitribai Phule Pune University, Nashik, and Divine College of Pharmacy, Satana. EDA thanks the Ministry of Tribal Affairs Govt. of India for providing financial assistance in the form of a fellowship (Award No.: 202021-NFST-MAH-01235).

KEYWORDS

- **functional food**
- **microbial infection**
- **prebiotic**
- **probiotic**
- **protein**

REFERENCES

Agrawal, R., (2005). Probiotics: An emerging food supplement with health benefits. *Food Biotechnology, 19*, 227–246.

Ahire, E. D., Talele, S. G., & Shah, H. S., (2020b). Nanoparticles as a promising technology in microbial pharmaceutics. *Applied Pharmaceutical Science and Microbiology*. Apple Academic Press.

Ahire, E., Sonawane, V., Surana, K., & Talele, G., (2021). Drug discovery, drug-likeness screening, and bioavailability: Development of drug-likeness rule for natural products. *Applied Pharmaceutical Practice and Nutraceuticals.* Apple Academic Press.

Ahire, E., Talele, S. G., & Talele, G. S., (2020a). Microbial pigments: A green microbial technology. *Biologically Active Natural Products.* Apple Academic Press.

Alissa, E. M., & Ferns, G. A., (2012). Functional foods and nutraceuticals in the primary prevention of cardiovascular diseases. *Journal Of Nutrition and Metabolism, 2012.*

Arya, S. S., Salve, A. R., & Chauhan, S., (2016). Peanuts as functional food: A review. *Journal Of Food Science and Technology, 53*, 31–41.

Balthazar, C., Pimentel, T., Ferrão, L., Almada, C., Santillo, A., Albenzio, M., Mollakhalili, N., et al., (2017). Sheep milk: Physicochemical characteristics and relevance for functional food development. *Comprehensive Reviews in Food Science And Food Safety, 16*, 247–262.

Bellavite, P., & Donzelli, A., (2020). Hesperidin and SARS-Cov-2: New light on the healthy function of citrus fruits. *Antioxidants, 9*, 742.

Bogdanov, S., (2012). Honey as nutrient and functional food. *Proteins, 1100*, 1400–2700.

Carvajal-Millan, E., Yaser, A. Z., & Haghi, A., (2021). *Natural Food Products and Waste Recovery: Healthy Foods, Nutrition Design, And Extraction of Valuable Compounds*. CRC Press.

Chang, R., (1996). Functional properties of edible mushrooms. *Nutrition Reviews, 54*, S91.

Chang, S. T., & Miles, P. G., (2004). *Mushrooms: Cultivation, Nutritional Value, Medicinal Effect, And Environmental Impact*. CRC Press.

Cross, M., Stevenson, L., & Gill, H., (2001). Anti-allergy properties of fermented foods: An important immunoregulatory mechanism of lactic acid bacteria? *International Immunopharmacology, 1*, 891–901.

De Renobales, M., Amores, G., Arranz, J., Virto, M., Barrón, L., Bustamante, M., De Gordoa, J. R., et al., (2012). Part-time grazing improves sheep milk production and its nutritional characteristics. *Food Chemistry, 130*, 90–96.

Earnshaw, R. G., (1992). The antimicrobial action of lactic acid bacteria: Natural food preservation systems. *The Lactic Acid Bacteria, 1.* Springer.

Elmer, G. W., (2001). Probiotics: Living drugs. *American Journal of Health-System Pharmacy, 58*, 1101–1109.

Fitzpatrick, M. C., Bauch, C. T., Townsend, J. P., & Galvani, A. P., (2019). Modelling microbial infection to address global health challenges. *Nature Microbiology, 4*, 1612–1619.

Florowska, A., Krygier, K., Florowski, T., & Dłużewska, E., (2016). Prebiotics as functional food ingredients preventing diet-related diseases. *Food & Function, 7*, 2147–2155.

Fooks, L., & Gibson, G. R., (2002). Probiotics as modulators of the gut flora. *British Journal of Nutrition,* 88, S39–S49.

Fuller, R., (1989). Probiotics in man and animals. *The Journal of Applied Bacteriology, 66*, 365–378.

Gobbetti, M., Cagno, R. D., & De Angelis, M., (2010). functional microorganisms for functional food quality. *Critical Reviews in Food Science and Nutrition, 50*, 716–727.

Graf, B. L., Rojas-Silva, P., Rojo, L. E., Delatorre-Herrera, J., Baldeón, M. E., & Raskin, I., (2015). Innovations in health value and functional food development of quinoa (*Chenopodium quinoa* Willd.). *Comprehensive Reviews in Food Science And Food Safety, 14*, 431–445.

Haghi, A., (2020). *Biologically Active Natural Products: Microbial Technologies and Phyto-Pharmaceuticals in Drug Development*. Apple Academic Press.

Hasler, C. M., (2002). Functional foods: Benefits, concerns and challenges—A position paper from the American council on science and health. *The Journal of Nutrition, 132*, 3772–3781.

Ivey, M., Massel, M., & Phister, T. G., (2013). Microbial interactions in food fermentations. *Annual Review of Food Science and Technology, 4*, 141–162.

Iyer, R., & Tomar, S., (2009). Folate: A functional food constituent. *Journal of Food Science, 74*, R114–R122.

Jarouliya, U., & Keservani, R. K., (2019a). "Pathways leading to child obesity. In: Debasis, B., (ed.), *Global Perspectives On Childhood Obesity: Current Status, Consequences And Prevention* (2nd edn., pp. 137–146). Academic Press, Elsevier, Chapter 12. ISBN: 9780128128404.

Jarouliya, U., & Keservani, R. K., (2019b). Protein function as cell surface and nuclear receptor in human diseases. In: Sanjay, B., & Debarshi, K. M., (eds.), *Medicinal Chemistry with Pharmaceutical Product Development* (pp. 1–32). Apple Academic Press, CRC Press, Taylor & Francis Group. Chapter 1. ISBN: 9781771887106.

Keservani, R. K., & Sharma, A. K., (2014). Flavonoids: Emerging trends and potential health benefits. *Journal Of Chinese Pharmaceutical Sciences, 23*(12), 815.

Keservani, R. K., Kesharwani, R. K., Sharma, A. K., Vyas, N., & Chadoker, A., (2010b). Nutritional supplements: An overview. *International Journal Of Current Pharmaceutical Review And Research, 1*(1), 59–75.

Keservani, R. K., Kesharwani, R. K., Vyas, N., Jain, S., Raghuvanshi, R., & Sharma, A. K., (2010a). Nutraceutical and functional food as future food: A review. *Der Pharmacia Lettre, 2*(1), 106–116.

Keservani, R. K., Sharma, A. K., & Kesharwani, R. K., (2016a). Nutraceutical and functional foods for cardiovascular health. In: *Food Process Engineering* (pp. 257–278, 291–312). Apple Academic Press, CRC Press. ISBN: 9781771884020.

Keservani, R. K., Sharma, A. K., & Kesharwani, R. K., (2016b). Medicinal effect of nutraceutical fruits for the cognition and brain health. *Scientifica*, 10. Article ID 3109254. doi: 10.1155/2016/3109254.

Keservani, R. K., Sharma, A. K., & Kesharwani, R. K., (2017). An overview and therapeutic applications of nutraceutical and functional foods. *Recent Advances in Drug Delivery Technology*, 160–201. ISBN:9781522507543.

Keservani, R. K., Sharma, A. K., & Kesharwani, R. K., (2020). *Nutraceuticals and Dietary Supplements: Applications in Health Improvement and Disease Management*. CRC Press. ISBN: 9781771888738.

Knorr, D., (1998). Technology aspects related to microorganisms in functional foods. *Trends In Food Science & Technology, 9*, 295–306.

Kubanek, J., Jensen, P. R., Keifer, P. A., Sullards, M. C., Collins, D. O., & Fenical, W., (2003). Seaweed resistance to microbial attack: A targeted chemical defense against marine fungi. *Proceedings of the National Academy of Sciences, 100*, 6916–6921.

Leroy, F., & De Vuyst, L., (2004). Lactic acid bacteria as functional starter cultures for the food fermentation industry. *Trends In Food Science & Technology, 15*, 67–78.

Licandro, H., Ho, P. H., Nguyen, T. K. C., Petchkongkaew, A., Van, N. H., Chu-Ky, S., Nguyen, T. V. A., et al., (2020). How fermentation by lactic acid bacteria can address safety issues in legumes food products? *Food Control, 110*, 106957.

Lu, X., Brennan, M. A., Narciso, J., Guan, W., Zhang, J., Yuan, L., Serventi, L., & Brennan, C. S., (2020). correlations between the phenolic and fiber composition of mushrooms and the glycaemic and textural characteristics of mushroom enriched extruded products. *LWT, 118*, 108730.

Manzi, P., Aguzzi, A., & Pizzoferrato, L., (2001). Nutritional value of mushrooms widely consumed in Italy. *Food Chemistry, 73*, 321–325.

Mizuno, T., (1995). Bioactive biomolecules of mushrooms: Food function and medicinal effect of mushroom fungi. *Food Reviews International, 11*, 5–21.

Moatsou, G., & Sakkas, L., (2019). Sheep milk components: Focus on nutritional advantages and biofunctional potential. *Small Ruminant Research, 180*, 86–99.

Mohapatra, A., Shinde, A. K., & Singh, R., (2019). Sheep milk: A pertinent functional food. *Small Ruminant Research, 181*, 6–11.

Nataraj, B. H., Ali, S. A., Behare, P. V., & Yadav, H., (2020). Postbiotics-parabiotics: The new horizons in microbial biotherapy and functional foods. *Microbial Cell Factories, 19*, 1–22.

Nwodo, U. U., Green, E., & Okoh, A. I., (2012). Bacterial exopolysaccharides: Functionality and prospects. *International Journal of Molecular Sciences, 13*, 14002–14015.

O'bryan, C., Crandall, P., Ricke, S., & Ndahetuye, J., (2015). Lactic acid bacteria (Lab) as antimicrobials in food products: Types and mechanisms of action. *Handbook of Natural Antimicrobials for Food Safety and Quality, 6*, 117–129.

Pang, G., Xie, J., Chen, Q., & Hu, Z., (2012). How functional foods play critical roles in human health. *Food Science and Human Wellness, 1*, 26–60.

Peter, K., (2004). *Handbook Of Herbs and Spices*. Wood Head publishing.

Rane, B. R., Bharath, M. S., Patil, R. R., Keservani, R. K., & Jain, A. S., (2020b). Novel approaches in nutraceuticals. In: Rajesh, K. K., Raj, K. K., & Anil, K. S., (eds.), *Enhancing the Therapeutic Efficacy of Herbal Formulations Through Novel Drug Delivery Systems* (pp. 241–266). IGI Global International Publisher, Pennsylvania, USA, Chapter 11. ISBN13: 9781799844532.

Rane, B. R., Patil, A. S., Keservani, R. K., & Jain, A. S., (2020a). Novel approaches in herbal formulation. In: Rajesh, K. K., Raj, K. K., & Anil, K. S., (eds.), *Enhancing the Therapeutic Efficacy of Herbal Formulations Through Novel Drug Delivery Systems* (pp. 43–68). IGI Global International Publisher, Pennsylvania, USA, Chapter 2. ISBN13: 9781799844532.

Rane, B. R., Tadavi, S. A., & Keservani, R. K., (2020c). Naturopathy. In: Anil, K. S., Raj, K. K., & Surya, P. G., (eds.), *Herbal Product Development* (pp. 321–347). Apple Academic Press, CRC Press, Taylor & Francis Group, Chapter 12. ISBN: 9781771888776.

Rautava, S., (2007). *Potential Uses of Probiotics in the Neonate* (pp. 45–53). Seminars in Fetal and Neonatal Medicine. Elsevier.

Raynal-Ljutovac, K., Lagriffoul, G., Paccard, P., Guillet, I., & Chilliard, Y., (2008). Composition of goat and sheep milk products: An update. *Small Ruminant Research, 79*, 57–72.

Reygaert, W. C., (2018). Green tea catechins: Their use in treating and preventing infectious diseases. *Biomed Research International, 2018.*

Roberfroid, M., (2004). *Inulin-Type Fructans: Functional Food Ingredients*. CRC Press.

Salminen, S., Bouley, C., Boutron, M. C., Cummings, J., Franck, A., Gibson, G., Isolauri, E., et al., (1998). Functional food science and gastrointestinal physiology and function. *British Journal Of Nutrition, 80*, S147–S171.

Salminen, S., Ouwehand, A., Benno, Y., & Lee, Y., (1999). Probiotics: How should they be defined? *Trends In Food Science & Technology, 10*, 107–110.

Sanders, M. E., & Veld, J. H., (1999). Bringing a probiotic-containing functional food to the market: Microbiological, product, regulatory and labeling issues. *Lactic Acid Bacteria: Genetics, Metabolism and Applications*, 293–315.

Sharma, A. K., Keservani, R. K., & Gautam, S. P., (2020). *Herbal Product Development.* Apple Academic Press, CRC Press, Taylor & Francis Group. ISBN: 9781771888776.

Singh, P., Kesharwani, R. K., & Keservani, R. K., (2017a). Antioxidants and vitamins: Roles in cellular function and metabolism. In: *Sustained Energy for Enhanced Human Functions and Activity* (pp. 385–407). Academic Press. ISBN:9780128054130.

Singh, P., Kesharwani, R. K., & Keservani, R. K., (2017b). Protein, carbohydrates, and fats: Energy metabolism. In: *Sustained Energy for Enhanced Human Functions and Activity* (pp. 103–115). Academic Press. ISBN:9780128054130.

Smilowitz, J. T., Lebrilla, C. B., Mills, D. A., German, J. B., & Freeman, S. L., (2014). Breast milk oligosaccharides: Structure-function relationships in the neonate. *Annual Review of Nutrition, 34*, 143–169.

Surana, K. R., Ahire, E. D., Sonawane, V. N., Talele, S. G., & Talele, G., (2021). Informatics and methods in nutrition design and development. *Natural Food Products and Waste Recovery: Healthy Foods, Nutrition Design, And Extraction of Valuable Compounds*, 33.

Syngai, G. G., Gopi, R., Bharali, R., Dey, S., Lakshmanan, G., & Ahmed, G., (2016). Probiotics-the versatile functional food ingredients. *Journal Of Food Science and Technology, 53*, 921–933.

Vanga, S. K., Singh, A., Orsat, V., & Raghavan, V., (2018). *Annex 2.5: Nutritional Comparison of Millets with Other Super Foods*, *1*, 1–18.

Walker, W. A., (2008). Mechanisms of action of probiotics. *Clinical Infectious Diseases, 46*, S87–S91.

Walls, I., & Scott, V. N., (1997). Use of predictive microbiology in microbial food safety risk assessment. *International Journal of Food Microbiology, 36*, 97–102.

Yan, F., & Polk, D. B., (2006). Probiotics as functional food in the treatment of diarrhea. *Current Opinion in Clinical Nutrition & Metabolic Care, 9*, 717–721.

Ziemer, C. J., & Gibson, G. R., (1998). An overview of probiotics, prebiotics and synbiotics in the functional food concept: Perspectives and future strategies. *International Dairy Journal, 8*, 473–479.

CHAPTER 10

APPROACHES FOR TREATING ULCER WITH FUNCTIONAL FOODS

MAYURI P. POL,[1] KHEMCHAND R. SURANA,[2] DHANANJAY M. PATIL,[1] EKNATH D. AHIRE,[3] and SANJAY J. KSHIRSAGAR[3]

[1]*Department of Pharmaceutics, Shreeshakti Shaikshanik Sanstha, Divine College of Pharmacy, Satana, Nashik, Maharashtra, India*

[2]*Department of Pharmaceutical Chemistry, Shreeshakti Shaikshanik Sanstha, Divine College of Pharmacy, Satana, Nashik, Maharashtra, India*

[3]*Department of Pharmaceutics, MET's Institute of Pharmacy, BKC, Affiliated to Savitribai Phule Pune University, Adgaon, Nashik, Maharashtra, India*

ABSTRACT

In recent years, functional food ingredients have been crucial for promoting health and lowering the risk of diseases like ulcers. Such foods are projected to give physiological advantages beyond their typical nutritional worth. Phytochemicals and bioactives must be present in foods, but their absorption, bioavailability, and metabolism are also crucial factors that affect how well they work. Processing, however, can also affect bioactive substances, therefore any changes to important food activities brought on by processing must be carefully considered. Key food ingredients and their benefits for health are provided. Functional foods are those whole, fortified, enriched, or more appropriate foods that provide health benefits beyond the availability of essential vitamins (e.g., nutrients and minerals), and this, Linking the

Applications of Functional Foods in Disease Prevention. Raj K. Keservani & Eknath D. Ahire, PhD (Eds.)

Intake of Useful Food or Food Functions Being Replicated, Randomized, Placebo-Controlled, Intervention Trials in Human Ulcers. This overview divides a large number of beneficial foods into categories based on the type of evidence supporting those functions, the strength of that evidence, and the recommended intakes. One of the areas of the foods and nutrient sciences that is currently being most actively researched and promoted is functional foods. It must be underlined, though, that these substances don't work like magic or are a cure-all for unhealthy behaviors. The simplest component of a comprehensive strategy to prevent ulcer problems is diet.

10.1 INTRODUCTIONS

Functional food is food claimed to have an additional function (often one associated with health promotion or disease prevention) by including new ingredients or extra existing ingredients. Functional foods can be "designed to have physiological benefits and/or lessen the threat of continual disease beyond simple nutritional functions and can be comparable in look to traditional food and fed on as a part of an ordinary diet" (Arai, 1996). The term may also apply to trends purposely bred into current fit for human consumption plants, consisting of red or gold potatoes having improved anthocyanin or carotenoid contents, respectively.

An Ulcer is a sickness of pores and skin, each that's lining the human frame externally and its inner viscera which may be demarcated as an open sore withinside the lining of epithelial cells or a deep lesion withinside the unique area which vitiates and might result in bleeding and flip out as bleeding sore or bleeding ulcers (Earl & Thomas, 1994).

10.2 HISTORY

Given that they all have flavor, scent, and nutritional value, all foods are functional to a certain extent. The concept was first put forth in Japan in the 1980s when, in response to rising healthcare expenses, the Ministry of Health and Welfare established a regulatory framework to certify specific meals with established health benefits in an effort to improve the nation's aging population's fitness (Kulshreshtha et al., 2017).

Certain foods are now designated as Foods for Specified Health Use, and thus may be eligible to get a special seal (FOSHU). Around 300 food products had been given FOSHU renown in Japan as of July 2002. Functional foods don't have a similar regulatory designation in the US. But many businesses have put forth definitions for this new food category. Functional ingredients were defined as "any modified meals or food element that can deliver a fitness gain beyond the standard vitamins it includes" by the National Academy of Sciences' Food and Nutrition Board in 1994 (Zojaji et al., 2009).

They are described as "ingredients that, through the distinctive property of inclusion of physiologically-energetic additions, offer a fitness advantage beyond plain nutrition" by the International Life Sciences Institute. The American Dietetic Association defined functional foods as those that are "complete, fortified, enriched, or enhanced" in a role paper published in 1999. More importantly, the paper states that these ingredients must be consumed as "part of a varied weight-reduction plan on a daily basis, at powerful levels" for clients to benefit from their potential fitness benefits. Although it is significantly less preferred by customers, the term "nutraceuticals," which was first introduced in 1991 by the Foundation for Innovation in Medicine to refer to virtually any bioactive component that can improve fitness, is frequently used interchangeably with functional foods. His definition of nutraceuticals reads, "the one's weight-reduction diet supplements that supply a focused shape of a presumed bioactive agent from a food, provided in a non-food matrix, and used to beautify fitness in dosages that exceed those that may be obtained from ordinary food," he astutely distinguished complete foods from the remoted additives derived from them (Faguet, 2015).

10.2.1 CHARACTERIZATION OF FUNCTIONAL FOODS

- Found in traditional food forms with inherent sensory characterizations.
- Containing physiological function additives that aren't fed on in medicinal/healing levels three.
- That confer scientifically validated secure for long-time period intake of the meant population.
- Containing functional additives (whether or not nutrients or phytochemicals) which can be brought to the food.
- It May be used to save you and deal with positive diseases.

10.3 FRUITS AS FUNCTIONAL FOODS

Best choices as functional foods used in ulcers.

Fruits: Any fresh or frozen fruits contain helpful fiber and antioxidants. Berries, apples, grapes, and pomegranates are among the best choices for healing polyphenols. If citrus fruits or juices like orange or grapefruit trigger reflux, avoid those (Taghipour et al., 2016).

10.3.1 BANANA

Bananas help the intestines work properly, regulate the body's acid-base balance, increase levels of the hormone that makes us happy, and improve heart health. Bananas also improve the way the nervous system works. They come in the form of mashed potatoes and can be consumed sparkling. Individuals with low acidity, who typically use this product, prefer to eat 2-3 culminations per day. But, because of the high levels of acidity, it is no longer larger than 1, according to today (Taghipour et al., 2016).

10.3.2 APPLE

Carrots and bitter cream are served with apples. In the oven, you could bake and create jelly and fruit beverages. Yet there is a guideline. Raw apples of confectionery varieties are preferable. Yet, the stressful effect of sour, young apples will harm the mucosal membrane. Apples are especially desired by those who have bleeding ulcers. Those whose hemoglobin is mixed up with blood. Apples, as you are aware, have a stronger effect on hemoglobin levels than other fruits or vegetables. Pomegranates include a range of vitamins C, so it is not always necessary to increase blood with them. Pears are also completely safe for the digestive system. They contain a significant amount of nutrients as well, including diets A, PP (nicotinic acid), K, and riboflavin, which is diet B2. This is an excellent way to obtain the necessary complex of minerals and nutrients without worrying about difficulties. With dried pears, you should be cautious and honest about your health (Floch, 2013).

10.3.3 VEGETABLES

Leafy greens, bright red and orange vegetables, and vegetables (like broccoli, cauliflower, and kale) are packed with vitamins and antioxidants that are especially good for your overall health and healing. Avoid spicy peppers and tomatoes, or products made with them, if they give you reflux. Limit raw vegetables because they are harder to digest (Mehta, 2016).

10.3.4 LEAN PROTEINS

Skinless poultry, lean beef like sirloin or tenderloin, fish, eggs, tofu, tempeh, dry beans, and peas are excellent sources of low-fat protein. Fatty fish like salmon, mackerel, and sardines provide omega-3 fats, which can reduce inflammation and may be helpful in preventing another ulcer (Ardakani et al., 2016).

10.3.5 FERMENTED DAIRY

Products such as kefir and Greek yogurt, which both include protein in addition to probiotics (beneficial bacteria), are examples of nutritious alternatives (Suerbaum & Michetti, 2002).

10.3.6 BREADS AND GRAINS

It includes foods like whole grain bread and grains that are either whole or cracked, such as oats, quinoa, farro, millet, or sorghum, in your diet to increase the amount of fiber you consume (Vomero & Colpo, 2014).

10.3.7 HERBS AND SPICES

Most mild herbs and spices are safe to use because they are sources of antioxidants. Turmeric, cinnamon, ginger, and garlic are the best options because they offer antibacterial and anti-inflammatory qualities. Try replacing sugar with honey as a sweetener (Li & Somerset, 2014).

10.4 FOODS TO TAKE IN LIMIT

1. **Alcohol:** All alcohol is a stomach irritant and will delay healing. Avoid wine, beer, and spirits (Fitzgerald et al., 2002).
2. **Caffeine:** You should cut back or stop drinking coffee, tea, and caffeinated sodas. They can increase stomach acid production (Zojaji et al., 2009).

3. **Milk:** There was a time when milk was used in treating ulcers, but research has found that it increases stomach acid. It's best to avoid it (Labov, 2006).
4. **Certain Meats:** Skip any highly seasoned meats, lunch meats, sausages, and fried or fatty meats and proteins (Zhang et al., 1997).
5. **High-Fat Foods:** Try to avoid large amounts of added fats, which can increase stomach acid and trigger reflux. You may need to avoid gravy, cream soups, and salad dressings, but healthy fats on the list are OK (Jarosz et al., 1998).
6. **Spicy Foods:** You may want to skip anything that is "hot," such as chili peppers, horseradish, black pepper, and sauces and condiments that contain them (Jarosz et al., 1998).
7. **Salty Foods:** Researchers have found that salty foods may promote the growth of *H. pylori*. Pickles, olives, and other brined or fermented vegetables are high in salt and linked to a higher risk of *H. pylori* ulcers (Podmore et al., 1998).
8. **Chocolate:** It can increase stomach acid production, and some people find that it triggers reflux symptoms (Green, 1941).
9. Recommended Timing: Try to eat five or six small meals each day, rather than three large ones. Stomach acid is produced every time you eat, but large meals require much more of it for digestion, which can be irritating (Barling, 1935).

10.5 WHAT IS NUTRITION?

Nutrition is readily consuming a healthful and balanced diet. Food and drink offer the strength and vitamins you want to be healthful. Understanding those vitamin phrases might also additionally make it simpler in order to make higher food choices (Davidson, 1928).

10.6 ULCER TYPES

Different forms of ulcers have been demarcated by scientists and are recognized through the starting place or region of incidence in the human body. Ulcer sorts recognized until now are Pressure ulcers, Genital ulcers, Ulcerative dermatitis (itis way inflammation), Anal fissures, Diabetic foot ulcers, Corneal ulcers (primarily based totally on the precise location of incidence), Mouth ulcers, additionally referred to as Aphthous ulcer (Canker sores), Peptic ulcer, Venous ulcer, Stress ulcer, Ulcerative sarcoidosis, Ulcerative lichen planus, Ulcerative colitis, Ulcerative disposition (Platt, 1936). However, the maximum not unusual place sorts of ulcers are:

1. **Peptic Ulcer:** This sort of ulcer is associated with pepsin and may arise in specific regions.
2. **Gastric Ulcer:** An ulcer takes place withinside the belly location and is frequently called a stomach ulcer.
3. **Duodenal Ulcers:** This sort of ulcer takes place withinside the duodenal location of the GIT tract or gastrointestinal tract.

What to eat when you have an ulcer? (Figure 10.1).

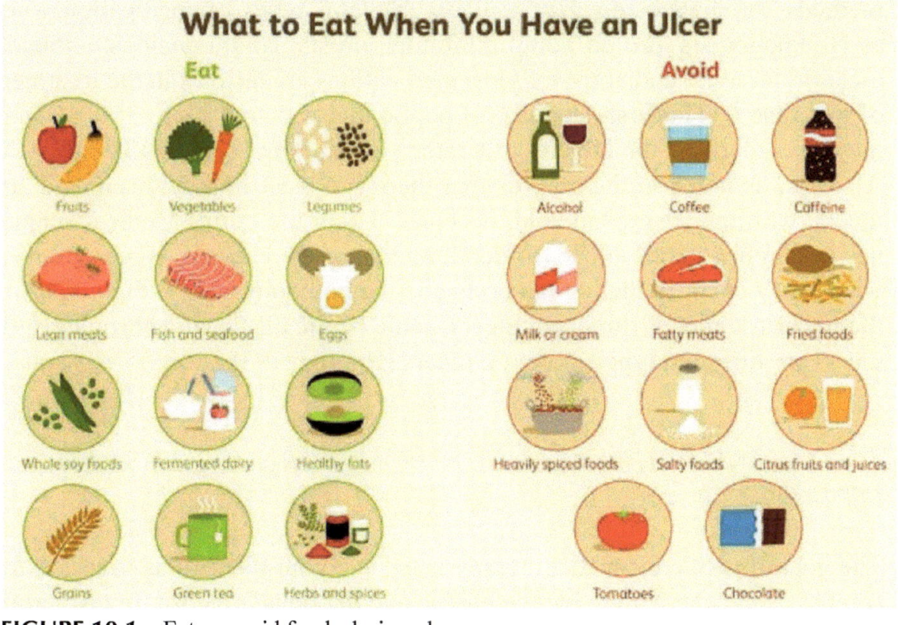

FIGURE 10.1 Eat or avoid foods during ulcers.

Source: Reprinted from Gillson, 2023. https://www.verywellhealth.com/what-can-i-eat-if-i-have-a-peptic-ulcer-1742154

10.6.1 PEPTIC ULCER

Peptic ulcers, also known as Peptic Ulcer Disorders (PUDs), are ulcers that can develop in either the gastric or duodenal regions and are linked to Pepsin, an enzyme that occurs in the stomach and breaks down proteins. The concept of acid suppression therapy was developed as a result of the discovery of H_2 receptor antagonists, which gave rise to the observed practice of surgical intervention for peptic ulcers in the form of stomach resection, vagotomy, etc. A triple-drug regimen was prescribed to treat the peptic condition after the discovery of the Helicobacter pylori organism as the causal factor, which was later adjusted to sequential treatment to prevent the development of antibiotic resistance (Archer & Graham, 1936). Peptic ulcer disease and accompanying headaches are still difficult to treat surgically. Due to an imbalance between competitive factors (gastric acid and pepsin) and protective factors, peptic ulcers or PUDs develop (gastric mucus, bicarbonate, prostaglandins). A quarter of those who suffer from this disorder has severe symptoms, such as bleeding, perforation, or obstruction of the gastric outlet. Both direct viewings using an endoscope and assessment radiography are effective methods for diagnosing peptic ulcers. Several types of medication, such as H_2 antagonists, proton pump inhibitors, prostaglandin analogs, antacids, ulcer protectives, and anti-H. Pylori medications are utilized in the treatment of this illness. The first accounts of peptic ulcers date back to 350 BC, and they were made by the Diocles of Craters, Celsius, and Galen (131-201 AD). Also, it was noted in the records that there is a clear link between acid and ulcers, with the adage "no acid, no ulcer" being the norm. Medical science has researched peptic ulcers in-depth and suggested several types based on where they occur in the human body and the pathophysiology that underlies them. Gastric ulcers, duodenal ulcers, acute peptic ulcers, and chronic peptic ulcers are different types of peptic ulcers (Bourne, 1938).

10.6.1.1 NUTRITIONAL SUGGESTIONS FOR SUFFERERS WITH PEPTIC ULCER

The goal of peptic ulcer diet therapy is to save your ulcer so as to lessen the soreness and ache withinside the gastric and duodenal mucosa. Investigation of dietary deficiencies is vital withinside the guidance of the perfect restoration diet. According to Marrota & Floch (Dalldorf, 1938), calorie distribution for sufferers with a peptic ulcer has to be normal, with values starting from 50–60% of carbohydrates, 10–15% of proteins, and 25–30% of lipids, with overall strength cost enough to keep or get better the dietary status. Reis (Lanman & Ingalls, 1937) advised that energy distribution must be adjusted in keeping with the patient's desires to normalize the dietary status, having as encouraged macronutrients a protein consumption of as much as 1.2 g/kg/

TABLE 10.1 Nutritional Recommendations for Patients with Peptic Ulcer to Accelerate the Healing Process

Characteristics	Recommendations	
Daily energy need (DEN)	Sufficient to maintain or recover the nutritional status. 20–25 kcal/kg: Weight loss 25–30 kcal/kg: Maintenance 30–35 kcal/kg: Weight gain	
	Acute Phase	Recovery Phase
Carbohydrate (%)	50–60	50-60
Proteins (g/kg/weight)	1.2	1.5
Lipids (%)	25–30	25–30
Vitamins A (μg)	900	3,000
Vitamin C (μg)	75	500
Vitamin B_{12} (μg)	2.4	2.4
Zinc (mg)	11	40
Selenium (μg)	55	400
Folic acid (μg)	400	400
Iron (mg)	45	45
Fiber (g)	20–30	20–30

Source: Adapted from Field et al., 1940

weight/day in the extreme stage (fifth to eighth week) and as much as 1.5 g/kg/weight/day in the restoration stage. Carbohydrates must be adjusted to the patient's wishes, without disaccharides' attention, so that you can keep away from fermentation and lipids without awareness of saturated fats (Field et al., 1940) (Table 10.1).

10.6.1.2 VITAMINS, FLAVONOIDS, AND PEPTIC ULCER

According to some authors, eliminating the bacteria H. pylori is the best treatment for PU. Likewise, certain human studies that used antioxidants to eradicate H. pylori found that vitamin C is crucial for the elimination of H. pylori in peptic ulcer patients. However, that study showed that when compared to greater doses, lower amounts of vitamin C were administered for longer periods of time (McCarrison, 1931).

Hence, it has been established that those who have peptic ulcers caused by *H. pylori* can consume up to 500 mg/day of vitamin C for a period of three months, which is less than the advised UL of 2,000 mg, in accordance with DRIs (Portnoy & Wilkinson, 1938). Large doses of vitamin C significantly reduced H pylori colonization, growth, or even eradication; however, the doses needed in those studies were astronomically high, usually 10 times or more than normal stages inside the stomach (Pohle et al., 2001).

Diet C functions as both an antioxidant and a pro-oxidant, and high levels may also cause kidney stones (Podmore et al., 1998). Having too much body iron storage makes vitamin C dangerous since iron encourages vitamin C's pro-oxidant properties (Brzozowski et al., 2001).

A recent study found that consuming more than 500 mg/day of vitamin C significantly increased its pro-oxidant interest and could actually trigger host cell DNA damage, increasing the risk of most malignancies. Hence, knowledge of the physiological levels of diet C inside the stomach is crucial for understanding how diet C protects against H. Pylori-caused PUD (Becker et al., 2003.).

It has long been established that the traditional diets recommended for treating gastric or duodenal ulcers frequently lack ascorbic acid (Aldoori et al., 1997). There have been cases of scurvy cases improving in PUD patients who followed the type of diet. Moreover, it is currently believed that early "pra scorbutic" moods are frequent in these situations (Mahmood et al., 1986).

Additionally, the link between ascorbic acid and collagen formation (Ankolekar et al., 2011), wound healing, and capillary fragility may also suggest that this diet is important for healing ulcers even though it is unlikely that ascorbic acid deficiency is a specific etiological factor in causing peptic ulcers in people (Lee et al., 2008).

Hence it could seem that the occurrence of a superimposed ascorbic acid deficit nation in people with peptic ulcer could be exceptionally damaging. The reduction of lipid peroxidation inside the gastric mucosa, which maintains the gastric microcirculation, is another aspect of vitamin C's protective effect. Intake of diet A seems to be protected as well, with an RR under 0.5 for those within the maximum as compared to the bottom quintile within the Health Professionals Follow-up Study. Vitamin C stimulates the expression of the antioxidant and vasodilator heme oxygenase enzyme within the gastric epithelium and inhibits the expression of the inducible nitric oxide synthase enzyme. Results from a prospective cohort look at what was covered by 47,806 males found that a diet high in diet A from all reassessments, as well as diets high in fruits and vegetables, may both slow the progression of DU, likely due to the fiber component of these foods (Ebringer et al., 2008).

In order to determine how certain factors affect the likelihood of a particular outcome, a potential cohort study tracks over time a group of people who are similar to one another but differ with respect to certain factors. However, larger studies are necessary to confirm results because there are so many different factors involved with this kind of study that could skew the data (Wang et al., 2004).

According to animal studies, diet A will cause the gastrointestinal tract to produce more mucus. Ulcers may form if mucosal protection is inadequate. Diet A may, therefore, also operate as a protective factor against the progression of ulcer disease. New research from China reveals that green tea and other substances high in flavonoids have the ability to protect against stomach cancer, *H. pylori* infection, and chronic gastritis (Kerstan et al., 2022).

Particularly, those components seem to stop *H. pylori* from growing. In addition, a recent laboratory study of green, white, oolong, and black teas found that they did not harm the good kinds of bacteria that are often found inside the stomach, including *L. acidophilus*, *L. plantarum*, and *B. lungum*, while inhibiting the growth of *H. pylori*. Nevertheless, since this was an in vitro experiment, we are unable to infer what may occur in a real body. Excellent results were obtained within the lab as the tea soaked for the full five minutes. Garlic, onions, colorful produce, and greens, including cranberries, strawberries, blueberries, broccoli, carrots, and snap peas, are among the items high in flavonoids (Chan & Woo, 2010).

10.6.2 MECHANISM ACTION OF USEFUL FOODS FOR PEPTIC ULCER

10.6.2.1 MILK OR MILK PRODUCTS AND PEPTIC ULCER

In population studies, associations between improved consumption of dairy products that have undergone fermentation (such as yogurt and cheese) and a reduced risk of PU were discovered, but consumption of unfermented milk is associated with an increased risk. The antibacterial properties of fermented dairy products, specifically lactobacillus and casein, are responsible for this effect; they prevent *H. pylori* from replicating. The elimination of *H. pylori* is regarded as an essential step in the management of PUD. Conventional triple-remedy eradication regimens, which include a proton pump inhibitor, clarithromycin, amoxicillin, or nitroimidazole, are inconvenient and frequently have uncertain outcomes. Furthermore, it is mentioned that as antibiotic resistance has increased over time, eradication costs have decreased. Several opportunities and complementary therapies were tried in this situation in an effort to get higher eradication without impacting compliance. In this search, fermented milk and several of the whey proteins that make it up have shown promise as potential candidates for supplementary medicine. They naturally have a higher level of impacted person acceptability. Casein and whey proteins, two key protein families, make up the protein structure of fermented milk. Casein is an insoluble protein that accounts for 80% of all available proteins. Bovine lactoferrin, -LA, glycomacropeptide, immunoglobulin, -lactoglobulin, and lactoperoxidase are all components of whey proteins, which are globular, water-soluble molecules. Whey may work as an antioxidant, immunological booster, anticancer, hypolipidemic, antiviral, antibacterial, and chelating agent, among other functions (Alanen & Muli, 2015). According to recent meta-evaluations and a number of randomized controlled studies, completely probiotic setups based on fermented milk increase *H. pylori* eradication rates by 10%. The efficacy of human whey proteins in the elimination of *H. pylori* has also been a subject of interest in recent studies. Yet, the role of whey proteins or fermented milk in medical exercise isn't always precisely characterized or widely addressed (Cox & Rasmussen, 2014).

10.6.3 USE OF PROBIOTICS IN PEPTIC ULCER

Probiotics are described as a meal complement primarily based totally on stay microorganisms, which have an effect on beneficially the human organism through supplying a microbial balance. There is a unique hobby in probiotics for treating contamination through *H. pylori*, as it performs an important function in the pathogenesis of chronic gastritis and peptic ulcer in adults.

Accordingly, a few research in humans used antioxidants to remove H. Pylori and determined that nutrition C has crucial results in microorganism eradication in sufferers of peptic ulcer. But that research confirmed that smaller doses of nutrition C for an extended time period had a higher reaction while as compared with better doses (Uyangaa et al., 2014).

Similarly, they acquired Bifidobacterium animal is and *L. acidophilus* (10^{1}0 CFU) two times an afternoon for the duration of six weeks, and concluded that ordinary consumption of yogurt containing Bifidobacterium animal is and *L. acidophilus* can successfully suppress the contamination due to *H. pylori* in humans. Use of antioxidants to remove Helicobacter pylori, and some authors display that the first-class remedy is the eradication of the bacteria (Hernández-Díaz et al., 2013).

10.7 DIABETIC FOOT ULCER

A diabetic foot ulcer is an ulcer of the foot location and is especially one of the problems of diabetes. During diabetes, many problems arise in the patient's body, like, mechanical modifications in the conformation of the bony structure of the foot, peripheral neuropathy, and atherosclerotic peripheral arterial disease. All those elements arise in excessive diploma among diabetes sufferers (Tables 10.2 and 10.3) (Wang et al., 2004).

TABLE 10.2 Functional Food Used in Diabetic Food Ulcer

Functional Foods	Examples
Yogurt and another dairy	Garlic lactobacilli, bifidobacteria
Onions, garlic	Sulfur compounds
Citrus fruit, carrots	β-Carotene
Rosemary	Squash, pumpkin carnosol
Teas, berries	Catechins
Onion, red grapes, citrus fruit, broccoli, Italian	Quercetin
Cabbage, broccoli, cauliflower, kale	Indoles
Grapes, strawberries	Ellagic acid
Whole grains, onions	Inulin, fructooligosaccharides (FOS)

TABLE 10.3 Functional Food Formats Availability of Nutraceuticals in Different

Nutraceuticals Ingredients	Functional Food Formats
Dietary lipids (eicosapentaenoic acid, docosahexaenoic acid, alpha-linoleic acid).	Fish, algae, krill, genetically modified seeds.
Phenolic compounds and polyphenols (phenolic acids, resveratrol, curcuminoids, catechins, flavanols, and flavones).	Tea, grape seed, olives, pomegranates, wine, cocoa, vegetables.
Carotenoids (beta-carotene, lycopene, lutein, astaxanthin).	Carrots, tomatoes, marigolds, green leafy vegetables, microalgae.
Bioactive peptides.	Milk, meat, fish, plants.
Minerals (Ca, Fe, Mg, Zn).	Usually available as soluble vitamins.

10.8 PRESSURE ULCER

Pressure ulcers are sort of ulcers at the derma place of the body. It is referred to as a stress ulcer because the sure diploma of stress at the pores and skin, which results in obstruction of blood floating to pores and skin, reasons an excessive situation of ulcer. The occurrence of stress ulcers is excessive for mattress-ridden sufferers and sufferers who're conversant in spinal damage and in a wheelchair. A pressure ulcer can occur at any location of the body that's under stress. The pathophysiology of stress ulcers lies in the reality that sure diploma of stress reasons obstruction of pores and skin nature. The blood vessels aren't capable of nourishing the pores and skin at the location of stress, and this results in the improvement of ulcers. Thus stress ulcer is a disorder of smooth tissue and primarily influences the bed ridden or wheel chair person who's managing a chronic disorder, coma, paralysis, or spinal cord damage. Pressure ulcers are excessive for hospitalized patients. Pressure ulcers also are liable for excessive rates (Wang et al., 2004).

10.9 NUTRITION AND WOUND RECOVERY

Although the proper nutrient consumption to inspire wound restoration is unknown, elevated desires for energy, protein, zinc, and Vitamins A, C, and E have been documented. High-protein oral dietary supplements are powerful in decreasing the occurrence of stress ulcers by 25% in patients at hazard. Energy, protein, arginine, and micronutrients (Vitamins A, C, and zinc) are all crucial in wound restoration. Proteins are the maximum critical macronutrient considering it's far essential for the restoration of tissues. Proteins are critical in preserving fantastic nitrogen stability and for all degrees of wound restoration consisting of fibroblast proliferation, collagen synthesis, angiogenesis, and immune function. In enteral formulations, protein may be in the shape of intact proteins, hydrolyzed proteins, or loose amino acids. The NPUAP/EPUAP recommends an international protein consumption for strain ulcer restoration of 1.25 to 1.5 g/kg of frame weight in step with day. For patients with Stage III/IV stress ulcers, the proposed stage is 1.5–2.0 g/kg, relying on the scale of the stress ulcer and the overall protein loss from draining wounds (Labov, 2006). In a study, the institution receiving better protein (1.8 g protein in step with kg frame weight) verified almost a two-fold extra charge of restoration than the ones randomized to decrease protein consumption (1.2 g protein in step with kg frame weight) (Uyangaa et al., 2014).

10.10 GENITAL ULCERS

Genital ulcers constitute ulcers in the genital regions, like scrotal ulcers, of each guy and female. It's regarded as a motive because of sexually transmitted sicknesses, inclusive of Genital herpes, Syphilis, Chancroid or *Chlamydia trachomatis*, Granuloma inguinale, and Lymphogranuloma venereum. The reasons can be attributed because of continual pores and skin sicknesses or pores and skin cancer. Around 70% to 80% of Genital ulcers are contamination by HSV-1 (Herpes simplex virus) or HSV-2 (Keri Marshall, 2004).

10.10.1 ULCERATIVE DERMATITIS

It is pores and skin sickness main to degradation of pores and skin region and might remodel to painful and instances bleeding. The principal organism causing ulcerative dermatitis is *Staphylococcus aureus*. Ulcerative colitis resulting from a microorganism may also end result in strain, and the affected person might also additionally expand strain ulcer and sickness complications. Stress Ulcer Stress ulcers are deep, focal mucosal harm penetrating the submucosa with excessive hazard for gastrointestinal bleeding. Stress can be a bodily strain or mental pressure. Stress ulcers especially arise in chronic sickness sufferers, like individuals who are tormented by Cancer, Spinal injury, and respiration sicknesses. Stress ulcers arise in nearly 75% of hospitalized patients (Table 10.4) (Alanen & Muli, 2015).

TABLE 10.4 Allowed Foods, Foods That Should be Consumed with Caution, and Foods That Must be Avoided

Food Groups	Allowed	Use with Caution	Prohibited
Dairy	Milk, low-fat cheese, yogurt, fermented milk	Fatty cheeses (mascarpone, cream cheese, gorgonzola)	–
Oilseeds	Flaxseed, Brazilian nut, Walnuts	–	–
Oils and olive oils	Vegetable oils, olive oil	–	Fried foods
Fruits	Apples, papaya, melon, banana	Orange, pineapple, acerola, passion fruits	Lemon
Vegetables	Leafy dark green vegetables, carrot, beet, green bean, spinach, kale, radish, zucchini, leek	Broccoli, cauliflower, cabbage, cucumber, onion, red pepper	Spicy pepper (black pepper, chilies)

TABLE 10.4 (Continued)

Food Groups	Allowed	Use with Caution	Prohibited
Legumes	Bean soup, lentils, chickpeas, soybean	Beans	–
Meats	Lean meat (beef, pork, chicken, fish)	Fatty meat, organ meats, sausages	–
Sweets	–	Concentrated sweets	Chocolates
Beverage	Natural juices	Citrus/acidic fruits juices	Coffee, black, tea, cola drink
Other foods	–	Industrialized seasonings, spices, and condiments	–

10.11 CONCLUSION

A significant part of individuals with PUD (either GU or DU) and functional dyspepsia are known to be affected by dietary variables, which are increasingly understood to play a significant role in both triggering and resolving symptoms. Research points to fatty foods as the primary trigger, but unfermented milk and dairy products, citrus fruits, spicy foods, coffee, alcohol, high salt intake, and soy sauce are other items that have been linked to the condition. The symptoms, however, seem to be improved by diets high in fiber, fermented milk, polyunsaturated fatty acids (particularly those in the n-6 family), and vitamins A and C. The outcomes of blind trial testing, however, are inconsistent. Furthermore, despite the fact that patients identify particular foods as PU symptom triggers, these patients frequently do not appear to alter their behavior in an effort to alleviate those symptoms; thus, any differences in dietary intake and lifestyle between patients and healthy individuals are minimal. As is true on a general level, emphasis should be placed on a fiber-rich diet. More dietary fiber consumption may help prevent DU from recurring if DU is linked to decreased gastrointestinal function in addition to the presence of a factor like H pylori. Several of the whey proteins found in fermented milk have shown promise as a supplemental treatment for PUD caused by *H. pylori*. Several studies have shown interest in the individual whey proteins' potential efficacy in the elimination of *H. pylori*. Unfortunately, the use of whey proteins or fermented milk in clinical practice is not yet commonly acknowledged or clearly defined. On the basis of the facts at hand, it is not possible to make a compelling case for a large change in dietary habits; further research is required in this area, as well as adjustments to alcohol and caffeine consumption, with the goal of addressing dyspepsia symptoms. Interventions backed by judgment and physiologic mechanisms, such as restricting alcohol or caffeine intake, are justified on a trial basis in individual patients despite the paucity of strong data in the literature. A common and sensible approach in the treatment of GERD includes avoiding heavy meals right before bedtime as well as making postural changes that may lessen reflux. Given that food can be used to prevent, treat, or even lessen the symptoms of PUD, a balanced diet is essential for its therapy. However, further research is needed to explicitly address diet therapy for the treatment of PUD and non-ulcer functional dyspepsia, which affects a sizable portion of the population.

ACKNOWLEDGMENT

We thank the METs, Institute of Pharmacy, BKC, affiliated under Savitribai Phule Pune University, Nashik, and Divine College of Pharmacy, Satana. EDA wants to thank the Ministry of Tribal Affairs Government of India for providing financial assistance in the form of a fellowship (Award No.: 202021-NFST-MAH-01235).

KEYWORDS

- **diabetic foot ulcer**
- **dyspepsia**
- **functional food**
- **genital ulcers**
- **nutraceuticals**
- **physiologic mechanisms**
- **ulcerative dermatitis**

REFERENCES

Alanen, J., & Muli, R., (2014/2015). *Assessment Tools for Pressure Ulcers* (Vol. 1, pp. 01–38). Centria University of applied science.

Aldoori, W. H., Giovannucci, E. L., Stampfer, M. J., Rimm, E. B., Wing, A. L., & Willett, W. C., (1997). Prospective study of diet and the risk of duodenal ulcer in men. *American Journal of Epidemiology, 145*(1), 42–50.

Ankolekar, C., Johnson, D., Pinto, M. D. S., Johnson, K., Labbe, R., & Shetty, K., (2011). Inhibitory potential of tea polyphenolics and influence of extraction time against *Helicobacter pylori* and lack of inhibition of beneficial lactic acid bacteria. *Journal of Medicinal Food, 14*(11), 1321–1329.

Arai, S., (1996). Studies on functional foods in Japan—State of the art. *Bioscience, Biotechnology, and Biochemistry, 60*(1), 9–15.

Archer, H. E., & Graham, G., (1936). The sub-scurvy state in relation to gastric and duodenal ulcer. *Lancet, 231*, 364–366.

Ardakani, M. J. E., Safaei, A., Oskouie, A. A., Haghparast, H., Haghazali, M., Shalmani, H. M., Peyvandi, H., et al., (2016). Evaluation of liver cirrhosis and hepatocellular carcinoma using protein-protein interaction networks. *Gastroenterology and Hepatology from Bed to Bench, 9*(1), S14.

Barling, B., (1935). Scurvy during treatment for gastric ulcer. *British Medical Journal, 1*, 358–359.

Becker, J. C., Grosser, N., Boknik, P., Schröder, H., Domschke, W., & Pohle, T., (2003). Gastroprotection by vitamin C – A heme oxygenase-1-dependent mechanism? *Biochemical and Biophysical Research Communications, 312*(2), 507–512.

Bourne, G., (1938). Vitamin C deficiency is estimated by capillary resistance. *British Medical Journal, 1*(4027), 560.

Brzozowski, T., Kwiecień, S., Konturek, P., Konturek, S., Ptak-Belowska, A., Mitis-Musioł, M., Duda, A., et al., (2001). Comparison of nitric oxide releasing NSAID and vitamin C with classic NSAID in healing of chronic gastric ulcers; involvement of reactive oxygen species. *Medical Science Monitor, 7*(4).

Chan, R. S., & Woo, J., (2010). Prevention of overweight and obesity: How effective is the current public health approach? *International Journal of Environmental Research and Public Health, 7*(3), 765–783.

Cox, J., & Rasmussen, L., (2014). Enteral nutrition in the prevention and treatment of pressure ulcers in adult critical care patients. *Critical Care Nurse, 34*(6), 15–27.

Dalldorf, G., (1938). The pathology of vitamin C deficiency. *Journal of the American Medical Association, 111*(15), 1376–1379.

Davidson, P. B., (1928). The development of deficiency disease during therapeutic diets. *Journal of the American Medical Association, 90*(13), 1014–1015.

Earl, R., & Thomas, P. R., (1994). *Opportunities in the Nutrition and Food Sciences: Research Challenges and the Next Generation of Investigators*. National Academies Press.

Ebringer, L., Ferenčík, M., & Krajčovič, J., (2008). Beneficial health effects of milk and fermented dairy products. *Folia Microbiologica, 53*(5), 378–394.

Faguet, G. B., (2015). A brief history of cancer: Age-old milestones underlying our current knowledge database. *International Journal of Cancer, 136*(9), 2022–2036.

Field, Jr. H., Robinson, W. D., & Melnick, D., (1940). Vitamins in peptic ulcer. *Annals of Internal Medicine, 14*(4), 588–592.

Fitzgerald, A. L., Maclean, D. R., & Veugelers, P. J., (2002). Dietary reference intakes: A comparison with the nova scotia nutrition survey. *Canadian Journal of Dietetic Practice and Research, 63*(4), 176–183.

Floch, M. H., (2013). Probiotic safety and risk factors. *Journal of Clinical Gastroenterology, 47*(5), 375–376.

Gillson, S. Following an Ulcer Diet: What to eat and what to avoid to manage symptoms. https://www.verywellhealth.com/what-can-i-eat-if-i-have-a-peptic-ulcer-1742154.

Green, C. M., (1941). The adequacy of diets used in the treatment of peptic ulcer. *Journal of the American Dietetic Association, 17*, 570–572.

Hernández-Díaz, S., Su, Y. C., Mitchell, A. A., Kelley, K. E., Calafat, A. M., & Hauser, R., (2013). Medications as a potential source of exposure to phthalates among women of childbearing age. *Reproductive Toxicology, 37*, 1–5.

Jarosz, M., Dzieniszewski, J., Dabrowska-Ufniarz, E., Wartanowicz, M., Ziemlanski, S., & Reed, P. I., (1998). Effects of high dose vitamin C treatment on *Helicobacter pylori* infection and total vitamin C concentration in gastric juice. *European Journal of Cancer Prevention*, 449–454.

Keri, M. N. D., (2004). Therapeutic applications of whey protein. *Alternative Medicine Review, 9*(2), 136–156.

Kerstan, A., Dieter, K., Niebergall-Roth, E., Dachtler, A. K., Kraft, K., Stücker, M., Daeschlein, G., et al., (2022). Allogeneic ABCB5+ mesenchymal stem cells for treatment-refractory chronic venous ulcers: A phase I/IIa clinical trial. *JID Innovations, 2*(1), 100067.

Keservani, R. K., & Sharma, A. K., (2014). Flavonoids: Emerging trends and potential health benefits. *Journal of Chinese Pharmaceutical Sciences, 23*(12), 815.

Keservani, R. K., Kesharwani, R. K., Sharma, A. K., Vyas, N., & Chadoker, A., (2010b). Nutritional supplements: An overview. *International Journal of Current Pharmaceutical Review and Research, 1*(1), 59–75.

Keservani, R. K., Kesharwani, R. K., Vyas, N., Jain, S., Raghuvanshi, R., & Sharma, A. K., (2010a). Nutraceutical and functional food as future food: A review. *Der Pharmacia Lettre, 2*(1), 106–116.

Keservani, R. K., Sharma, A. K., & Kesharwani, R. K., (2017). An overview and therapeutic applications of nutraceutical and functional foods. *Recent Advances in Drug Delivery Technology*, 160–201. ISBN: 9781522507543.

Keservani, R. K., Sharma, A. K., & Kesharwani, R. K., (2020). *Nutraceuticals and Dietary Supplements: Applications in Health Improvement and Disease Management*. CRC Press. ISBN: 9781771888738.

Kulshreshtha, M., Srivastava, G., & Singh, M. P., (2017). Pathophysiological status and nutritional therapy of peptic ulcer: An update. *Environmental Disease, 2*(3), 76.

Labov, J. B., (2006). National and state standards in science and their potential influence on undergraduate science education. *CBE – Life Sciences Education, 5*(3), 204–209.

Lanman, T. H., & Ingalls, T. H., (1937). Vitamin C deficiency and wound healing: An experimental and clinical study. *Annals of Surgery, 105*(4), 616.

Lee, S. Y., Shin, Y. W., & Hahm, K. B., (2008). Phytoceuticals: Mighty but ignored weapons against *Helicobacter pylori* infection. *Journal of Digestive Diseases, 9*(3), 129–139.

Li, L., & Somerset, S., (2014). Digestive system dysfunction in cystic fibrosis: Challenges for nutrition therapy. *Digestive and Liver Disease, 46*(10), 865–874.

Mahmood, T., Tenenbaum, S., Niu, X. T., Levenson, S. M., Seifter, E., & Demetriou, A. A., (1986). Prevention of duodenal ulcer formation in the rat by dietary vitamin A supplementation. *Journal of Parenteral and Enteral Nutrition, 10*(1), 74–77.

McCarrison, R., (1931). Some surgical aspects of faulty nutrition. *The Lancet, 217*(5621), 1151–1154.

Mehta, D., (2016). Ulcer-review on types, anti-ulcer drugs, anti-ulcer medicinal plants, anti-ulcer drug market, diagnostics, and current global clinical trials status. *Invent Rapid Pharm Pract., 2*, 1–8.

Platt, R., (1936). Scurvy as the result of dietetic treatment. *Lancet, 281*, 366–367.

Podmore, I. D., Griffiths, H. R., Herbert, K. E., Mistry, N., Mistry, P., & Lunec, J., (1998). Vitamin C exhibits pro-oxidant properties. *Nature, 392*(6676), 559–559.

Pohle, T., Brzozowski, T., Becker, J. C., Van, D. V. I. R., Markmann, A., Konturek, S. J., Moniczewski, A., et al., (2001). Role of reactive oxygen metabolites in aspirin-induced gastric damage in humans: Gastroprotection by vitamin C. *Alimentary Pharmacology and Therapeutics, 15*(5), 677–687.

Portnoy, B., & Wilkinson, J. F., (1938). Vitamin C deficiency in peptic ulceration. *British Medical Journal, 1*(4027), 554.

Singh, P., Kesharwani, R. K., & Keservani, R. K., (2017a). Antioxidants and vitamins: Roles in cellular function and metabolism. In: *Sustained Energy for Enhanced Human Functions and Activity* (pp. 385–407). Academic Press. ISBN: 9780128054130.

Singh, P., Kesharwani, R. K., & Keservani, R. K., (2017b). Protein, carbohydrates, and fats: Energy metabolism. In: *Sustained Energy for Enhanced Human Functions and Activity* (pp. 103–115). Academic Press. ISBN: 9780128054130.

Suerbaum, S., & Michetti, P., (2002). *Helicobacter pylori* infection. *New England Journal of Medicine, 347*(15), 1175–1186.

Taghipour, N., Molaei, M., Mosaffa, N., Rostami-Nejad, M., Aghdaei, H. A., Anissian, A., Azimzadeh, P., & Zali, M. R., (2016). An experimental model of colitis induced by dextran sulfate sodium from acute progresses to chronicity in C57BL/6: Correlation between conditions of mice and the environment. *Gastroenterology and Hepatology from Bed to Bench, 9*(1), 45.

Uyangaa, E., Patil, A. M., & Eo, S. K., (2014). Prophylactic and therapeutic modulation of innate and adaptive immunity against mucosal infection of herpes simplex virus. *Immune Network, 14*(4), 187–200.

Vomero, N. D., & Colpo, E., (2014). Nutritional care in peptic ulcer. *ABCD: Arquivos Brasileiros de Cirurgia Digestiva (São Paulo), 27*, 298–302.

Wang, K. Y., Li, S. N., Liu, C. S., Perng, D. S., Su, Y. C., Wu, D. C., Jan, C. M., et al., (2004). Effects of ingesting *Lactobacillus*-and *Bifidobacterium*-containing yogurt in subjects with colonized *Helicobacter pylori*. *The American Journal of Clinical Nutrition, 80*(3), 737–741.

Zhang, H. M., Wakisaka, N., Maeda, O., & Yamamoto, T., (1997). Vitamin C inhibits the growth of a bacterial risk factor for gastric carcinoma: *Helicobacter pylori*. *Cancer: Interdisciplinary International Journal of the American Cancer Society, 80*(10), 1897–1903.

Zojaji, H., Talaie, R., Mirsattari, D., Haghazali, M., Molaei, M., Mohsenian, N., Derakhshan, F., & Zali, M. R., (2009). The efficacy of *Helicobacter pylori* eradication regimen with and without vitamin C supplementation. *Digestive and Liver Disease, 41*(9), 644–647.

CHAPTER 11

FUNCTIONAL FOODS IN ACIDITY MANAGEMENT AND PREVENTION

KHEMCHAND R. SURANA and SUNIL K. MAHAJAN

Department of Pharmaceutical Chemistry, MGV's Pharmacy College, Panchvati, Nashik, Maharashtra, India

ABSTRACT

Herbs, veggies, and fruits are being taken into consideration to be the ability supply for treating and controlling numerous ailments. "Let food be thy medication and medication thy food," the guiding principle via Hippocrates, is currently receiving excellent interest. In this context, the idea of functional food is enormously appreciated. Functional foods are "meals or food factor which gives each health and dietary benefits." This chapter includes functional food for prevention and remedy of acidity. Factors like *H. pylori* infection, intake of NSAID's, alcohol, tumors of the acid generating cells, smoking, chewing tobacco contributes for acidity. It is affecting hundreds of thousands of people worldwide. Conventional remedy consists of the medication which reduces the rate of gastric acid secretion, drugs which protects the mucus membrane that lines the GIT or the medication which removes Helicobacter pylori. In maximum of those instances, destructive reactions and occurrence of relapse has been observed. The difficulty is to introduce more secure tablets with most therapeutic effect, without destructive reactions. In the prevailing have a look at a formulation (PHF-1) containing decided on functional foods which acts via way of means of numerous mechanisms to prevent and treat ulcers become advanced and evaluated for its efficacy against ethanol caused gastric ulcers.

Applications of Functional Foods in Disease Prevention. Raj K. Keservani & Eknath D. Ahire, PhD (Eds.)

11.1 INTRODUCTION

Acidity occurs when there is excess secretion of acids in the gastric glands of the stomach, producing gas, bad breath, stomach ache and other symptoms. It triggers an unpleasant reaction that also often makes you feel sick and miserable (Petriccione et al., 2015). "Acidity may be triggered because of a protracted gap among meals, empty stomach or immoderate consumption of tea, coffee, smoking or alcohol." When the secretion of acids is greater than typical as experience heartburn or acid reflux disease disorder or GERD (gastroesophageal reflux disease), that's generally precipitated while we devour a heavy meal or highly spiced foods. Acidity can emerge as pretty annoying. In order to dispose of the same, we frequently motel to numerous treatments (Sampaio et al., 2017). However, when you have failed at doing so, then worry not; we have got were given your back. In this chapter, we've blanketed a listing of a few domestic treatments for acidity that can be of excellent assist in doing away with acidity problems (Yoshime et al., 2016). Acidity happens whilst there's extra secretion of acids withinside the gastric glands of the stomach. When the secretion is greater than usual, we feel, what's usually called heartburn, which is typically brought on off via way of means of intake of highly spiced foods. Here are a few domestic treatments to remedy acidity (Neves et al., 2015):

- Skip the aerated beverages in addition to the caffeine. Opt for herbal tea instead – have a pitcher of lukewarm water each day.
- Include banana, watermelon and cucumber to your day-by-day diet. Watermelon juice is amazing for curing acidity.
- Nariyalpaani is thought to assuage the system in case you be afflicted by acidity.
- Drink a pitcher of milk – everyday. Jaggery, lemon, banana, almonds and yogurt are all regarded to provide you immediate comfort from acidity. Ginger aids in digestion. Either purchase powdered ginger in tablet bureaucracy or add the herb in your recipes.
- An easy guidance of lemon water with sugar may be sipped on an hour earlier than lunch to lessen uneasiness. Have vegetables like drumsticks, beans, pumpkin, cabbage, carrot and spring onions.

Acidity is a better threat for the ones who've an imbalance in Pitta. An already irritated Pitta may be in addition imbalanced through greater highly spiced and fried meals inflicting in addition problems. The irritated Pitta prevents agni (digestive fire) from digesting the meals. This outcomes withinside the build-up of ama (toxins) (Canani and Terrin, 2010). Therefore, you must additionally remember of ingredients that could calm irritated Pitta and preserve your agni in check (Canani et al., 2006). In order to digest meals, the gastric glands in our stomach secrete acid. However, whilst those gastric glands secrete extra acid, we enjoy acidity (Risch, 2003).

Also called acid reflux, acidity may be because of different factors including abnormal consuming habits, extra consumption of highly spiced meals, normal smoking, or intake of alcohol (Holbrook et al., 2009). Heartburn, a painful, burning sensation withinside the chest or throat, is the most common symptom of acidity. Also, problems in swallowing, common burping, hiccups, or indigestion are different signs and symptoms of acidity. It may be a touch irritating at times, because it prevents you from playing your food and reasons a whole lot of discomfort too. So, right here are 20 easy and powerful domestic treatments to help you get remedy from acidity (Ephron, 2012). Here's a quick list of food you should stock up on to ease the symptoms associated with acidity (Fass and Tougas, 2002).

11.2 FUNCTIONAL FOODS FOR ACID REFLUX OR HEARTBURN

11.2.1 FRIED, FATTY FOODS

Talk about a triple whammy. First, foods that are high in fat and drenched in oil (bacon and sausage, French fries, potato chips and doughnuts, to name a few) cause the lower esophageal sphincter the ring of muscle connecting the esophagus and stomach to relax, allowing acid to leak out and up into the esophagus (Dağlı, and Kalkan, 2017; Surdea-Blaga et al., 2019). Also, they are digested slowly and sit in your stomach longer, giving gastric acid more time to work its way up into your esophagus, increasing your risk of heartburn. And excessive-fats meals can purpose weight gain. That more poundage will increase stress for your abdomen, making it less complicated for fluid to journey upward, into the esophagus. Opt alternatively for lean cuts of meat, chook and fish, cooked at the grill or with inside the oven (Asl et al., 2015; Jarosz and Taraszewska, 2014).

11.2.2 CITRUS FRUITS

Their vitamin C can be the draw; however those fruits' excessive acid content material makes them unstable for reflux (Valussi, 2012; Keservani et al., 2010a, 2017, 2020) (our stomachs clearly produce loads of acid; why make matters worse). Lemon is and limes are the largest culprits; however, oranges, grapefruit, grapes, and blueberries make the list, too. Offset their acidity by buffering them with more alkaline (less acidic) foods, such as low-fat yogurt (Cencic and Chingwaru, 2010; Etienne et al., 2013; Keservani et al., 2010b; Juansah, et al., 2012).

11.2.3 TOMATOES

Also relatively acidic, this common base for plenty food is likewise a culprit in causing gastric acid to again up into the esophagus and no, cooking tomatoes won't help (Rani et al., 2016). If you can't face up to the strong flavor of marinara sauce, neutralize the acid via way of means of including one-quarter teaspoon of baking soda on your recipe (Anthon and Barrett, 2012).

11.2.4 CHOCOLATE

Its high fat content, caffeine and theobromine, a compound found in the cacao bean, can all trigger acid reflux. Chocolate causes the intestines to release serotonin, which causes the decrease esophageal sphincter to loosen up and open, permitting that undesirable flow returned up the esophagus. Dark chocolate, which incorporates much less fats than milk chocolate, may be much less probable to purpose acid reflux (Jinap and Hollender, 1995; Harasym et al., 2020).

11.2.5 PEPPERMINT

A cup of refreshing, mint-infused tea can calm a turbulent tummy. But peppermint, in any form, can also cause gastric reflux and heartburn (Valussi, 2012; Croyle and Talarczyk, 2019). Savor a cup of soothing chamomile tea or suck on liquorice flavored hard candy to soothe your stomach (Necheles and Meyer, 1934).

11.2.6 CARBONATED BEVERAGES

The bubbles in carbonated drinks are like air pockets, increasing on your belly. This will increase strain at the esophageal sphincter, growing the danger for acid to reflux into the esophagus. Try decaf iced tea or get dressed up tap water with slices of cucumber (Johnson et al., 2010; Wang and Lussi, 2012; Urala and Lähteenmäki, 2007).

11.2.7 COFFEE

While research at the issue is mixed, a few have proven that caffeinated liquids might also additionally impair the characteristic of the muscle groups that separate the esophagus and stomach. What's more, consuming caffeinated coffee may additionally boom the quantity of acid that your stomach produces (Moon et al., 2009; Pérez-Martínez et al., 2010). If a decaf brew isn't for you, strive for a darkish roast coffee (it has a tendency to have much less caffeine in step with bean that lighter roasts). Or comfortable as much as a cup of natural tea (Rao and Fuller, 2018).

11.2.8 ALCOHOL

More reasons to drink in moderation: Alcohol can irritate the esophagus and make the stomach produce more acid. It also relaxes that gateway of the lower esophageal sphincter muscle (Hine and Hine, 1952). When it involves wine, white is greater acidic than purple and can purpose greater heartburn (Prasad and Sivamani, 2021). Beer is acidic, and it's carbonated each terrible news. But it has a decrease alcohol content material as compared to different alcoholic beverages, and can be higher for heartburn. Darker drinks (brandy, whisky and darkish rum) incorporate compounds known as congeners, which could growth belly infection and heartburn symptoms (Luy and Parera, 1986).

11.2.9 MILK

Alkaline in nature, milk is the correct meals to appease the indignant esophagus and stomach lining. Rich in pitta pacifying properties, milk at room temperature reduces the build-up of acid, facilitates calm the acid reflux disease disorder and neutralizes the extra acids present withinside the stomach. Warm milk must be avoided (Najafi et al., 2009). Cold milk: Milk has an excessive quantity of calcium, which facilitates to save you acid build-up via way of means of absorbing the extra acid produced. Also with if the milk is cold it positive does offers immediate comfort from the burning sensation that one feels in the course of reflux. Avoid including sugar (Steuart, 1920).

11.2.10 TULSI

One of the most secure ayurvedic products, Tulsi enables in stimulating the stomach to provide extra mucous. It includes antiulcer residences and decreases the consequences of gastric acids (Kumar et al., 2013). One of the very best approaches to conquer the soreness felt for the duration of acidity is with the aid of using chewing 4–6 Tulsi leaves to discover immediately relief (Rai et al., 2018).

11.2.11 FENNEL SEEDS

The reason one can find fennel as the base ingredient of many mouth refreshing products is that this small little gem helps your stomach to cool down (Valussi, 2012; Corbo et al., 2014). Chew fennel after every meal as it has many gastrointestinal benefits. Once every few days you can also have fennel tea as it keeps your digestive tract healthy and helps in indigestion and bloating (Dubrovskii et al., 2019).

11.2.12 BUTTERMILK

One of the best drinks to be enjoyed in the hot scorching weather is a glass of chilled buttermilk. Considered to be a sattvic food as per Ayurveda, buttermilk or chaas is known for its nutrients and soothing effect on the stomach. It contains lactic acid that normalizes the acidity in the stomach. If you don't like drinking it just like that, try adding a dash of black pepper, a pinch of salt, a pinch of asafetida and few freshly chopped coriander leaves (Trachoo and Mistry, 1998).

11.2.13 NON-ACIDIC FRUITS

While fruits like lemon, orange and grapefruit cause acidity, alkaline fruits like banana, apples and pears, peaches, melons, berries, avocados have been shown to neutralize the acidity through pacifying the aggravated Pitta, hence providing relief (Wang, et al., 2021).

11.2.14 HONEY

Honey is an elixir for those who suffer from hyperacidity. It stops the overproduction of acid in the stomach and protects the stomach lining from its damaging effects, and calms those bouts of acidity (Naydenova et al., 2014).

Drink a glass of lukewarm water with a spoon of honey and lemon juice every morning on an empty stomach (Bogdanov et al., 1999).

11.2.15 BOILED VEGETABLES

While they may not sound tempting, boiled veggies are a treat for your flaming stomach. Not only do they assimilate easily, but they also need much less acid to be digested. So the next time you feel acidity flaring up, add a bowl of boiled veggies to your lunch (Najgebauer-Lejko and Tabaszewska, 2014).

11.2.16 AMLA POWDER

Amla may be very powerful in neutralizing stomach acids. It is a Pittashamak and subsequently it regulates manufacturing of extra Pitta withinside the body. It additionally heals the internal lining of the stomach and esophagus, which reduces the burning sensation and discomfort (Raaf et al., 2021).

11.2.17 AMLA FIGHTS WITH ACIDITY

The abnormal meals conduct and unusual consumption of sweet, sour, highly spiced and oily meals may also reason acidity, and additionally, tea, coffee, and smoking are inflicting that trouble. The physiological elements are anger, grief and depression. This trouble is conquered through taken of 1 gram of amla powder and a small quantity of sugar combined with milk or water two times a day. Extract of Amla confirmed the capacity to lessen appreciably the severity of ulceration of stomach brought on through 80% ethanol and sturdy alkalies at each dose used (Mirunalini and Krishnaveni, 2013.). Amla is extraordinarily nutritious and can be critical nutritional reassets of Vitamin C, minerals and amino acids. It additionally carries tannins, phyllembelic acid, phyllemblin, rutin, curcuminoides and emblicol and phenolic compounds. That is what the proof suggests, and as part of a balanced, plant-primarily based totally weight-reduction plan that rations highly spiced ingredients and fried ingredients and emphasizes leafy and starchy vegetables, amla may be an easy answer for each lingering and remoted instances of acid reflux disease disorder (Goraya and Bajwa, 2015). Long defined as a "wonder berry" in Ayurvedic medicine, amla (additionally called Indian gooseberry or *Emblica officinalis*) has been verified in current years to have an super array of fitness benefits. This numerous choice of proof extends to acid reflux disease disorder and its associated issues. In one take a look at of topics with gastroesophageal reflux disease (GERD), amla turned into proven to restriction heartburn, regurgitation, and indigestion (Alkandari and Sidhu, 2019).

11.2.18 BANANAS

This low-acid fruit can assist in neutralizing stomach acid via way of means of coating an indignant esophageal lining. And now no longer most effective are bananas alkaline, they're additionally wealthy in pectin, a soluble fiber that enables continues meals flowing properly via the digestive tract (Akubor et al., 2003). This allows you to sense complete longer, so you're much less in all likelihood to overeat. Other alkaline ingredients consist of melons (especially cantaloupe and honeydew), cauliflower and almonds. Bananas include herbal antacids that act as a thing to combat acidity. And have few slices of apple earlier than bedtime to alleviate heartburn or reflux (Sanaeifar et al., 2016). Banana pulp carries antioxidants, inclusive of, vitamins, carotenoids, and phenolic compounds which includes catechin, epicatechin, lignin, tannins, and flavonoids (Keservani and Sharma, 2014) in addition to anthocyanins. Ulcer has been a first-rate assignment of growing nations. It is a lesion as a result of elements inclusive of the microorganism Helicobacter pylori and extra acidity. Researchers examined the efficiency of banana towards this worry and it changed into proven to have substantial antiulcer activity via way of means of who evaluated the ulcer index, gastric wall mucus, gastric juice pH, and quantity of ulcer-brought about animals (Kittur and Tharanathan, 2001).

11.2.19 ALMONDS

It allows for heartburn because it neutralizes the juices to your stomach, relieving and stopping acidity overall. Almonds and different dry culmination also can be taken as a munching in case you are in an assembly and now no longer capable of taking your food on time in order that fasting does now no longer create hyperacidity. Have 3–4 almonds after eating (Christian and Ukhun, 2006).

11.3 CONCLUSION

From the present study, it can be concluded that the functional food significantly in secretion of protect gastric acidity in stomach which is induced by number of contributing factors. The protection was found to be dose dependent. Since the ingredients used acts by various mechanisms and also possess nutritional value, the present formulation can be considered as an alternative, cost effective, safer and potential medicine to treat gastric ulcers. Due to the increasing popularity with health-conscious consumers, functional food science is becoming popular and creating new interest in marketing the products.

ACKNOWLEDGMENT

We thank the Divine College of Pharmacy Satana and MGVs, College of Pharmacy, Panchvati, Nashik, for their constant support and for providing all facilities to complete this work.

KEYWORDS

- **acid reflux**
- **acidity**
- **functional food**
- **heat burn**

REFERENCES

Akubor, P. I., Obio, S. O., Nwadomere, K. A., & Obiomah, E., (2003). Production and quality evaluation of banana wine. *Plant Foods for Human Nutrition, 58*(3), 1–6.

Alkandari, D., Sarfraz, H., & Sidhu, J. S., (2019). Development of a functional food (pan bread) using amla fruit powder. *Journal of Food Science and Technology, 56*(4), 2287–2295.

Anthon, G. E., & Barrett, D. M., (2012). Pectin methyl esterase activity and other factors affecting pH and titratable acidity in processing tomatoes. *Food Chemistry, 132*(2), 915–920.

Asl, S. F., Mansour-Ghanaei, F., Samadi, H., & Joukar, F., (2015). Evaluations of life style factors and the severity of gastroesophageal reflux disease; a case-control study. *International Journal of Molecular Epidemiology and Genetics, 6*(1), 27.

Bogdanov, S., Lüllmann, C., Martin, P., Von, D. O. W., Russmann, H., Vorwohl, G., Oddo, L. P., et al., (1999). Honey quality and international regulatory standards: Review by the international honey commission. *Bee World, 80*(2), 61–69.

Canani, R. B., & Terrin, G., (2010). Gastric acidity inhibitors and the risk of intestinal infections. *Current Opinion in Gastroenterology, 26*(1), 31–35.

Canani, R. B., Cirillo, P., Roggero, P., Romano, C., Malamisura, B., Terrin, G., Passariello, A., et al., (2006). Therapy with gastric acidity inhibitors increases the risk of acute gastroenteritis and community-acquired pneumonia in children. *Pediatrics, 117*(5), e817–e820.

Cencic, A., & Chingwaru, W., (2010). The role of functional foods, nutraceuticals, and food supplements in intestinal health. *Nutrients, 2*, 611–625.

Christian, A., & Ukhun, M. E., (2006). Nutritional potential of the nut of tropical almond (*Terminalia catappia* L.). *Pakistan Journal of Nutrition, 5*(4), 334–336.

Corbo, M. R., Bevilacqua, A., Petruzzi, L., Casanova, F. P., & Sinigaglia, M., (2014). Functional beverages: The emerging side of functional foods: Commercial trends, research, and health implications. *Comprehensive Reviews in Food Science and Food Safety, 13*, 1192–1206.

Croyle, J. R., & Talarczyk, P., (2019). The relationship between acidity and short-term student performance on exams. *Journal Of Student Research, 8*.

Dağlı, Ü., & Kalkan, İ. H., (2017). The role of lifestyle changes in gastroesophageal reflux diseases treatment. *Turk J. Gastroenterol., 28*(Suppl 1), S33–S37.

Dubrovskii, I. I., Arseneva, T. P., Evstigneeva, T. N., Gorshkova, S. B., Bazarnova, Y. G., & Iakovchenko, N. V., (2019). Development of formulation and technology of yogurt with prolonged shelf life enriched with biologically active substances from fennel seed extract. *Agronomy Research, 17*(S2), 1313–1323.

Ephron, N., (2012). *Heartburn*. Hachette UK.

Etienne, A., Génard, M., Lobit, P., Mbeguié-A-Mbéguié, D., & Bugaud, C., (2013). What controls fleshy fruit acidity? A review of malate and citrate accumulation in fruit cells. *Journal of Experimental Botany, 64*(6), 1451–1469.

Fass, R., & Tougas, G., (2002). Functional heartburn: The stimulus, the pain, and the brain. *Gut, 51*(6), 885–892.

Goraya, R. K., & Bajwa, U., (2015). Enhancing the functional properties and nutritional quality of ice cream with processed amla (Indian gooseberry). *Journal of Food Science and Technology, 52*(12), 7861–7871.

Harasym, J., Kaim, U., Bogacz-Radomska, L., & Olędzki, R., (2020). Development of functional foods by traditional food processes. *Sustainability of the Food System*. Elsevier.

Hine, J., & Hine, M., (1952). The relative acidity of water, methanol and other weak acids in isopropyl alcohol solution. *Journal of the American Chemical Society, 74*(21), 5266–5271.

Holbrook, W. P., Furuholm, J., Gudmundsson, K., Theodors, A., & Meurman, J. H., (2009). Gastric reflux is a significant causative factor of tooth erosion. *Journal of Dental Research, 88*(5), 422–426.

Jarosz, M., & Taraszewska, A., (2014). Risk factors for gastroesophageal reflux disease: The role of diet. *Przeglad Gastroenterologiczny, 9*, 297.

Jinap, S., Dimick, P. S., & Hollender, R., (1995). Flavor evaluation of chocolate formulated from cocoa beans from different countries. *Food Control, 6*(2), 105–110.

Johnson, T., Gerson, L., Hershcovici, T., Stave, C., & Fass, R., (2010). Systematic review: The effects of carbonated beverages on gastro-oesophageal reflux disease. *Alimentary Pharmacology & Therapeutics, 31*(6), 607–614.

Juansah, J., Budiastra, I. W., Dahlan, K., & Seminar, K. B., (2012). The prospect of electrical impedance spectroscopy as non-destructive evaluation of citrus fruits acidity. *Int. J. Emerg. Technol. Adv. Eng, 2*(11), 58–64.

Keservani, R. K., & Sharma, A. K., (2014). Flavonoids: Emerging trends and potential health benefits. *Journal of Chinese Pharmaceutical Sciences, 23*(12), 815.

Keservani, R. K., Kesharwani, R. K., Sharma, A. K., Vyas, N., & Chadoker, A., (2010b). Nutritional supplements: An overview. *International Journal of Current Pharmaceutical Review and Research, 1*(1), 59–75.

Keservani, R. K., Kesharwani, R. K., Vyas, N., Jain, S., Raghuvanshi, R., & Sharma, A. K., (2010a). Nutraceutical and functional food as future food: A review. *Der Pharmacia Lettre, 2*(1), 106–116.

Keservani, R. K., Sharma, A. K., & Kesharwani, R. K., (2017). An overview and therapeutic applications of nutraceutical and functional foods. *Recent Advances in Drug Delivery Technology,* 160–201. ISBN:9781522507543.

Keservani, R. K., Sharma, A. K., & Kesharwani, R. K., (2020). *Nutraceuticals and Dietary Supplements: Applications in Health Improvement and Disease Management*. CRC Press. ISBN: 9781771888738.

Kittur, F. S., Saroja, N., & Tharanathan, R., (2001). Polysaccharide-based composite coating formulations for shelf-life extension of fresh banana and mango. *European Food Research and Technology, 213*(4), 306–311.

Kumar, S., Rai, D., & Singh, D., (2013). The functional, rheological and sensory attributes of tulsi (Holy Basil, *Ocimum sanctum*) extract based herbal ice-cream. *The Bioscan, 8*, 77–80.

Luy, J. C., & Parera, J. M., (1986). Acidity control in alcohol dehydration. *Applied Catalysis, 26*, 295–304.

Mirunalini, S., Vaithiyanathan, V., & Krishnaveni, M. A. N. I., (2013). Amla: A novel ayurvedic herb as a functional food for health benefits-a mini. *Int. J. Pharma. Pharmaceut Sci, 5*.

Moon, J. K., Yoo, H. S., & Shibamoto, T., (2009). Role of roasting conditions in the level of chlorogenic acid content in coffee beans: Correlation with coffee acidity. *Journal of Agricultural and Food Chemistry, 57*, 5365–5369.

Najafi, M. N., Mortazavi, S. A., Koocheki, A., Khorami, J., & Rekik, B., (2009). Fat and protein contents, acidity and somatic cell counts in bulk milk of Holstein cows in the Khorasan Razavi province, Iran. *International Journal of Dairy Technology, 62*(1), 19–26.

Najgebauer-Lejko, D., Grega, T., & Tabaszewska, M., (2014). Yoghurts with addition of selected vegetables: Acidity, antioxidant properties and sensory quality. *Acta Scientiarum Polonorum Technologia Alimentaria, 13*(1), 35–42.

Naydenova, N., Mihaylova, G., Ilieva, T., & Boycheva, S., (2014). Milk products as functional foods: Production of yoghurt enriched with herbal extracts and honey. *Ecology And Future-Scientific Journal of Ecology and Environment, 13*, 71–75.

Necheles, H., & Meyer, J., (1934). On the inhibition of gastric secretion by oil of peppermint. *American Journal of Physiology-Legacy Content, 110*(3), 686–691.

Neves, L. C., Tosin, J. M., Benedette, R. M., & Cisneros-Zevallos, L., (2015). Post-harvest nutraceutical behavior during ripening and senescence of 8 highly perishable fruit species from the Northern Brazilian Amazon region. *Food Chemistry, 174*, 188–196.

Pérez-Martínez, M., Caemmerer, B., De Peña, M. P., Cid, C., & Kroh, L. W., (2010). Influence of brewing method and acidity regulators on the antioxidant capacity of coffee brews. *Journal Of Agricultural and Food Chemistry, 58*, 2958–2965.

Petriccione, M., De Sanctis, F., Pasquariello, M. S., Mastrobuoni, F., Rega, P., Scortichini, M., & Mencarelli, F., (2015). The effect of chitosan coating on the quality and nutraceutical traits of sweet cherry during postharvest life. *Food and Bioprocess Technology, 8*(2), 394–408.

Prasad, B. N., & Sivamani, S., (2021). Fate of acidity and alcohol during fermentation of Salalah and Sri Lanka coconut water. *European Journal of Engineering and Technology*, 9.

Raaf, A., Suriaini, N., Djafar, F., Syamsuddin, Y., & Supardan, M. D., (2021). Effect of drying temperature on the moisture loss, acidity and characteristics of Amla fruit. In: *IOP Conference Series: Earth and Environmental Science* (Vol. 667, No. 1, p. 012047). IOP Publishing.

Rad, A. H., Mehrban, M., Nejhad, V. G., & Khazani, S., (2014). *Evaluation of Physicochemical, Rheological and Sensory Properties of Probiotic Chocolate*, *4*(1), 01–13.

Rai, H. K., Rai, D. C., Singh, A. K., & Kumar, S., (2018). To study the effect of Tulsi addition on chemical and textural property of Shrikhand. *Journal of Pharmacognosy and Phytochemistry, 7*(3), 2866–2870.

Rani, K., Parvez, N., & Sharma, P. K., (2016). Nutraceutical as functional foods. *World Journal of Pharmaceutical Research, 5*, 452–468.

Rao, N. Z., & Fuller, M., (2018). Acidity and antioxidant activity of cold brew coffee. *Scientific Reports, 8*(1), 1–9.

Risch, H. A., (2003). Etiology of pancreatic cancer, with a hypothesis concerning the role of N-nitroso compounds and excess gastric acidity. *Journal of the National Cancer Institute, 95*(13), 948–960.

Sampaio, K. A., Ayala, J. V., Van, H. V., Monteiro, S., Ceriani, R., Verhé, R., & Meirelles, A. J., (2017). Impact of crude oil quality on the refining conditions and composition of nutraceuticals in refined palm oil. *Journal of Food Science, 82*(8), 1842–1850.

Sanaeifar, A., Bakhshipour, A., & De La Guardia, M., (2016). Prediction of banana quality indices from color features using support vector regression. *Talanta, 148*, 54–61.

Steuart, D. W., (1920). The development of acidity in milk. *Journal of Dairy Science, 3*(1), 52–59.

Surdea-Blaga, T., Negrutiu, D. E., Palage, M., & Dumitrascu, D. L., (2019). Food and gastroesophageal reflux disease. *Current Medicinal Chemistry, 26*, 3497–3511.

Trachoo, N., & Mistry, V. V., (1998). Application of ultrafiltered sweet buttermilk and sweet buttermilk powder in the manufacture of nonfat and low-fat yogurts. *Journal of Dairy Science, 81*(12), 3163–3171.

Urala, N., & Lähteenmäki, L., (2007). Consumers' changing attitudes towards functional foods. *Food Quality and Preference, 18*, 1–12.

Valussi, M., (2012). Functional foods with digestion-enhancing properties. *International Journal of Food Sciences and Nutrition, 63*, 82–89.

Wang, L., Jiang, X., Zhao, L., Wang, F., Liu, Y., Zhou, H., He, H., & Han, Y., (2021). A candidate PpRPH gene of the D locus controlling fruit acidity in peach. *Plant Molecular Biology, 105*(3), 321–332.

Wang, X., & Lussi, A., (2012). Functional foods/ingredients on dental erosion. *European Journal of Nutrition, 51*, 39–48.

Yoshime, L. T., De Melo, I. L. P., Sattler, J. A. G., De Carvalho, E. B. T., & Mancini-Filho, J., (2016). Bitter gourd (*Momordica charantia* L.) seed oil as a naturally rich source of bioactive compounds for nutraceutical purposes. *Nutrire, 41*(1), 1–7.

CHAPTER 12

ROLE OF FUNCTIONAL FOODS IN PSYCHOTIC DISORDERS

AFSAR S. PATHAN,[1] KAJAL A. BENDALE,[1] PRATIKSHA M. PANDAGALE,[2] CHANDRASHEKHAR D. PATIL,[3] and EKNATH D. AHIRE[4]

[1]*Department of Pharmacology, R. C. Patel Institute of Pharmaceutical Education and Research, Shirpur, Maharashtra, India*

[2]*Department of Pharmacology, Oriental College of Pharmacy, Navi Mumbai, Maharashtra, India*

[3]*Department of Pharmacology, Shreeshakti Shaikshanik Sansthas, Divine College of Pharmacy, Satana, Nashik, Maharashtra, India*

[4]*Department of Pharmaceutics, METs Institute of Pharmacy, BKC, Affiliated to Savitribai Phule University, Adgoan, Nashik-3, Maharashtra, India*

ABSTRACT

We all know about nowadays lifestyles and workloads and which brought major impact on mental health, functional disorder which mainly effect on the brain anatomy and physiology, and change the behavioral status. Psychosis is defined as a group of mental disorders characterized by symptoms such as fixed false beliefs, hallucinations, disorganized thoughts, and clang associations, treatment available on psychosis disorder is only for symptomatic not cure types. It is necessary to develop the treatment which counteract and improve the disease condition and functional food are the good candidate for the psychosis disorders, which has variety of activity by which they act on pathophysiology of disease. In this chapter, we focus

Applications of Functional Foods in Disease Prevention. Raj K. Keservani & Eknath D. Ahire, PhD (Eds.)

on the role of different functional food in psychotic disorders and different types of psychotic disorders.

12.1 INTRODUCTION

The Food and Drug Administration defines "food" as "articles used primarily for taste, aroma or nutritive value" (Hasler, 1996). Further, foods that provide an additional physiological advantage that may help prevent disease or improve health are known as Functional foods (Keservani et al., 2010a, b, 2017, 2020). Such foods include categories of nutraceuticals and designer foods. Designer food is defined as food that is created with the intention of providing health advantages in addition to its typical nutritional value. Designer foods are frequently used interchangeably with functional/fortified foods, which refer to the nutrients already present in them as well as additional nutrients (Rajasekaran and Kalaivani, 2013). Nutraceuticals, on the other hand, are substances extracted from herbal products, dietary supplements (nutrients), specific diets, and processed foods like cereals are utilized as medicine (Nasri et al., 2014). The Institute of Medicine of the U.S. National Academy of Sciences has defined Functional foods as "foods that encompass potentially healthful products including any modified food or food ingredient that may provide a health benefit beyond the traditional nutrients it contains" (Hasler, 1996). Furthermore, Health Canada's definition of functional foods is as follows: "Similar in appearance to conventional food, consumed as part of the usual diet, with demonstrated physiological benefits and/or to alleviate the risk of chronic disease beyond basic nutritional functions" (Henry, 2010). Functional foods are a marketing word that is used around the world but is not legally defined. Japan is the only country that recognizes functional foods as a separate category. Functional foods are not possible to discriminate based on age, country, and gender. Foods like low-fat and skim milk, fruit juice, cereals, and yogurt, for example, are consumed by people of all ages, but tea, cholesterol-lowering products, and soy products are mostly consumed by adults and the elderly (Ozen, Pons and Tur, 2012). As a result, functional food (Keservani et al., 2020) supplies the body with the necessary vitamins, minerals, lipids, proteins, carbs, and other nutrients for a healthy existence (Cencic and Chingwaru, 2010). Food can become functional by using any of the following ways:

- Getting rid of a component that could be dangerous if taken;

- Increasing the concentration of a component up to the point when it can induce a beneficial effect;
- The addition of a component that is not normally present in foods but when added can enhance the benefits;
- Replacing a component that is in excess with a beneficial one;
- Increasing a component's bioavailability or stability.

The functional food business has grown dramatically in recent years due to advances in science and technology. Functional foods are a hot topic in the food industry right now, and they're going to revolutionize health and well-being through diet.

12.2 TYPES OF FUNCTIONAL FOODS

12.2.1 FUNCTIONAL FOODS OF ANIMAL ORIGIN

The biologically active animal-derived component which has gained popularity in recent years is "probiotics" (Hasler, 2002). They are defined as live micro-organisms that when administered in adequate amounts, confer a health benefit on the host" (Reid, Gadir and Dhir, 2019). Probiotics are a type of functional food since they provide additional nutritional advantages in a variety of situations. Lactic-acid-producing bacteria are found in the majority of probiotics currently on the market (Lin, 2003). These products help in restoring gut microflora and also have the ability to block the invasion of certain pathogenic micro-organisms into the gut (Lin, 2003). There are various products derived from animals that play a significant role as functional foods. Some of them are described as follows. Calcium, an essential vitamin that can help prevent osteoporosis and colon cancer, is best found in dairy products (Mestre Prates and Mateus, 2002). Also, fermented dairy products (Probiotics) are found to be effective in gut disorders. In addition to probiotics, prebiotics are a type of fermentable carbohydrate that is also considered to be a functional food. Prebiotics are defined as nondigestible food ingredients that beneficially affect the host by selectively reducing the growth and/or activity of one or a limited number of bacteria in the colon and thus improve host health (Mestre Prates and Mateus, 2002). Prebiotics include starch, dietary fibers, oligosaccharides, and sugars. Dairy consumers are more likely to be underweight, have a lower risk of high blood pressure, and are less likely to suffer from strokes, colon cancer, or osteoporosis (MAA, 2017), thus having a huge impact on health benefits. One of the most important functional foods is fish since it improves functioning and lowers disease risk. Fish contains compounds that may reduce the risk of diseases like heart disease, cancer, diabetes, dementia, and other chronic and inflammatory disorders (Sarojnalini and Hei, 2019). They are also recognized as a nutrient-dense food including omega-3-fatty acids, trace minerals, vitamins as well as polyunsaturated fatty acids (Ramalho Ribeiro et al., 2019).

Another important component is the Meat. Anserine, carnosine, glutathione, L-carnitine, creatine, and taurine are all bioactive substances found in meat and meat products that are beneficial to one's health (Khan et al., 2011). Glutathione has been shown to play a critical role in the preservation of a variety of vital processes, including redox equilibrium, oxidative stress reduction, metabolic detoxification enhancement, and immune system control. According to data, it also appears to have an essential role in the treatment of a range of chronic and age-related illnesses (Minich and Brown, 2019). Due to its high protein content and amino acid levels, rabbit meat has recently earned a lot of popularity as a functional diet (Dalle Zotte and Szendro, 2011).

Eggs are also considered to be a good source of proteins, microproteins, lipids, and cholesterol. Lysozime, ovomucid, ovoinhibitor, and cystanin are physiologically active proteins found in eggs. It is also a good source of lecithin and one of the few foods that contains choline (Miranda et al., 2015). Choline is now officially recognized as a functional food, and it plays a wide range of roles in human metabolism, from cell structure development to neurotransmitter synthesis and release (Zeisel and Da Costa, 2009). As a result, choline's position as a functional food is beneficial, as its lack can cause a variety of neurological illnesses, including psychotic disorder.

12.2.2 FUNCTIONAL FOODS OF PLANT ORIGIN

Numerous functional foods derived from plant origin exist. But only a handful of them has scientific evidence required for approval by the FDA. Oat soluble fiber from psyllium seed husk, soy protein, and sterol are examples of such foods (Hasler, 2002). Some foods do not have scientific evidence yet are undergoing various trials to prove their claim as functional food. Such foods are cranberries, garlic, nuts, grapes, and chocolate (Ahire et al., 2021; Hasler, 2002).

The cranberry (*Vacciunium macrocarpon*) is considered one of America's first functional foods (McKay and Wilson, 2016). The consumption of cranberry juice has been linked to the prevention of Urinary Tract Infections (UTI). Cranberries include A-type Proanthocyanidins (PAC), a flavonoid (Keservani and Sharma, 2014) class that inhibits bacteria's ability to cling to and invade the bladder lining, protecting the bladder against UTI. They are also good for enhancing oral and gastrointestinal health, as well as promoting cardiovascular function and reducing cancer formation (Ahire et al., 2020; McKay and Wilson, 2016).

Garlic (*Allium savitum*) is a fragrant herbaceous plant that is used as a functional food to prevent a wide range of illnesses. Allicin, a key organo-sulfur molecule found in garlic, is the biologically active ingredient (Rahman, 2007). It is the active ingredient in garlic's antibacterial properties. Allicin (diallylthiosulfinate) reacts with the thiol groups of Gram-positive and Gram-negative bacteria, thus inactivating essential enzymes for their growth (Leontiev et al., 2018). Garlic's anticancer, antioxidant, ability to alleviate cardiovascular disorders (Keservani et al., 2016a), boost immune systems, and anti-diabetic properties are just a few of its many benefits (Rahman, 2007; Surana et al., 2021).

Nuts are a great example of functional foods as they are loaded with essential vitamins and minerals along with omega fatty acids. Peanuts have been popular as a functional food because they include proteins, fiber, polyphenols, and antioxidants that may be used in a number of foods (Arya, Salve and Chauhan, 2016; Singh et al., 2017a). Peanuts also include resveratrol, phenolic acids, flavonoids, and phytosterols, which have been shown to inhibit the absorption of cholesterol from the diet (Arya, Salve and Chauhan, 2016).

Grapes are regarded to be a high-nutrient and antioxidant-rich food (Seccia, Viscecchia and Nardone, 2019). These antioxidants aid in the prevention of oxidation and free radical damage by neutralizing excess free radicals (Mohamed, 2015). Polyphenols such as anthocyanins, flavonoids, and resveratrol are among the other substances found in grapes. These exert various effects such as antioxidant, cardioprotective, anticancer, anti-inflammatory, anti-aging, and anti-microbial properties (Seccia, Viscecchia and Nardone, 2019).

Humans have been consuming cocoa and chocolate for thousands of years. Chocolate's components have been discovered to provide a number of health benefits. It includes antioxidant flavanol compounds such as catechins (Koca and Tasci, 2016). Catechins act as an antioxidant via two mechanisms:

i. **Direct Mechanism:** Chelating metal ions, scavenging ROS.

ii. **Indirect Mechanism:** Generating Phase II detoxification enzymes, inducing antioxidant enzymes, blocking pro-oxidant enzymes, and inducing antioxidant enzymes (Bernatoniene and Kopustinskiene, 2018).

Other minerals found in chocolate products include magnesium and calcium.

12.3 PSYCHOTIC DISORDERS

Psychosis is defined as a group of mental disorders characterized by symptoms such as fixed false beliefs, hallucinations, disorganized thoughts, clang associations, word salad, echolalia (Meaningless repetition of another person's spoken words), and abnormal motor behavior (Rosenthal and Khandai, 2019). The clinical diagnosis of psychosis is mainly based on two main symptoms, delusions and hallucinations. Delusions are fixed, incorrect beliefs that are in contradiction with reality. On the contrary, hallucinations are defined as the perception of a non-existing object or event. Hearing voices or seeing objects even when no one has spoken or is there is an example. Such symptoms mainly cause the disabled functioning of the daily activities and also negatively impact functional outcomes. Schizophrenia is the most prevalent type of psychotic disease. According to studies, patients with schizophrenia have three times the risk of dying young than the general population (Murray et al., 2021). Also, the relapse may occur due to positive symptoms such as suspiciousness, delusions, and hallucinations (Patel et al., 2014).

12.4 TYPES OF PSYCHOTIC DISORDERS

Psychotic disorders are classified as given in subsections.

12.4.1 SCHIZOPHRENIA

Schizophrenia is a chronic and severe mental illness with a large global disease burden, ranking among the top 15 major causes of disability in 2016 (Moreno-ku and Martı, 2018). It has affected nearly 1% of the population worldwide (Gonz et al., 2016). Globally more than 21 million people are affected by it (Kumbol et al., 2018). Genetic susceptibility, prenatal and early childhood stress, and environmental factors all have a role in the development of schizophrenia. The initial symptoms of schizophrenia usually appear in early adulthood, and the severity worsens with age (Deng, Xu and Yeung, 2017). They are mainly divided into three types: positive, negative, and cognitive symptoms. Positive symptoms include hallucinations and delusions. Negative symptoms include blunted effects, a deficit in social functioning, and avolition (Kumbol et al., 2018). Cognitive defects cannot be considered a symptom but an outcome of a chronic state of illness.

12.4.2 SCHIZOAFFECTIVE DISORDER (SAD)

Schizoaffective disorder represents a relationship between schizophrenic disorder and affective symptoms (Cheniaux et al., 2008). Affective symptoms include prolonged sadness, irritability or anxiety, lethargy, and lack of energy. Schizoaffective disorders can be interpreted as an atypical form of schizophrenia with affective symptoms (Cheniaux et al., 2008). Schizoaffective disorders involve major depression, mania, or mixed episodes. It is characterized as acute relapse with the need for hospitalization (Stip and Tourjman, 2010).

12.4.3 SCHIZOPHRENIFORM DISORDER

Schizophreniform disorder is a mental ailment that falls between schizophrenia and brief psychosis (Arciniegas, 2015). The presence of two or more psychotic symptoms, at least one of which may be delusions, hallucinations, or disorganized speech, during a period of 1–6 months, defines the condition (Arciniegas, 2015). Delusions, hallucinations, disorganized speech, unusual behavior, lack of efficacy, and lack of interest in life are all symptoms that are similar to schizophrenia.

12.4.4 BRIEF PSYCHOTIC DISORDER (BPD)

BPD is defined by the presence of delusions, hallucinations, formal thought disorder, or abnormal psychomotor behavior (Arciniegas, 2015). The medical condition lasts at least one day but not more than one month (Arciniegas, 2015). It is a brief episode of psychotic behavior that is usually brought on by a stressful situation.

12.4.5 DELUSIONAL DISORDER

In the absence of hallucinations, one or more delusions lasting at least one month are required for diagnosis (Arciniegas, 2015). Other than the delusions, the functional aspects are not disturbed. The following are some of the most common types of delusions:

- Delusional jealousy;
- Bizzare;
- Erotomanic;
- Grandiose;
- Persecutory;
- Somatic;
- Mixed;
- Thought broadcasting;
- Thought insertion.

12.4.6 SHARED PSYCHOTIC DISORDER (PSD)

It is a fascinating and poorly understood psychotic disorder. This is an uncommon condition in which a person develops delusions after being in a close connection with someone who has strong delusions (Arciniegas, 2015). It is defined by the transmission of delusions from a person (the primary patient) who is already suffering from a psychotic disease to another person (the secondary patient) who is in close proximity to them ('Psychiatry & Psychosomatic Medicine,' 2016).

12.4.7 SUBSTANCE INDUCES PSYCHOTIC DISORDER (SIPD)

It is defined as a psychiatric disease having symptoms of delusions and/or hallucinations during or soon after substance intoxication or withdrawal (Fiorentini et al., 2021). These symptoms are distinct from primary psychotic disorders among substance users in which the symptoms:

- Preceded the onset of substance use;
- Continued substantially after the cessation of severe intoxication or acute withdrawal;
- Significantly more than one would expect based on the kind, amount, or duration of substance use;

- There is a history of psychotic episodes that occurred independently of substance use (David and Mcd, 2015).

12.5 EPIDEMIOLOGY OF FUNCTIONAL FOODS USE IN PSYCHOTIC DISORDERS

Nutritional intervention in the treatment and prevention of mental disease is becoming increasingly important. Nutrient supplementation and dietary techniques for overcoming unpleasant symptoms and cognitive behavior have received attention. The pathways that link diet with psychoses are as follows.

Nutrient deficiency has increased the incidence of mental diseases due to a lack of diet management and poor understanding (Teasdale, Mörkl and Müller-Stierlin, 2020). Vitamin B6, B9, and B12 deficiency causes an interruption in the one-carbon metabolic cycle. This further causes an increase in homocysteine levels. This disorder causes oxidative stress and vascular damage, which can cause neurodegeneration and schizophrenia by disrupting the blood-brain barrier (Teasdale, Mörkl and Müller-Stierlin, 2020). Also, lack of vitamin D is linked to the development of schizophrenic conditions. A recent study concluded that people with neonatal vitamin D deficiency had a 44% increased risk of developing schizophrenia later in their life (Teasdale, Mörkl and Müller-Stierlin, 2020). As a result of low nutritional quality, inflammation has been related to immunological dysregulation and mental diseases.

People living with schizophrenia or at least first-episode psychosis are found to have altered gut microbial composition than people with non-psychosis (Teasdale, Mörkl and Müller-Stierlin, 2020). A study concluded that the use of functional foods, especially those rich in anti-oxidant and anti-inflammatory properties has helped to treat schizophrenic symptoms (Matrisciano and Pinna, 2019). Furthermore, a balanced dietary pattern that includes functional foods like vegetables and fruits can help to alleviate schizophrenia symptoms.

12.6 ROLE OF FUNCTIONAL FOODS IN PSYCHOTIC DISORDERS

According to the research, almost 1 in 150 people will be diagnosed with psychotic diseases at some point during their lives (Moreno-ku and Martı, 2018). These numbers suggest the need to focus on the treatment and prevention of mental diseases. The etiology of such disorders is likely to be unknown as a complex mechanism is involved in it. Despite the lack of clinical trials examining the impact of functional foods on psychosis, nutritional interventions are becoming more common among schizophrenia patients (Teasdale, Mörkl and Müller-Stierlin, 2020). Several ingredients, including vitamin C, milk proteins, and herbal extracts (Sharma et al., 2020; Rane et al., 2020a–c) including ginkgo biloba, lemon balm, valerian, and kava, have been shown to improve mood (Hamer, Owen and Kloek, 2005) that can certainly help in conditions such as psychotic depression. Certain lifestyle and nutritional factors have a significant part in the disease's genesis. The role of various functional foods is elaborated below.

12.6.1 NUTRIENTS

Nutrients include compounds or substances with nutritional value, i.e., amino acids, fatty acids, vitamins, and minerals (Kumar, Patel and Seshadri, 2019; Singh et al., 2017b). Recent research suggests that amino acid imbalances are linked to the pathophysiology of schizophrenia (a disease with further aggravates to cause psychosis). Glycine, cysteine, glutamate, and serine are amino acids that are involved in the metabolism of homocysteine and the production of glutathione (Saleem, no date). Homocysteine is thought to cause vascular damage by causing oxidative stress, apoptosis, mitochondrial malfunction, and apoptosis. Additionally, increased levels cause cognitive deficits in persons with emotional disorders (Neuroscience, 2014). Glutathione, on the other hand, is a tripeptide that is the most abundant anti-oxidant in brain cells. Glutathione shortage causes an increase in reactive oxygen species (ROS), which leads to DNA damage. As a result, there is a link between amino acids and psychotic illnesses. Therefore, adding nutrients such as amino acids to the daily diet can turn out to be beneficial in the long run.

Eicosapentaenoic acid (EPA) and Docosahexaenoic acid (DHA) are polyunsaturated fatty acids that play a vital role in neural cells. Their availability determines phospholipid structure and structure, in turn, affects function (Mossaheb et al., 2012). In animal research, PUFA-deficient diets resulted in altered monoaminergic transmission. Also, clinical studies indicate that supplementation with PUFA enhances serotonin responsivity (Mossaheb et al., 2012). Reduced PUFA levels have been linked to schizophrenia in studies. Omega-3 polyunsaturated fatty acids (PUFAs) have also been shown to be beneficial in the treatment of schizophrenia symptoms (Hoenders et al., 2018). With an increase in Phospholipase A2 (PLA2) concentration in the brain, arachidonic acid is released, which in turn results in inflammation. PLA2 activity is dramatically reduced by omega-3 fatty acid intake, preventing brain degeneration (Hsu and Huang, 2020) and having a good impact on schizophrenia symptom prevention.

A deficiency in particular vitamins has been linked to the beginning of schizophrenia symptoms. Thiamine (Vitamin B1) is responsible for the production of confusion, psychotic symptoms, and neurological signs. Vitamin B12 deficiency leads to symptoms of disorientation and depression, whereas pyridoxine (Vitamin B6) deficiency results in seizures (Petrie and Ban, 1985; Keservani et al., 2016b). Supplementing with Vitamin B6 has also been shown to help schizophrenics with psychotic symptoms (Toriumi et al., 2021). Pyridoxine phosphate and pyridoxamine phosphate are both transformed to Pyridoxal Phosphate (PLP), an active co-factor. This PLP is an important co-factor for the formation of neurotransmitters. It is involved in the decarboxylation process of glutamine to Gamma amino butyric acid (GABA) (Tong, 2014). As a result, a decrease in PLP concentration causes a decrease in GABA concentration, which in turn increases the likelihood of seizures. Vitamin C insufficiency has been associated with cognitive impairment, abnormal behavior, and motor impairments, and vitamin C supplementation has been shown to be effective in the prevention and treatment of schizophrenia in patients. It is found that Vitamin C is involved in processes such as protection of the neuron against oxidative stress, alleviating inflammation, regulating neurotransmission, and affecting neuronal development ('Preventive and therapeutic potential of vitamin C.pdf,' no date). All of these processes are linked with the pathophysiology of psychosis.

Minerals such as zinc, copper, and magnesium play an important modulatory role in controlling NMDA receptors, under excitation which can produce specific forms of memory dysfunction. Low levels of these nutrients may produce abnormal NMDA activity thus leading to abnormal behavior (Cornish and Mehl-Madrona, 2008). Zinc deficiency has been linked to changes in behavior and cognition, as well as a decreased ability to learn and depression. Current research suggests that zinc supplementation along with therapy may reduce the number of medications required, reduce the cost involved in it and thus lead to favorable outcomes (Petrilli et al., 2017). An increase in magnesium concentration has also been observed to reduce anxiety and hallucinations, whilst a decrease in magnesium concentration can intensify the same symptoms.

12.6.2 HERBAL PRODUCTS

Evidence was found that several Chinese medicines and ginkgo biloba were added to antipsychotic medication to alleviate psychotic symptoms, according to a report of 110 controlled randomized clinical studies. In addition, when glycine, sarcosine, N-acetyl cysteine, estradiol, and vitamin B6 were added to antipsychotic medication therapy, symptoms improved (Hoenders et al., 2018).

Glycine, when taken alone as a medicine, led to improved symptom outcomes that were absent in placebo-treated patients, according to a pilot investigation of short-term therapy of risk syndrome patients for psychosis (Manuscript, 2014). In addition, when used with non-clozapine anti-psychotics, sarcosine was found to alleviate symptoms of schizophrenia (Bruggeman, Knegtering, and Jong, 2018).

N-acetyl cysteine, another chemical, has been found to help people with schizophrenia with negative symptoms and functioning (Dodd et al., 2017). Gingko biloba has been shown to help with akathisia and tardive dyskinesia (Bruggeman, Knegtering, and Jong, 2018). Ginkgo biloba extract improved the efficacy of the antipsychotic medicine haloperidol in individuals with persistent schizophrenia on positive symptoms, according to a study (Atmaca, Tezcan and Kuloglu, 2005). Another study concluded that the use of ginkgo biloba in schizophrenic patients helped reduce positive symptoms. In schizophrenic patients, the extract had no effect on negative symptoms. These findings support the use of ginkgo biloba as an additional therapy in schizophrenia patients (Brondino, De Sivestri, Re, Lanati, Thiemann, 2013). Gingko biloba has also been regarded as a neuroprotective agent in mental illness. The factors that affect neuroprotection include oxidative stress, neuroinflammation, neurogenesis, and an increase in vulnerability to schizophrenia. Evidence suggests a close link between neuroprotection and the onset of symptoms of schizophrenia (Kim and Na, 2017).

According to a recent study, kava (*Piper methysticum*) extract reduced the stereotyped behavior caused by amphetamine in mice (Molz et al., 2021). This may help to alleviate psychotic symptoms. Another study revealed several isolated cases of psychotic reactions following kava use and suggested that it possesses anti-psychotic activities, yet there is no conclusive evidence to it (Cairney, Maruff and Clough, 2002). Due to the increase in the use of kava, further investigations are needed to unwind its therapeutic efficacy in psychotropic disorders. In male ICR mice, the root extract of *Albizia zygia* was found to exhibit antipsychotic effects, including reductions in positive, negative, and cognitive symptoms of schizophrenia (Kumbol et al., 2018). In traditional medicine, the leaves of the plant have been used in the management of mental troubles. In addition, an extract of *Albizia zygia* leaves was found to have antipsychotic-like effects in mice (Amoateng et al., 2017).

12.7 FUTURE PROSPECT AND CHALLENGE

Functional foods and their usefulness in the prevention of psychotic illnesses are currently the subject of substantial investigation. Efforts are being made by the research institutes to identify substances having the potential in reducing risks of chronic diseases. There is now a clear explanation for functional foods, and there is a link between them and the prevention of psychotic symptoms. However, comprehensive research is needed for linking such foods to psychotic diseases. The complexity of the pathophysiology of psychotic diseases, as well as unraveling them to uncover treatment elements, is a key problem in the area. In such cases, the research methodology should be well defined. Natural products need to be studied more for the treatment and prevention of psychiatric illnesses.

12.8 CONCLUSION

Functional foods are defined as foods that have components that can alter physiological function and aid in the prevention of chronic diseases. In recent years, functional food has attracted a lot of attention. Consumption of functional foods has increased as a result of the need for a healthy lifestyle. The usefulness of such meals in the prevention and treatment of psychosis has been highlighted in different ways. Many studies are further needed to optimize the dose of functional foods and their interaction with other drugs. Furthermore, the benefits of functional meals should be thoroughly investigated.

ACKNOWLEDGMENT

We thank the MET's, Institute of Pharmacy, BKC, affiliated under Savitribai Phule Pune University, Nashik, and R. C. Patel Institute of Pharmaceutical Education and Research, Shirpur. EDA thanks the Ministry of Tribal Affairs

Govt. of India for providing financial assistance in the form of a fellowship (Award No.: 202021-NFST-MAH-01235).

KEYWORDS

- **functional food**
- **management**
- **nutrition**
- **psychosis**

REFERENCES

Ahire, E. D., Surana, K. R., Patil, C. D., Shah, H. S., Sonwane, G. B., & Talele, S. G., (2020). Role of omega-3 fatty acids in different neurodegenerative disorders. *Applied Pharmaceutical Science and Microbiology: Novel Green Chemistry Methods and Natural Products*, 173.

Ahire, E., Sonawane, V., Surana, K., & Talele, G., (2021). Drug discovery, drug-likeness screening, and bioavailability: Development of drug-likeness rule for natural products. *Applied Pharmaceutical Practice and Nutraceuticals.* Apple Academic press.

Amoateng, P., et al., (2017). *Psychotropic Effects of an Alcoholic Extract from the Leaves of Albizia zygia (Leguminosae-Mimosoideae), 2017.*

Arciniegas, D. B., (2015). Psychosis. *Continuum Lifelong Learning in Neurology, 21*(3), 715–736. doi: 10.1212/01.CON.0000466662.89908.e7.

Arya, S. S., Salve, A. R., & Chauhan, S., (2016). Peanuts as functional food: A review. *Journal of Food Science and Technology, 53*(1), 31–41. doi: 10.1007/s13197-015-2007-9.

Atmaca, M., Tezcan, E., & Kuloglu, M., (2005). *The Effect of Extract of Ginkgo Biloba Addition to Olanzapine on Therapeutic Effect and Antioxidant Enzyme Levels in Patients with Schizophrenia*, 652–656.

Bernatoniene, J., & Kopustinskiene, D. M., (2018). The role of catechins in cellular responses to oxidative stress. *Molecules, 23*(4), 1–11. doi: 10.3390/molecules23040965.

Brondino, N., De Sivestri, A., Re, S., Lanati, N., & Thiemann, P., (2013). A Systematic review and meta-analysis of *Ginkgo biloba* in neuropsychiatric disorders. *Evidence-Based Complementary and Alternative Medicine, 2013*, 915691.

Bruggeman, R., Knegtering, H., & Jong, M. De., (2018). *Natural Medicines for Psychotic Disorders a Systematic Review, 206*(2), 81–101. doi: 10.1097/NMD.0000000000000782.

Cairney, S., Maruff, P., & Clough, A. R., (2002). The neurobehavioural effects of kava. *Australian & New Zealand Journal of Psychiatry, 36*(5), 657–662.

Cencic, A., & Chingwaru, W., (2010). *The Role of Functional Foods, Nutraceuticals, and Food Supplements in Intestinal Health*, 611–625. doi: 10.3390/nu2060611.

Cheniaux, E., et al., (2008). Does schizoaffective disorder really exist? A systematic review of the studies that compared schizoaffective disorder with schizophrenia or mood disorders. *Journal of Affective Disorders, 106*(3), 209–217. doi: 10.1016/j.jad.2007.07.009.

Cornish, S., & Mehl-Madrona, L., (2008). *The Role of Vitamins and Minerals in Psychiatry*, 33–42.

Dalle, Z. A., & Szendro, Z., (2011). The role of rabbit meat as functional food. *Meat Science, 88*(3), 319–331. doi: 10.1016/j.meatsci.2011.02.017.

David, J., & Mcd, R., (2015). *Primary and Substance-Induced Psychotic Disorders in Methamphetamine Users.* Author Griffith Research Online.'

Deng, H., Xu, J., & Yeung, W. F., (2017). Ginkgo biloba versus placebo for schizophrenia. *Cochrane Database of Systematic Reviews, 2017*(1). doi: 10.1002/14651858.CD012524.

Dodd, S., et al., (2017). *Cognitive Effects of Adjunctive N – Acetyl Cysteine in Psychosis*, 1–11. doi: 10.1017/S0033291716002932.

Fiorentini, A., et al., (2021). Substance-induced psychoses: An updated literature review. *Frontiers in Psychiatry, 12*, 1–15. doi: 10.3389/fpsyt.2021.694863.

Gonz, J., et al., (2016). *HHS Public Access, 16*(10), 2131–2144. doi: 10.1017/S1461145713000606.Preclinical.

Hamer, M., Owen, G., & Kloek, J., (2005). The role of functional foods in the psychobiology of health and disease. *Nutrition Research Reviews, 18*(1), 77–88. doi: 10.1079/nrr2005103.

Hasler, C. M., (1996). Functional foods: The Western perspective. *Nutrition Reviews*. doi: 10.1111/j.1753-4887.1996.tb03810.x.

Hasler, C. M., (2002). Functional foods: Benefits, concerns and challenges – A position paper from the American council on science and health. *Journal of Nutrition, 132*(12), 3772–3781. doi: 10.1093/jn/132.12.3772.

Henry, C. J., (2010). Functional foods. *European Journal of Clinical Nutrition, 64*(7), 657–659. doi: 10.1038/ejcn.2010.101.

Hoenders, H. J. R., et al., (2018). Natural medicines for psychotic disorders: A systematic review. *Journal of Nervous and Mental Disease, 206*(2), 81–101. doi: 10.1097/NMD.0000000000000782.

Hsu, M., & Huang, Y., (2020). *Beneficial Effects of Omega-3 Fatty Acid Supplementation in Schizophrenia: Possible Mechanisms*, 1–17.

Keservani, R. K., & Sharma, A. K., (2014). Flavonoids: Emerging trends and potential health benefits. *Journal of Chinese Pharmaceutical Sciences, 23*(12), 815.

Keservani, R. K., Kesharwani, R. K., Sharma, A. K., Vyas, N., & Chadoker, A., (2010b). Nutritional supplements: An overview. *International Journal of Current Pharmaceutical Review and Research, 1*(1), 59–75.

Keservani, R. K., Kesharwani, R. K., Vyas, N., Jain, S., Raghuvanshi, R., & Sharma, A. K., (2010a). Nutraceutical and functional food as future food: A review. *Der Pharmacia Lettre, 2*(1), 106–116.

Keservani, R. K., Sharma, A. K., & Kesharwani, R. K., (2016a). Nutraceutical and functional foods for cardiovascular health. In: *Food Process Engineering* (pp. 257–278, 291–312). Apple Academic Press, CRC Press. ISBN: 9781771884020.

Keservani, R. K., Sharma, A. K., & Kesharwani, R. K., (2016b). Medicinal effect of nutraceutical fruits for the cognition and brain health. *Scientifica*, 10. Article ID 3109254. doi: 10.1155/2016/3109254.

Keservani, R. K., Sharma, A. K., & Kesharwani, R. K., (2017). An overview and therapeutic applications of nutraceutical and functional foods. *Recent Advances in Drug Delivery Technology*, 160–201. ISBN:9781522507543.

Keservani, R. K., Sharma, A. K., & Kesharwani, R. K., (2020). "*Nutraceuticals and Dietary Supplements: Applications in Health Improvement and Disease Management*. CRC Press. ISBN: 9781771888738.

Khan, M. I., et al., (2011). Meat as a functional food with special reference to probiotic sausages. *Food Research International, 44*(10), 3125–3133. doi: 10.1016/j.foodres.2011.07.033.

Kim, Y. K., & Na, K. S., (2017). Neuroprotection in schizophrenia and its therapeutic implications. *Psychiatry Investigation, 14*(4), 383–391. doi: 10.4306/pi.2017.14.4.383.

Koca, I., & Tasci, B., (2016). Garlic as a functional food. *Acta Horticulturae, 1143*, 139–146. doi: 10.17660/ActaHortic.2016.1143.20.

Kumar, S., Patel, R., & Seshadri, S., (2019). The role of nutraceuticals in preventing and managing mental disorders. *Functional Foods and Mental Health*, pp. 50–71.

Kumbol, V. W., et al., (2018). *Albizia zygia* root extract exhibits antipsychotic-like properties in murine models of schizophrenia. *Biomedicine & Pharmacotherapy, 106*, 831–841. doi: 10.1016/j.biopha.2018.07.030.

Leontiev, R., et al., (2018). A comparison of the antibacterial and antifungal activities of thiosulfinate analogues of allicin. *Scientific Reports, 8*(1), 1–19. doi: 10.1038/s41598-018-25154-9.

Lin, D. C., (2003). Probiotics as functional foods. *Nutrition in Clinical Practice, 18*(6), 497–506. doi: 10.1177/0115426503018006497.

MAA, G., (2017). Health benefits of milk and functional dairy products. *MOJ Food Processing & Technology, 4*(4), 108–111. doi: 10.15406/mojfpt.2017.04.00099.

Manuscript, A., (2014). *NIH Public Access, 23*(8), 931–940. doi: 10.1016/j.euroneuro.2012.09.008.Glycine.

Matrisciano, F., & Pinna, G., (2019). Functional food and nutrition for the management of autism spectrum disorders and schizophrenia. *Functional Foods and Mental Health Publisher* (Vol. 1, No. 1, pp. 307–330). Food Science, Publisher 9781708767006.

McKay, D. L., & Wilson, T., (2016). Cranberry juice: Effects on health. *Beverage Impacts on Health and Nutrition*, 101–113. doi: 10.1007/978-3-319-23672-8_7.

Mestre, P. J. A., & Mateus, C. M. R. P., (2002). Functional foods from animal sources and their physiologically active components. *Revue de Medecine Veterinaire, 153*(3), 155–160.

Minich, D. M., & Brown, B. I., (2019). A review of dietary (Phyto)nutrients for glutathione support. *Nutrients, 11*(9), 1–20. doi: 10.3390/nu11092073.

Miranda, J. M., et al., (2015). Egg and egg-derived foods: Effects on human health and use as functional foods. *Nutrients, 7*(1), 706–729. doi: 10.3390/nu7010706.

Mohamed, S., (2015). Antioxidants as functional foods in health and diseases. *Austin Journal of Nutrition and Food Sciences, 3*(3), 1–8. Available at: www.austinpublishinggroup.com (accessed on 22 February 2023).

Molz, C., et al., (2021). *Kava Decreases the Stereotyped Behavior Induced by Amphetamine in Mice, 265*. doi: 10.1016/j.jep.2020.113293.

Moreno-ku, B., & Martı, C., (2018). Prevalence of psychotic disorders and its association with methodological issues. *A Systematic Review and Meta-Analyses*, 1–25.

Mossaheb, N., et al., (2012). *Polyunsaturated Fatty Acids in Emerging Psychosis*, 576–591.

Murray, A. J., et al., (2021). Oxidative stress and the pathophysiology and symptom profile of schizophrenia spectrum disorders. *Frontiers in Psychiatry, 12*, 1–16. doi: 10.3389/fpsyt.2021.703452.

Nasri, H., et al., (2014). New concepts in nutraceuticals as alternative for pharmaceuticals. *International Journal of Preventive Medicine, 5*(12), 1487–1499.

Neuroscience, B., (2014). *Homocysteine Levels in Schizophrenia and Affective Disorders —Focus on Cognition, 8*, 1–10. doi: 10.3389/fnbeh.2014.00343.

Online, F. A., et al., (n.d). *International Journal of Nutrition ISSN No: 2379-7835 Classification of Functional Foods*, (3), 7–16. doi: 10.14302/issn.2379.

Ozen, A. E., Pons, A., & Tur, J. A., (2012). Worldwide consumption of functional foods: A systematic review. *Nutrition Reviews, 70*(8), 472–481. doi: 10.1111/j.1753-4887.2012.00492.x.

Patel, K. R., et al., (2014). Schizophrenia: Overview and treatment options. *P and T, 39*(9), 638–645.

Petrie, W. M., & Ban, T. A., (1985). *Vitamins in Psychiatry Do They Have a Role ?, 65*, 58–65.

Petrilli, M. A., et al., (2017). *The Emerging Role for Zinc in Depression and Psychosis, 8*, 1–12. doi: 10.3389/fphar.2017.00414.

Rahman, M. S., (2007). Allicin and other functional active components in garlic: Health benefits and bioavailability. *International Journal of Food Properties, 10*(2), 245–268. doi: 10.1080/10942910601113327.

Rajasekaran, A., & Kalaivani, M., (2013). Designer foods and their benefits: A review. *Journal of Food Science and Technology, 50*(1), 1–16. doi: 10.1007/s13197-012-0726-8.

Ramalho, R. A., et al., (2019). Farmed fish as a functional food: Perception of fish fortification and the influence of origin – insights from Portugal. *Aquaculture, 501*, 22–31. doi: 10.1016/j.aquaculture.2018.11.002.

Rane, B. R., Bharath, M. S., Patil, R. R., Keservani, R. K., & Jain, A. S., (2020b). Novel approaches in nutraceuticals. In: Rajesh, K. K., Raj, K. K., & Anil, K. S., (eds.), *Enhancing the Therapeutic Efficacy of Herbal Formulations through Novel Drug Delivery Systems* (pp. 241–266). IGI Global International Publisher, Pennsylvania, USA, Chapter 11. ISBN13: 9781799844532.

Rane, B. R., Patil, A. S., Keservani, R. K., & Jain, A. S., (2020a). Novel approaches in herbal formulation. In: Rajesh, K. K., Raj, K. K., & Anil, K. S., (eds.), *Enhancing the Therapeutic Efficacy of Herbal Formulations through Novel Drug Delivery Systems* (pp. 43–68). IGI Global International Publisher, Pennsylvania, USA, Chapter 2. ISBN13: 9781799844532.

Rane, B. R., Tadavi, S. A., & Keservani, R. K., (2020c). Naturopathy. In: Anil, K. S., Raj, K. K., & Surya, P. G., (eds.), *Herbal Product Development* (pp. 321–347). Apple Academic Press, CRC Press, Taylor & Francis Group, Chapter 12. ISBN: 9781771888776.

Reid, G., Gadir, A. A., & Dhir, R., (2019). Probiotics: Reiterating what they are and what they are not. *Frontiers in Microbiology, 10*, 1–6. doi: 10.3389/fmicb.2019.00424.

Rosenthal, L. J., & Khandai, A. C., (2019). Psychotic disorders. *Fundamentals of Consultation Liaison Psychiatry: Principles and Practice*, 109–132. doi: 10.1007/978-3-319-57111-9_2084.

Saleem, S., Shaukat, F., Gul, A., Arooj, M., & Malik, A., (2017). Potential role of amino acids in pathogenesis of schizophrenia. *International Journal of Health Sciences*, *11*(3), 63.

Sarojnalini, C., & Hei, A., (2019). Fish as an important functional food for quality life. *Functional Foods*, 1–19. doi: 10.5772/intechopen.81947.

Seccia, A., Viscecchia, R., & Nardone, G., (2019). Table grapes as functional food: Consumer preferences for health and environmental attributes. *BIO Web of Conferences, 15*, 03011. doi: 10.1051/bioconf/20191503011.

Sharma, A. K., Keservani, R. K., & Gautam, S. P., (2020). *Herbal Product Development*. Apple Academic Press, CRC Press, Taylor & Francis Group. ISBN: 9781771888776.

Shaw, R. J., Pao, M., Holland, J. E., & DeMaso, D. R., (2016). Practice patterns revisited in pediatric psychosomatic medicine. *Psychosomatics. 57*(6), 576–585.

Singh, P., Kesharwani, R. K., & Keservani, R. K., (2017a). Antioxidants and vitamins: Roles in cellular function and metabolism. In: *Sustained Energy for Enhanced Human Functions and Activity* (pp. 385–407). Academic Press. ISBN:9780128054130.

Singh, P., Kesharwani, R. K., & Keservani, R. K., (2017b). Protein, carbohydrates, and fats: Energy metabolism. In: *Sustained Energy for Enhanced Human Functions and Activity* (pp. 103–115). Academic Press. ISBN:9780128054130.

Stip, E., & Tourjman, V., (2010). Aripiprazole in schizophrenia and schizoaffective disorder: A review. *Clinical Therapeutics, 32*(SUPPL. 1), S3–S20. doi: 10.1016/j.clinthera.2010.01.021.

Surana, K. R., Ahire, E. D., Sonawane, V. N., Talele, S. G., & Talele, G., (2021). Informatics and methods in nutrition design and development. *Natural Food Products and Waste Recovery: Healthy Foods, Nutrition Design, and Extraction of Valuable Compounds*, 33.

Teasdale, S., Mörkl, S., & Müller-Stierlin, A. S., (2020). Nutritional psychiatry in the treatment of psychotic disorders: Current hypotheses and research challenges. *Brain, Behavior, & Immunity – Health, 5*, 100070. doi: 10.1016/j.bbih.2020.100070.

Tong, Y., (2014). *Seizures Caused by Pyridoxine (Vitamin B6) Deficiency in Adults: A Case Report and Literature Review, 3*(2), 52–56. doi: 10.5582/irdr.2014.01005.

Toriumi, K., et al., (2021). *Vitamin B6 Deficiency Hyperactivates the Noradrenergic System, Leading to Social Deficits and Cognitive Impairment*. doi: 10.1038/s41398-021-01381-z.

Zeisel, S. H., & Da Costa, K. A., (2009). Choline: An essential nutrient for public health. *Nutrition Reviews, 67*(11), 615–623. doi: 10.1111/j.1753-4887.2009.00246.x.

CHAPTER 13

FUNCTIONAL FOODS FOR VIRAL INFECTIONS

KHUSHALI PAGAR,[1] SACHIN GUNJAL,[1] KHEMCHAND R. SURANA,[2] TUSHAR LOKHANDE,[3] SUNIL MAHAJAN,[3] and EKNATH D. AHIRE[4]

[1]*Department Pharmaceutics, Shreeshakti Shaikshanik Sanstha, Divine College of Pharmacy, Satana, Nashik, Maharashtra, India*

[2]*Department of Pharmaceutical Chemistry, Shreeshakti Shaikshanik Sanstha, Divine College of Pharmacy, Satana, Nashik, Maharashtra, India*

[3]*Department of Pharmaceutical Chemistry, MGV's Pharmacy College, Panchvati, Nashik, Maharashtra, India*

[4]*Department of Pharmaceutics, MET's Institute of Pharmacy, Affiliated to Savitribai Phule Pune University, BKC, Adgaon, Nashik, Maharashtra, India*

ABSTRACT

Functional food is very important to control of viral infections. An array of nutrients like macro and micro-nutrients, vitamins, etc., is related to boosting the host's immune responses in opposition to intracellular pathogens. The functional food has immunomodulatory results in managing the infection and inflammation processes. They also have results in managing dietary deficiency of any form, i.e., malnutrition. The maximum vital additives that may be brought to meals are probiotics, prebiotics, symbiotic, nutrient, and polyphenols. The elements consist of anorexia, conventional behaviors,

Applications of Functional Foods in Disease Prevention. Raj K. Keservani & Eknath D. Ahire, PhD (Eds.)

reduction of intestinal absorption, metabolic damage, sickness metabolism of lipids and carbohydrates, discount of vitamins, iron, zinc, and copper.

13.1 INTRODUCTION

The term 'functional food' was born in Japan. The Japanese were the first to look at that food should have a function beyond gastronomic pleasure and strength and nutrient supply to the human organism. Following this, the United States of America legislates those merchandise inside the FOSHU (Foods of Specified Health Use) legislation. Only the products which meet certain requirements can have the FOSHU stamp. The idea of a fixed cost of food was later adopted by countries in Europe and North America. There is no consensus between Europe and the United States concerning a concrete definition, leading to a series of different terms: nutraceutical, designer food, pharma food, etc., which has contributed to increasing the confusion among professionals and consumers. Whereas, the united states decide upon the term nutraceutical, European experts determined in the FUFOSE (Functional Food Science in Europe) project, under the auspices of the European Union and ILSI Europe (International Life Sciences Institute Europe, 1999) to adopt the time period 'purposeful meals' with a consensus definition (López-Varela et al., 2002). It is essential to focus on that purposeful meals need to be a meals, now no longer a drug. Beneficial effects need to be received with the aid of using consuming normal amounts of purposeful meals inside the 'regular' diet. As it's been currently stated, the ultimate purpose of the scientific network and food enterprise has to be to expand practical ingredients for enhancing existence quality (Alkhatib, 2020). Therefore, as Western civilization is going through as much as an innovative boom in immune-mediated and intestine-associated fitness problems, which includes autoimmune and inflammatory sicknesses, it appears ok to examine the consequences of practical ingredients at the immune machine (Banik et al., 2021).

13.2 FUNCTIONAL FOOD

Basically, practical ingredients are used to decorate positive physiological capabilities, so that it will save you or maybe to therapy sicknesses. Although herbal ingredients (Sharma et al., 2020; Rane et al., 2020a–c), which include strawberries or onions, had been defended as protected as practical ingredients, best the ones ingredients which can be submitted to positive technique are taken into consideration practical ingredients (Konstantinopoulos et al., 2020.). Thus, numerous strategies to acquire practical ingredients consist of the addition or elimination of an aspect, change of the meals processing, genetic engineering, etc., that's permitting the meals enterprise to expand new merchandise with extra fee for the market (Ling et al., 2015).

13.3 OBJECTIVES OF FUNCTIONAL FOODS

The consensus statement of the FUFOSE project (International Life Sciences Institute Europe, 1999) establishes that basic research in biological sciences is relevant in order to establish the mechanisms which modulate the function of interest in the maintenance of health and in the reduction of disease. This can also be sustained by means of a significant, epidemiologically proven association between a marker of the amount ingested of a specific component and the specific beneficial effect which is attributed to it. One of the clearest objectives of the study of functional foods is to identify and validate biological markers in order to understand the interaction mechanisms between the food component and a body function (Reygaert, 2018):

1. **Probiotics:** Living micro-organisms which while ingested in positive amounts, have an advantageous effect on host fitness, which is going past traditional dietary consequences. The microorganism most usually used as probiotics are *Lactobacilli* and *Bifidobacteria.* They may be given with fermented ingredients which include yogurt, fermented greens or meats and they'll in short set up with inside the intestine (Hodgkinson et al., 1993; Pannu and Bhatnagar 2019).
2. **Prebiotics:** Ingredients or compounds which have an useful impact at the microflora with inside the host itself, which includes fiber, fructo-oligosaccharides, inulin, lactulose, and sugar alcohols. They are short-chain carbohydrates that can be fermented withinside the big bowel and stimulate the increase of probably useful bifidobacteria (Dang and Marsland, 2019; Jayawardena et al., 2020).
3. **Synbiotics:** A combination of prebiotics and probiotics (Wintergerst et al., 2007).
4. **Nutrients:** Minerals, nutrients, fatty acids or nutritional fiber, for example, which can be particular and feature a totally focused movement. Objectives of practical ingredients The consensus declaration of the FUFOSE assignment establishes that simple studies in

organic sciences is applicable so that it will set up the mechanisms which modulate the characteristic of hobby with inside the protection of fitness and withinside the discount of sickness (Martineau et al., 2017). This also can be sustained with the aid of an extensive, epidemiologically tested affiliation among a marker of the quantity ingested of a particular aspect and the particular useful impact that's attributed to it. One of the maximum clean goals of the have a look at of practical ingredients is to become aware of and validate organic markers so that it will apprehend the interplay mechanisms among the meals aspect and a frame characteristic (Bikle et al., 2002).

5. **Polyphenols:** These are plant secondary metabolites offering quite a number protective capabilities towards ultraviolet radiations, infectious sicknesses, and oxidative stress. Most of them have planar and fused phenolic earrings with hydroxyl agencies as substituents. They may be categorized into phenolic acids, flavonoids, polyphenolic amides, and different polyphe-nol compounds (which include stilbenes or lignans) (Pinto et al., 2020). Polyphenols are tested inhibitors of irritation and prevent expression of pro-inflammatory genes. Polyphenols had been observed to be powerful towards a vast variety of viruses which include Epstein-Barr virus, enterovirus, herpes simplex virus (HSV), influenza virus and different virus-inflicting respiration tract-associated infections (Scholz-Ahrens et al., 2007). Most of the polyphenols exert their interest mainly as anti-virals in a synergistic manner. For example, a crude extract organized from Sambucas nigra changed into capable of inhibiting viral replication effectively in a Vero cellular line and brought about alteration of viral structure (Huang et al., 2021).

Quite some mechanisms to provide an explanation for the antiviral residences of those compounds had been proposed. For example, polyphenols from *Broussonetia papyrifera* have been observed to inhibit the proteases of MERS/SARS-CoV2 thereby efficiently blockading the unfold of the virus (Wilson et al., 2019). This polyphenol is famous as an effective antioxidant and forestalls improvement of inflammatory responses. It has been observed to be up-regulated underexposure to UV radiation, ozone, and in reaction to fungal infections In addition, resveratrol has been confirmed to be an amazing inhibitor of higher respiratory tract viral infections (Taliun et al., 2021). Resveratrol without delay interferes with the replication of influenza virus by means of blockading the nuclear to cytoplasmic translocation of Ribonucleoprotein particles. In case of contamination by means of coronaviruses, resveratrol has been observed to noticeably boom the survival fee of virus inflamed cells withinside the frame by means of decreasing their frequency of apoptosis the antiviral residences of resveratrol are augmented by means of its robust immune modulating capabilities. In addition, resveratrol management has additionally been found to bring about upregulation of the IFNß gene for arrest of viral replication. However, no matter such a lot of fitness augmenting roles of resveratrol, its healing utility has been restrained because of negative solubility and coffee bioavailability (Marjani et al., 2017).

13.4 THE FOUNDATION OF ANTI-OXIDATIVE CAPACITY OF GREEN TEA

Consumption of green tea has come to be famous globally over the previous couple of a long time to its tested position as a robust antioxidant (Rahman and Idid, 2021). Green tea is crafted from the leaves of *Camellia sinensis* and the additives liable for its anti-oxidative assets are catechins. Catechins also are plant secondary metabolites belonging to the elegance of hydroxyflavans which have a flavan-3-ol skeleton and its substituted derivatives. Catechins incorporate approximately 10% of the dry weight of green tea leaves and consist of (-)-epigallocatechin gallate (EGCG), (-)-epigallocatechin (EGC), (-)-epicatechin gallate (ECG) and (-)-epicatechin (EC) because the primary elements with EGCG accounting for approximately 50% of the full quantity of catechins in inexperienced tea. EGCG has been stated to own many different pharmacotherapeutic residences which include antioxidative, anti-bacterial, and antiviral roles. EGCG serves as an effective inhibitor of HIV replication in addition to an agglutinating agent of influenza viruses. Antiviral interest of catechins is believed to be because of their inhibition of host-virus membrane fusion by means of prevention of acidification of the endosomal compartment (Vitapole, 2000; Kesharwani et al., 2016).

13.5 PROBIOTICS

The position of intestinal microflora in preserving the general fitness of the man or woman has been set up past arguments. However, just like our personal organs, the consortium of intestine microbes also are prone to occasional dysbiosis ensuing in altered functionalities with unwanted consequences (Roberfroid, 2000). Probiotics are stay microbes which can complement and refill the ill intestine vegetation and repair homeostasis. Gut commensals join and talk with primary organs of the human frame to make sure protection of physiological homeostasis, with the mind through the intestine-mind axis and with the lung through the intestine-lung axis. Probiotics can control a variety of host immune responses, similar to the resident gut bacteria, which reduces the infectivity of some viral infections, primarily those in Journal of Food Science. Functional components of earlier vitamins changed as a result of respiratory tract illnesses. (Mortaz et al., 2013). Probiotics have already been advocated as a mitigation approach for a weakened immune machine following COVID-19 contamination wherein intestine microfloral dysbiosis with extensive discounts in *Lactobacillus* and *Bifidobacterium* had been seen. They had been related to the expanded expression of mucous glycoproteins to reinforce the epithelial barrier towards viral invasion. Probiotics also are implicated in the resurrection of tight junction complexes following virus inflicted harm through expanded expression and redistribution of zonula occludent (ZO-2) proteins and protein kinase C (PKC) (Englyst et al., 2000).

13.5.1 MOLECULAR MECHANISM OF THE ACTION OF PROBIOTICS

Although the precise mechanism of probiotic primarily based totally prophylaxis of viral infections nonetheless stays to be delineated, those microbial lines are believed to exert their immunomodulatory (Keservani et al., 2015) consequences through interplay of the bacterial elements like peptidoglycan, lipoteichoic acid, and nucleic acid with their cognate Toll-like receptors in addition to muramyl dipeptide, which binds to the nod-like receptors and next modulation of Treg cells, TH17 cells, and type 3 innate lymphoid cells. NK cells, T-cells, and stages of circulating antibodies make up the immune system's immune arsenal, which probiotic lines also strengthen for increased viral contamination warning and boom (Roberfroid, 2002). These also are implicated in modulating the stages of proinflammatory and immunoregulatory cytokines leading to stronger viral clearance with minimum bystander harm to lung cells. Collectively, the probiotics limitation viral replication by means of:

- enhancement of the intestinal epithelial barrier;
- attaching to and sequestering viruses;
- by means of secreting antiviral additives which includes bacteriocins and hydrogen peroxide;

- by means of induction of basal stage nitric oxide (NO) technology by means of host cells;
- by means of activation of macrophages and dendritic cells;
- by means of inducing CD^{8+} T cells into cytotoxic T lymphocytes; and
- by means of inducing differentiation of B cells by means of TH2 cells.

Although uncommon incidents of probiotic-related bacteremia and fungemia had been stated in some immunocompromised sufferer's probiotics are commonly safe.

13.6 PREBIOTICS

To maintain our intestine microbiota glad and rejuvenated, they ought to be fed with particular vitamins that aren't absorbable by means of our intestine, however tailormade for those microbes. The intestine microbiota prospers on positive plant derived nutritional fibers, which human enzymes are in-able to digesting. These nutritional fibers are together referred to as prebiotics. Numerous not unusual place fruits and greens which include apples, artichokes, bananas, berries, legumes, tomatoes, asparagus, etc., are enriched in prebiotic fibers which include fructans, oligosaccharides, arabino-oligosaccharides, isomalto-oligosaccharides, xylo-oligosaccharides, resistant starch, lactosucrose, lactobionic acid, galactomannan, psyllium, polyphenols, and polyunsaturated fatty acids, amongst others (López-Varela et al., 2002). In addition to offering critical vitamins to the intestine microbes, the prebiotics additionally gain our frame in a number of different methods which include facilitating absorption of calcium, regulating the glycemic index, boosting digestion, and preserving the epithelial cellular lining healthful. New insights have additionally cautioned the useful position of prebiotics in control of persistent intestine problems which includes irritable bowel syndrome (Ross et al., 2000).

13.7 ANTIOXIDANT NUTRIENTS

Micronutrients, comprising nutrients and minerals, are named so because of the reality that necessities by means of our frame are in fairly smaller or less amounts. However, they may be vital because the number one nutrients and their deficiency results in impairment of many key metabolic reactions and biochemical pathways in addition to malfunctioning of the host immune reaction. They have an extensive position to play in preserving the capability of each innate and adaptive immune responses and micronutrient homeostasis is imperative to the sustenance of a healthful immune machine (Rayman, 2000). Some nutrients like A, C, E, flavonoids (Keservani and Sharma, 2014), and carotenoids are readily available from the food regimen and consequently can characteristic relentlessly in neutralizing oxidative unfastened radicals (Keservani et al., 2010a, b, 2017, 2020). A loss in the stability of those factors could make some on eat risk of contamination and sicknesses, and supplementation, on the alternative hand, can increase the restoration process, mainly from viral infections Vitamin D as a version nutrition with a practical meals tag a multidimensional nutrition in phrases of healing capacity is Vitamin D, additionally cited as "calciferol." It is a fat-soluble nutrition present in exclusive bureaucracy this is clearly found in some ingredients, however by and large produced endogenously by means of decrease layers of pores and skin dermis by means of a chemical response pushed by means of the response of photons of daylight with LDL cholesterol Hence, the primary bureaucracy observed in human flow are Vit D3 (cholecalciferol) and D2 (ergocalciferol). Vitamin D acquired from meals reassets or synthesized in presence of daylight isn't biologically energetic. It is activated by means of successive enzymatic hydroxylation occasions taking place respectively withinside the kidney and the liver, to yield 1, 25-dihydroxyvitamin D. Vitamin D is involved regulating many physiological methods in the frame including reabsorption of calcium and magnesium, which can be additionally concerned as pH buffers, mediating bone and extra-skeletal metabolism, regulating cardiovascular homeostasis, and most importantly modulating immunological responses of our bodies. Vitamin D is likewise implicated in a marked discount in pro-inflammatory cytokines for that reason reducing the chance of a cytokine hurricane thereby decreasing the want for extended inflammatory occasions. An advocated dose of nutrition D3 tiers from 2,000 to 10,000 IU/day (Fearne and Hughes 1999).

13.8 FUNCTIONAL FOODS MECHANISMS OF OPTIMIZING THE IMMUNE SYSTEM ANTIVIRAL

Defense enhancing the antiviral immune defense can benefit from the practical meals consumption of a great form of plant, animal, and fungi species, ate up throughout exclusive diets and cultural practices which include conventional natural remedy which includes teas, roots, mushrooms, and fermented flowers and leaves; MD additives which includes olive-primarily based totally merchandise, oily fish, seeds, culmination, and greens; famous drinks which includes espresso; and protein-wealthy ingredients which includes hen extract and soybean peptides (Moriguchi and Muraga, 2000). The majority of such ingredients comprise clearly taking placenutrients and minerals (e.g., nutrients C, D, B6, B12, A, E, and minerals of zinc, copper, iron, and selenium), and different phenolic compounds which can be immunoprotective especially through anti-oxidation and anti-irritation residences. Other ingredients which includes oily fish omega-three fatty acids comprise monounsaturated fatty acids which includes omega-three fatty acids (EPA and DHA) in oily fish, which may be enzymatically transformed to specialized pro-resolving mediators (SPMs) referred to as resolvins, protectins, and maresins, which can be molecules that assist inflammatory decision and recuperation of inflamed sites which include the respiration tract, that can save you acute lung damage. Fermented food products (e.g., yogurt, pickles, fermented culmination, greens, plant, and drinks) comprise probiotics, and feature additionally been proven to decorate intestine microorganism profile and intestine-lung axis-associated respiration fitness (Surana et al., 2021). A précis of systematic evaluations and randomized managed trials stated decreased prevalence and severity of higher and decrease respiration tract infections (odds ratio ~ 0.8–0.5) by means of the usage of exclusive probiotics, mainly lactobacilli and bifidobacterial. The efficacy of probiotics in decreasing COVID-19 (Ahire and Kshirsagar, 2021) inflamed sufferers has now no longer but been set up, however the prophylactic blessings for reinforcing the immune machine are supportive in their long time use, mainly thinking about that enhancing intestine microbiota profile has been lately implicated in stopping COVID-19 (Ahire et al., 2020a) in older and high-hazard people with compromised immune systems. Selected meals dietary supplements and micronutrient nutrients and hint factors had been reviewed somewhere else in phrases of optimizing the immune responses (Grimble, 1997). Given the promising position of famous practical ingredients, which includes the ones in the MD which include olive oil, and Asian and African natural teas and fermented ingredients and famous drinks as a part of way of life sickness prevention, it's miles vital to contextualize the antiviral mechanisms of such functional foods. Below is an assessment of famous ingredients inside numerous nutritional patterns, which include olive oil nutraceuticals, famous nutrients which includes nutrition D, conventional medicinal herbs and roots, and protein peptides for stopping viral infections which include COVID-19, mainly while they may be followed as a part of an lively way of life (Douglas and Ohlstein, 2000).

13.9 OLIVE OIL ANTIVIRAL POSITION

Olive oil (OO), and its elements (leaves and bark), shape and vital immune-improving practical meals because of the extensive NCD preventative blessings, mainly of cardiovascular sickness, diabetes, and cancer, that have been reviewed somewhere else. OO, mainly extra-virgin OO (EVOO) carries monounsaturated fatty acids, and numerous polyphenols which include oleuropein and hydroxytyrosol, that have numerous antioxidative and anti-inflammatory residences, which may be related with extensive anti-viral and antibacterial capacity. Oleuropein has proven a capacity antiviral interest towards respiration syncytial virus (RSV), a not unusual place to perspiration contamination (URI) virus. This impact has been attributed to the antioxidative assets of elenolic acid as a primary fragment in oleuropein, which has lengthy been proven to have amazing antiviral sports towards herpes, influenza A and B, and parainfluenza 1, 2, and 3 viruses. Antioxidant potential of OO changed into later proven to be unbiased of the dimensions of the antiviral impact, with oleuropein displaying advanced antiviral consequences in comparison with different secoiridoid glycosides remoted from *Ligustrum lucidum* (Vickers, 2017; Mvubu et al., 2016). The anti-oxidation defensive blessing of OO, mainly EVOO, which has a better phenolic content material, promotes its position for reinforcing the immune machine defense towards viruses. Hydroxytyrosol antiviral mechanisms have been confirmed through its inactivation consequences on influenza A viruses, mainly throughout the virus morphological changes, which includes the presence of a viral envelope that's an imperative membrane protein concerned in numerous elements of the virus existence cycle which include its assembly, budding, and pathogenesis. The mechanisms of which OO nutraceuticals defend against viral infections have regularly centered at the hydroxytyrosol preventative consequences on HIV from coming into the host cellular and binding the catalytic site of the HIV-1, and its inhibitory impact on each viral access and integration. Regular consumption of olive leaf extracts, wealthy in polyphenol flavonoids, had been proven to be liable for a 33% discount in URI. Such promising antiviral capacity changed into attributed to the subsequent anti-oxidation actions of oleuropein with dose-based inhibition of the copper sulfate-precipitated oxidation of low-density lipoproteins (LDLs), and precipitated boom in nitric oxide manufacturing in macrophages and practical interest (Yurie and Ikeguchi, 2017). In every other have a look at amongst high-faculty athletes who had been at risk of URI, olive leaf extract supplementation (20 g, carrying a 100 mg oleuropein) turned into proven to reduce the length of contamination (28% discount in sick days) yet now no longer the prevalence fee. Thus, olive polyphenols (each in OO and leaf), mainly oleuropein and hydroxytyrosol, appear to sell antiviral defense and may be an adjoining prevention to govern URIs. Exploring OO mechanisms for shielding onwards novel viruses which includes COVID-19, mainly its protein viral envelop characteristic and interplay with host cells might additionally be vital. The blessings of OO consumption, mainly as a part of a balanced food regimen which includes the MD may be similarly augmented through bodily interest, mainly power and resistance kind workout. Such a method is probably to be an powerful prevention of viral infections. In phrases of the advocated OO dose, a slight dose of 20–30 g/day (mainly polyphenol-wealthy EVOO) in mixture with different nutritional practical ingredients may be advocated for reinforcing the immune machine, that's in step with latest NCD prevention recommendations (Martirosyan and Singh, 2015; Xia et al., 2020).

13.10 VITAMIN D ANTIVIRAL ROLE

The position of nutrition D in ameliorating the consequences of each NCDs and CDs is now properly accepted. Vitamin D reduces acute respiration tract contamination, and its deficiency is related with susceptibility to viral infections, and additionally with numerous cancers, diabetes, and CVD (Alyane et al., 2008). Research into the position of vitamin D in stopping COVID-19 and influenza viral infections has received latest momentum through numerous evaluations and meta-analyzes, mainly for the reason that those who are prone to COVID-19 infections are especially high-risk people with numerous NCDs which includes diabetes and CVD, suggesting that nutrition D might be the lacking hyperlink among NCDs and viral CDs (Syngai et al., 2016). Based at the latter, numerous mechanisms of the way nutrition D availability can lessen the hazard of viral infections and COVID-19 through the subsequent mechanisms:

- Lowering viral replication prices through cathelicidins and defensins, and stopping lung injures that result in pneumonia through its anti-inflammatory cytokines;

- Capacity for nutrition D supplementation effectiveness in decreasing the hazard of influenza mainly in poor people;
- Documented decreased hazard of COVID-19 throughout the summertime season indicated by means of a decrease wide variety of instances with inside the southern hemisphere, in comparison with better wide variety of instances with inside the wintry weather months while 25-hydroxy Vitamin D concentrations are lowest;
- Vitamin D deficiency has been observed to make a contribution to acute respiration misery syndrome; and that case-fatality prices boom with age and with persistent sickness comorbidity, each of which can be related to decrease 25-hydroxyvitamin D concentration (Shui and Leong, 2006).

This can be completed mostly by sun exposure, and to a lesser extent, through dietary intake or supplementation. Vitamin D performs a vital position in regulating sleep patterns, circadian rhythm, improving sleep quality, and in a roundabout way ameliorating sleep apnea. For example, nutrition D receptors and the enzymes that manipulate its activation and degradation had been observed in mind areas concerned in sleep regulation; and vitamin D is likewise concerned in melatonin manufacturing pathways, and may have an effect on stressed legs syndrome and obstructive sleep apnea syndrome (López-Varela et al., 2002).

The exceptional supply of nutrition D is sunshine exposure, however it's miles considerable in numerous ingredients which include oily fish, tuna, dairy, and egg yolk. Grant et al. advocated supplementation in people with maximum hazard of COVID-19 contamination or nutrition D deficiency (10,000 IU/d of nutrition D3) to elevate 25(OH) D concentrations above 40–60 ng/mL (100–150 nmol/L). For remedy of those who come to be infected with COVID-19, higher vitamin D3 doses are advocated (Aparicio-Soto et al., 2016).

13.11 ANTIVIRAL ROLE OF POPULAR TRADITIONAL MEDICINE

Traditional antiviral medicinal treatments throughout exclusive cultures are basically primarily based totally on a mixture of numerous practical ingredients and nutraceuticals with lively immunomodulators, polyphenols, anti-inflammatory and anti-oxidation additives. *Armeniacae semen* (apricot seeds), *Cinnamomi Cortex* (Chinese cinnamon), *Glycyrrhizae radix* (liquorice root), and *Ephedrae herba* form a Japanese conventional remedy called "Maoto," which is regularly administered orally as granules to adults with seasonal influenza. It has been proven to be nicely tolerated and related to equal medical and virological efficacy to neuraminidase inhibitors in supporting progeny influenza viruses to go away without re-infecting the host cellular. Licorice roots which comprise the lively aspect. Glycyrrhizin, had been proven to inhibit influenza A virus uptake into the cellular, and decreased CCID50 by means of 90%. Licorice and curcumin have lately been reviewed for his or her postulated antiviral capacity (Stark and Madar, 2002). Other not unusual place conventional natural treatments for respiration viruses which have been supported by means of medical proof consist of berries' extracts, *Echinacea*, *Clinacanthus siamensis*, *Punica granatum* (pomegranate), *Psidium guajava* Linn. (guava tea), *Epimedium koreanum* Nakai; *Scutellaria baicalensis* Georgi (Baicalin), and *Peonia lactiflora* Pall. (Bai Shao). Examples in their antiviral position consist of discount in viral replication, enhancement of anti-influenza virus IgG and IgA antibody manufacturing, development of T-cellular characteristic (e.g., stimulation of interferon-gamma production through T-cells), neuraminidase inhibitor, virus budding prevention, inhibition of viral RNA and viral protein synthesis, viral hemagglutination, viral binding to and penetration into host cells. Fungi also are typically utilized in Asian and Chinese remedy to elevate the immune machine (Ahire et al., 2021). For example, *Cordyceps militaris* is a mushroom historically used for numerous pharmaceutical functions in East Asia, which include China, for reinforcing immunity. In a human have a look at 1.5 g/day of *Cordyceps militaris* for four weeks more advantageous the NK cellular interest and lymphocyte proliferation and in part expanded cytokine secretion. Immune improving antiviral mechanisms of conventional remedy mainly roots and fungi are vital preventing and controlling novel influenza viruses, which include COVID-19 (Ahire et al., 2020b; Perona and Botham, 2013).

13.11.1 SELENIUM

Selenium (Se) has additional important health effects, particularly in relation to immune response, viral disease and cancer prevention. Numerous studies suggest that deficiency of Se is accompanied by a loss of immunocompetence. In fact, both cell-mediated immunity and B-cell function can be impaired (Alkhatib, 2020). This failure in the immune system is probably not unconnected with the fact that Selenium is normally found in significant amounts in immune tissues such as liver, spleen and lymph nodes. On the other hand, supplementation with Se, even in 'selenium-replete' individuals, has marked immunostimulant effects, including an enhancement of proliferation of activated T cells and an improvement of NK-cell activity. Selenium also appears to abrogate the age-related deficiency of lymphocytes from the aged host to respond to stimulation by proliferation and differentiation into cytotoxic effector cells (Leporatti and Ivancheva, 2003).

13.11.2 PEPTIDE AND AMINO ACIDS

Promising proof has been proven approximately the restorative and antioxidative position of conventional medicinal herbs and peptides put up trauma or bodily challenges, which can be vital in lung damage pathology. GSH is taken into consideration vital in immune modulation, reworking of the extracellular matrix, apoptosis, and mitochondrial respiratory through its gamma-glutamyl cysteine synthetase heavy and mild subunits oxidant/antioxidant reaction to phenolic antioxidants, and is taken into consideration key to the improvement of an oxidant/antioxidant imbalance in lung irritation (Gobbetti et al., 2010). In every other have a look at it changed into proven that anserine, beta-alanyl-three-methyl-l-histidine, a dipeptide, replenished the unfastened radical scavenging enzymes SOD (superoxide dismutase) and preserved catalase and GSH cofactors, whilst maintaining cellular integrity and homeostasis, collectively with a hematological boom in crimson blood cellular volume-to-attention and an attenuated white blood cell elevation following muscular assignment in healthful men. Dietary consumption of anserine, carnosine dipeptides, and different animal-primarily based totally amino acids which include taurine, creatine, and 4-hydroxyproline sell the immunological defense of human beings towards infections by means of microorganism, fungi, parasites, and viruses (which include coronavirus) through improving the metabolism and capabilities of monocytes, macrophages, and different cells of the immune machine. Plant-primarily based totally peptides from soybean have additionally been proven to modulate cell immune systems (expanded lymphocytes and granulocytes wide variety, expanded CD11b(+) cells and CD56(+) herbal killer cells), alter neurotransmitters (reduced adrenaline and expanded dopamine), and improve mind characteristic (Weidmann and Craik, 2016). However, fish, meat, and chicken are the number one reasserts of immunomodulatory peptides and amino acids, and for this reason they've long been taken into consideration practical ingredients taken to relieve fatigue, respiration, and blood less signs in older people, mainly in Asia.

13.11.3 PLANT CYCLOTIDES

Plant cyclotides are nicely-studied anti-virals, because they may be mimicked for antiviral drug improvement, given their solid chemical structure. They were initially derived from African tea, which was used in traditional African medicine to promote childbearing because to their uteroactive antiviral HIV properties. The defensive mechanisms of plant cyclotides towards infections and pathogens are postulated through stopping malfunctioning of the immune cells by means of increase-inhibiting increase consequences at the human immune machine mainly on lymphocytes (e.g., T-cells), which can motive an over-reactivity of this defense equipment throughout infections. Cyclotides may be acquired from numerous flowers which include Violaceae and Rubiaceae, however are plentiful in numerous different plant families, mainly Cucurbitaceae (e.g., squash, pumpkin, zucchini), Fabaceae (legumes, peas, beans), and Solanaceae (eggplant, tomato, potato, pepper). Therefore, it's miles probably that intake of such ingredients, mainly seasonal consumption performs a defensive position in improving the immune machine and complements antiviral defense mechanisms (Dórea and da Costa, 2005).

13.11.4 COFFEE AND CAFFEINE

Coffee, caffeine, and clearly caffeinated drinks are widely recognized to set off numerous fitness blessings and prevention of sickness. All sorts of espresso intake (otherwise roast beans, fermented or non-fermented leaves) are not unusual place throughout numerous cultures internationally for centuries. Epidemiological proof shows that eating 2–3 cups of espresso every day is related to decreased prevalence of metabolic sicknesses which can be regularly concurrent with a compromised immune machine which includes diabetes. Therefore, its miles viable to mean an advantageous position for caffeine as a beneficial immunomodulator. Nutraceuticals inside espresso have proven exclusive antiviral outcomes, with caffeic acid inhibiting the multiplication of influenza A virus in vitro, while caffeine, quinic acid, and chlorogenic acid do now no longer. Caffeic acid has also been shown to exhibit antiviral activity against herpes simplex virus (DNA virus) and polio virus (RNA virus), as well as to reduce progeny virus output (mostly within three hours of contamination) and to decrease the degeneration of virus-inflamed cells. However, caffeine stated immuno-defensive mechanisms from laboratory *in vivo* and *in vitro* trials had been equivocal. Positive caffeine consequences on innate immunity contain suppression of neutrophil and monocyte chemotaxis, and pro-inflammatory cytokines (which includes TNF-a) from human blood, however caffeine has additionally been stated to suppress antibody manufacturing and human lymphocyte characteristic as indicated by means of decreased T-cellular proliferation and impaired manufacturing of Th1 (IL-2 and interferon-gamma), Th2 (IL-4, IL-5), and Th3 (IL-10) cytokines. Caffeine suppression of LPS has additionally been stated in a latest human have a look at in women with obesity. The latter have a look at additionally observed that caffeine ameliorates the obesity-precipitated metabolic side-consequences following extreme workout way of life intervention which include improved LPS, insulin movement, glucose homeostasis, and androgen stages. This shows that caffeine optimizes the metabolic and immunoprotective blessings while blended with different way of life additives, mainly workout (Dziki et al., 2015).

13.12 CONCLUSION

It can be concluded that there was a significant change towards natural functional foods due to viral infection. The majority of the consumers intensified their consumption of for their perceived medicinal properties. The selection of natural functional foods to prevent or treat viral infection is significantly influenced by gender, age and employment status. It is recommended that studies be done to gather empirical evidence on the efficacy of natural functional foods against viral infection.

ACKNOWLEDGMENT

We thank the MET's, Institute of Pharmacy, BKC, affiliated under Savitribai Phule Pune University, Nashik, and Divine College of Pharmacy, Satana. EDA thanks the Ministry of Tribal Affairs Govt. of India for providing financial assistance in the form of a fellowship (Award No.: 202021-NFST-MAH-01235).

KEYWORDS

- **functional food**
- **human body immunity**
- **malnutrition**
- **nutrition**
- **viral infection**

REFERENCES

Ahire, E. D., & Kshirsagar, S. J., (2021). *Immune Responses Induced by Different Vaccine Platforms Against Coronavirus Disease-19,* 243–257.

Ahire, E. D., Sonawane, V. N., & Surana, K. R., (2020a). Role of drug repurposing in current treatment strategies against COVID-19; systemic review. *Pharm Reson.*, 24–29.

Ahire, E. D., Talele, S. G., & Shah, H. S., (2020b). Nanoparticles as a promising technology in microbial pharmaceutics. *Applied Pharmaceutical Science and Microbiology.* Apple Academic Press.

Ahire, E., Sonawane, V., Surana, K., & Talele, G., (2021). Drug discovery, drug-likeness screening, and bioavailability: Development of drug-likeness rule for natural products. *Applied Pharmaceutical Practice and Nutraceuticals.* Apple Academic Press.

Alkhatib, A., (2020). Antiviral functional foods and exercise lifestyle prevention of coronavirus. *Nutrients, 12*(9), 2633.

Alyane, M., et al., (2008). Cardioprotective effects and mechanism of action of polyphenols extracted from propolis against doxorubicin toxicity. *Pakistan Journal of Pharmaceutical Sciences, 21*(3).

Aparicio-Soto, M., et al., (2016). Extra virgin olive oil: A key functional food for prevention of immune-inflammatory diseases. *Food & Function, 7*(11), 4492–4505.

Banik, S. P., et al., (2021). Functional foods beyond nutrition: Therapeutic interventions to combat COVID-19 and other viral diseases. *Journal of Food Bioactives, 15.*

Bikle, D. D., et al., (2002). Insulin-like growth factor I is required for the anabolic actions of parathyroid hormone on mouse bone. *Journal of Bone and Mineral Research, 17*(9), 1570–1578.

Dang, A. T., & Marsland, B. J., (2019). Microbes, metabolites, and the gut-lung axis. *Mucosal Immunology, 12*(4), 843–850.

Dórea, J. G., & Da Costa, T. H. M., (2005). Is coffee a functional food? *British Journal of Nutrition, 93*(6), 773–782.

Douglas, S. A., & Ohlstein, E. H., (2000). Human urotensin-II, the most potent mammalian vasoconstrictor identified to date, as a therapeutic target for the management of cardiovascular disease. *Trends in Cardiovascular Medicine, 10*(6), 229–237.

Dziki, D., et al., (2015). Ground green coffee beans as a functional food supplement – Preliminary study. *LWT – Food Science and Technology, 63*(1), 691–699.

Englyst, K. N., et al., (2000). Starch analysis in food. *Encyclopedia of Analytical Chemistry*, 4246–4262.

Fearne, A., & Hughes, D., (1999). Success factors in the fresh produce supply chain: Insights from the UK. *Supply Chain Management: An International Journal.*

Gobbetti, M., et al., (2010). Functional microorganisms for functional food quality. *Critical Reviews in Food Science and Nutrition, 50*(8), 716–727.

Grimble, R. F., (1997). Effect of antioxidative vitamins on immune function with clinical applications. *International Journal for Vitamin and Nutrition Research (Internationale Zeitschrift fur Vitamin-und Ernahrungsforschung. Journal International de Vitaminologie et de Nutrition), 67*(5), 312–320.

Hodgkinson, C. A., et al., (1993). Mutations at the mouse microphthalmia locus are associated with defects in a gene encoding a novel basic-helix-loop-helix-zipper protein. *Cell, 74*(2), 395–404.

Huang, S., et al., (2021). fMRI-based decoding of visual information from human brain activity: A brief review. *International Journal of Automation and Computing, 18*(2), 170–184.

Jayawardena et al., (2020). Enhancing immunity in viral infections, with special emphasis on COVID-19: A review. *Diabetes & Metabolic Syndrome: Clinical Research & Reviews, 14*(4), 367–382.

Keservani, R. K., & Sharma, A. K., (2014). Flavonoids: Emerging trends and potential health benefits. *Journal of Chinese Pharmaceutical Sciences, 23*(12), 815.

Keservani, R. K., Kesharwani, R. K., Sharma, A. K., & Jarouliya, U., (2015). Dietary supplements, nutraceutical and functional foods in immune response (immunomodulators). In: Debasis, B., Harry, G. P., & Anand, S., (eds.), *Nutraceutical and Functional Foods in Human Life and Disease Prevention* (pp. 343–358). Boca Raton, Fl. CRC Press, Taylor and Francis, Chapter 20. ISBN: 9781482237214.

Keservani, R. K., Kesharwani, R. K., Sharma, A. K., Vyas, N., & Chadoker, A., (2010b). Nutritional supplements: An overview. *International Journal of Current Pharmaceutical Review and Research, 1*(1), 59–75.

Keservani, R. K., Kesharwani, R. K., Vyas, N., Jain, S., Raghuvanshi, R., & Sharma, A. K., (2010a). Nutraceutical and functional food as future food: A review. *Der Pharmacia Lettre, 2*(1), 106–116.

Keservani, R. K., Sharma, A. K., & Kesharwani, R. K., (2017). An overview and therapeutic applications of nutraceutical and functional foods. *Recent Advances in Drug Delivery Technology*, 160–201. ISBN:9781522507543.

Keservani, R. K., Sharma, A. K., & Kesharwani, R. K., (2020). "*Nutraceuticals and Dietary Supplements: Applications in Health Improvement and Disease Management*. CRC Press. ISBN: 9781771888738.

Kesharwani, R. K., Singh, P., & Keservani, R. K., (2016). Green coffee bean extract and chlorogenic acids: Chemistry and novel antioxidant benefits. In: Bagchi, D., Moriyama, H., & Swaroop, A., (eds.), *Green Coffee Bean Extract in Human Health* (pp. 1–17). Boca Raton, Taylor and Francis, USA. Chapter 1. ISBN: 9781498716376.

Konstantinopoulos, P. A., et al., (2020). Germline and somatic tumor testing in epithelial ovarian cancer: ASCO guideline. *Journal of Clinical Oncology, 38*(11), 1222–1245.

Leporatti, M. L., & Ivancheva, S., (2003). Preliminary comparative analysis of medicinal plants used in the traditional medicine of Bulgaria and Italy. *Journal of Ethnopharmacology, 87*(2–3), 123–142.

Ling, W., et al., (2015). Not all contexts are created equal: Better word representations with variable attention. In: *Proceedings of the 2015 Conference on Empirical Methods in Natural Language Processing* (pp. 1367–1372).

López-Varela, S., et al., (2002). Functional foods and the immune system: A review. *European Journal of Clinical Nutrition, 56*(3), S29–S33.

Marjani, M., et al., (2017). Big IoT data analytics: Architecture, opportunities, and open research challenges. *IEEE Access, 5*, 5247–5261.

Martineau, A. R., et al., (2017). Vitamin D supplementation to prevent acute respiratory tract infections: Systematic review and meta-analysis of individual participant data. *BMJ, 356*.

Martirosyan, D. M., & Singh, J., (2015). A new definition of functional food by FFC: What makes a new definition unique? *Functional Foods in Health and Disease, 5*(6), 209–223.

Moriguchi, S., & Mikako, M., (2000). *Vitamin E and Immunity*, *1*(1), 305–336.

Mortaz, E., et al., (2013). Probiotics in the management of lung diseases. *Mediators of Inflammation, 2013*.

Mvubu, N. E., et al., (2016). Canonical pathways, networks and transcriptional factor regulation by clinical strains of *Mycobacterium tuberculosis* in pulmonary alveolar epithelial cells. *Tuberculosis, 97*, 73–85.

Pannu, N., & Bhatnagar, A., (2019). Resveratrol: From enhanced biosynthesis and bioavailability to multitargeting chronic diseases. *Biomedicine & Pharmacotherapy, 109*, 2237–2251.

Perona, J. S., & Botham, K. M., (2013). Olive oil as a functional food: Nutritional and health benefits. In: *Handbook of Olive Oil* (pp. 677–714). Springer, Boston, MA.

Pinto, D., et al., (2020). Cross-neutralization of SARS-CoV-2 by a human monoclonal SARS-CoV antibody. *Nature, 583*(7815), 290–295.

Rahman, M. T., & Idid, S. Z., (2021). Can Zn be a critical element in COVID-19 treatment? *Biological Trace Element Research, 199*(2), 550–558.

Rane, B. R., Bharath, M. S., Patil, R. R., Keservani, R. K., & Jain, A. S., (2020b). Novel approaches in nutraceuticals. In: Rajesh, K. K., Raj, K. K., & Anil, K. S., (eds.), *Enhancing the Therapeutic Efficacy of Herbal Formulations Through Novel Drug Delivery Systems* (pp. 241–266). IGI Global International Publisher, Pennsylvania, USA, Chapter 11. ISBN13: 9781799844532.

Rane, B. R., Patil, A. S., Keservani, R. K., & Jain, A. S., (2020a). Novel approaches in herbal formulation. In: Rajesh, K. K., Raj, K. K., & Anil, K. S., (eds.), *Enhancing the Therapeutic Efficacy of Herbal Formulations Through Novel Drug Delivery Systems* (pp. 43–68). IGI Global International Publisher, Pennsylvania, USA, Chapter 2. ISBN13: 9781799844532.

Rane, B. R., Tadavi, S. A., & Keservani, R. K., (2020c). Naturopathy. In: Anil. K. S., Raj, K. K., & Surya, P. G., (eds.), *Herbal Product Development* (pp. 321–347). Apple Academic Press, CRC Press, Taylor & Francis Group, Chapter 12. ISBN: 9781771888776.

Rayman, M. P., (2000). The importance of selenium to human health. *The Lancet, 356*(9225), 233–241.

Reygaert, W. C., (2018). An overview of the antimicrobial resistance mechanisms of bacteria. *AIMS Microbiology, 4*(3), 482.

Roberfroid, M. B., (2000). Prebiotics and probiotics: Are they functional foods? *The American Journal of Clinical Nutrition, 71*(6), 1682S–1687S.

Roberfroid, M. B., (2002). Global view on functional foods: European perspectives. *British Journal of Nutrition, 88*(S2), S133–S138.

Ross, S. E., et al., (2000). Inhibition of adipogenesis by Wnt signaling. *Science, 289*(5481), 950–953.

Scholz-Ahrens et al., (2007). Prebiotics, probiotics, and synbiotics affect mineral absorption, bone mineral content, and bone structure. *The Journal of Nutrition, 137*(3), 838S–846S.

Sharma, A. K., Keservani, R. K., & Gautam, S. P., (2020). *Herbal Product Development.* Apple Academic Press, CRC Press, Taylor & Francis Group. ISBN: 9781771888776.

Shui, G., & Leong, L. P., (2006). Residue from star fruit as valuable source for functional food ingredients and antioxidant nutraceuticals. *Food Chemistry, 97*(2), 277–284.

Stark, A. H., & Madar, Z., (2002). Olive oil as a functional food: Epidemiology and nutritional approaches. *Nutrition Reviews, 60*(6), 170–176.

Surana, K. R., Ahire, E. D., Sonawane, V. N., Talele, S. G., & Talele, G., (2021). Informatics and methods in nutrition design and development. *Natural Food Products and Waste Recovery: Healthy Foods, Nutrition Design, and Extraction of Valuable Compounds,* 33.

Syngai, G. G., et al., (2016). Probiotics-the versatile functional food ingredients. *Journal of Food Science and Technology, 53*(2), 921–933.

Taliun, D., et al., (2021). Sequencing of 53, 831 diverse genomes from the NHLBI TOPMed program. *Nature, 590*(7845), 290–299.

Vickers, N. J., (2017). Animal communication: When I'm calling you, will you answer too? *Current Biology, 27*(14), R713–R715.

Vitapole, D., (2000). *Functional Dairy Products*. France: John Libbey Eurotext.

Weidmann, J., & Craik, D. J., (2016). Discovery, structure, function, and applications of cyclotides: Circular proteins from plants. *Journal of Experimental Botany, 67*(16), 4801–4812.

Wilson, M. R., et al., (2019). Clinical metagenomic sequencing for diagnosis of meningitis and encephalitis. *New England Journal of Medicine, 380*(24), 2327–2340.

Wintergerst, E. S., et al., (2007). Contribution of selected vitamins and trace elements to immune function. *Annals of Nutrition and Metabolism, 51*(4), 301–323.

Xia, T., et al., (2020). Nutrients and bioactive components from vinegar: A fermented and functional food. *Journal of Functional Foods, 64*, 103681.

Yurie, H., Ikeguchi et al., (2017). The efficacy of a scaffold-free bio 3D conduit developed from human fibroblasts on peripheral nerve regeneration in a rat sciatic nerve model. *PLOS One, 12*(2), e0171448.

CHAPTER 14

ROLE OF FUNCTIONAL FOODS IN GROWTH AND DEVELOPMENT

SHANTI BHUSHAN MISHRA,[1] SHRADHANJALI SINGH,[2] AMIT KUMAR SINGH,[3] ANIL KUMAR SINGH,[3] and DIVYA RANI SHARMA[4]

[1]*Department of Pharmacognosy, United Institute of Pharmacy, Prayagraj, Uttar Pradesh, India*

[2]*Department of Pharmaceutical Chemistry, United Institute of Pharmacy, Prayagraj, Uttar Pradesh, India*

[3]*Department of Pharmaceutics, United Institute of Pharmacy, Prayagraj, Uttar Pradesh, India*

[4]*Department of Pharmacology, United Institute of Pharmacy, Prayagraj, Uttar Pradesh, India*

ABSTRACT

The value of functional foods was ascertained by the nutritional value derived from food which was known to prevent several diseases. Functional foods by diet can maintain the health determinants and hence lessen the possibility of diseases. Several physiological systems including the antioxidative system and gastrointestinal system, have been involved in the utilization of functional foods. As per the published and explored scientific work on functional food, health claims can be claimed by the consumers in terms of reduced risk associated with any disease. This chapter illustrates insight related to the various categories of functional foods involved in the growth and development of human being.

Applications of Functional Foods in Disease Prevention. Raj K. Keservani & Eknath D. Ahire, PhD (Eds.)

14.1 INTRODUCTION

The idiom 'functional food' was devised in the 1980s in Japan at the early stage. The market of functional food augments steadily, getting around $21.3 billion in the United States and $8 billion in Europe (Kobayashi et al., 2017).

Despite of many definitions of functional food, the most promising definition is "a food considered for preventing disease along with their nutritive value." Nutraceuticals, probiotics and functional foods have common characteristics with respect to essential nutrients. These fortified, enriched foods can be consumed on a regular basis through diet (Keservani et al., 2010a, b, 2017, 2020).

The food which has certain physiologic effect is considered to be a macronutrient and if the food is taken more than the given recommendation then it is called micronutrient. Apart from providing nutritional value several food components including oligosaccharides capable of modulating body functions and lessens the risk associated with several diseases.

14.2 ROLES OF FUNCTIONAL FOODS

Health-related facets can be linked with developed facets related with nutrition considers (Figure 14.1) (Kumar et al., 2015; Georgieva et al., 2018):

- Maintenance of health determinants including immune system, gastrointestinal health, mental health, physical health, age-related health conditions;
- Lower risk linked with obesity;
- Diminished risk associated with diet-related manifestations including cardiovascular disease, metabolic disease, and musculoskeletal disease.

14.3 EARLY DEVELOPMENT AND GROWTH

14.3.1 MOTHER'S DIET

Short and long period progress of the child depends upon the intake of nutrient in pregnancy and lactation. Despite of the nutrition provided certain diseases. Factors influencing the time course of pregnancy, childbirth and components of breast milk depend upon intake of energy and proteins,

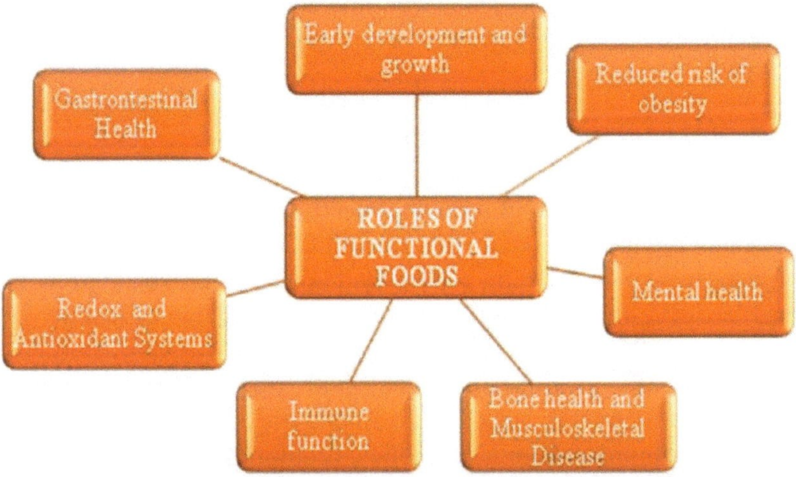

FIGURE 14.1 Roles of functional foods.

including amino acids, polyunsaturated fatty acids (PUFAs) micronutrients considering, zinc, folic acid, iodine, and iron.

14.3.2 DIET OF CHILDREN AND ADOLESCENTS

Behavioral functions and cognitive activities considering sensory and visual abilities may be affected by intake of diet in childhood and adolescence, specifically polyunsaturated fatty acids, and trace minerals like zinc, iron, and iodine. Functional food will provide the possible components considering calcium, vitamin D, vitamin K, fluoride, and prebiotic fructans, which emphasizes on the skeletal development at the adolescence stage. Growth of the gastrointestinal system in children and adolescents can get triggered by food components considering probiotics, prebiotics, oligosaccharides, and inulin.

14.3.3 IMMUNE FUNCTION

Immune function is considered to be a vital part of the biological system and can be affected by improper nutrition. For optimal immune response possible components which can contribute includes vitamins, antioxidants, copper, zinc and manganese, PUFAs, nucleotides, L-arginine, nucleosides, and probiotics.

14.3.4 GASTROINTESTINAL STATUS

The edge between intake of food and metabolic functions offers a targetive approach in the progressive improvement of functional foods. The ability of GI system to work efficiently can be maintained by preventing the invasion of bad bacteria by maintaining an appropriate balance of healthy bacteria (Roberfroid, 1996).

The improvement in the functional food is positioned as per the use of components which was considered to amend the composition and metabolic action of the gut microflora is prebiotics, probiotics, and synbiotics.

Feasible health profit of probiotics considers:

- Declined prevalence and relentlessness of GI illness;
- Largely enhancement in gut function, considering declined in constipation and diarrhea.

In view of improving the condition associated with the gut, growth stimulators suggesting prebiotics and undigestible foods or certain bacterial species which can modify the metabolic actions can be considered. These improvements will lead in enhancement of immune system and hence declining the risk associated with colon cancer. Since the risk associated with colon cancer is being decreased the production of small chain fatty acids will augment the absorption of magnesium and calcium.

The food items which contain prebiotics are dairy products, breakfast cereals and bars, baked goods, meat products, and breads salad dressings.

14.3.5 MENTAL HEALTH CONDITION

Promotion of optimum health determinants including mental health condition by using functional foods could affect the mental act and persuades several behavior suggesting cognitive performance, vitality and mood, short-term memory, and reaction to stress.

With respect to mental ability suggesting memory and decision time, glucose has some beneficial influence. In view of reducing the pain sucrose can be benefitted. In view of augmenting the cognitive performance, caffeine can augment vigilance, psychomotor ability, memory, and reaction time in adolescents and vitamin B does in older patients.

Meals associated with high amounts of carbohydrate are prone to exhibit sleep and calmness. In view of overcoming this issue, tryptophan, a category of amino acid can decline the time required for sleep, along with tyrosin and tryptophan can solve the jet lag issue. Another problem associated with less amount of sleep is depression. In view of improving the condition of depression folic acid, n-3 fatty acids and S-adenosylmethionine (SAMe) has marked effect on augmenting the condition.

14.3.6 REDOX AND ANTIOXIDANT SYSTEM

Human system is incorporated with antioxidant system which entailed intake of nonvitamin foods including polyphenols and erstwhile natural antioxidants. Various activities and protection of cells inside the body depend upon these systems. These systems are associated with pathologies of cells. Scientific explorations were well established and considers as hypotheses in which exact mechanism of dietary antioxidants, their beneficial effects were unjustified and hence need to be explored rationally. But if the antioxidants were consumed as in the form of fresh fruits and vegetables (Ziegler, 1991), they have a marked increase in their antioxidant potential towards protecting cells in the body.

Aging is the major factor which governs the physical, metabolic activities in the human body. It is related with the chronic conditions suggesting Parkinson's disease, cardiovascular disease, cataract, cancer, age-related macular degeneration, osteoarthritis, and Alzheimer's disease. These conditions may exaggerate the conditions associated with enhanced oxidative stress.

Together with oxidative stress, some defense components include:

- Antioxidant related enzymes;
- Manganese, copper, selenium, minerals, and trace elements are co-factors for antioxidant enzymes;
- Glutathione, which is considered a reducing agent, is oxidized and reduced reversibly.

The useful and potential components for functional food considers vitamins C as antioxidants (Keservani and Sharma, 2014). Effective antioxidants suggesting plant related foods that consists of components like cranberry, goji, blueberry, acai, pomegranate, mangosteen, grapes, tomato are now being surveyed with the help of the functional food industry.

Challenging aspects related with variation in mental ability is one of the prime focus nowadays especially for older peoples. Prevention of these challenges can be done by considering a variety of food components. This food component including fatty acids diminishes the risk of depression. The problem of memory and mental concentration can be improved by the use of gingko biloba (Keservani et al., 2016b).

14.4 PHYSICAL PERFORMANCE

Physically active is the most productive approach for functional food. Athletes being physically active can get help from liquid food containing suitable content of energy substrates, electrolytes and fluid balance. This type of food is appropriate and digestible also. To ascertain the nutritional desires this liquid food can help people to consume food appropriately.

Potential for functional foods can be gained by influencing active performance and recovery on account of usage of different types of carbohydrates considering a range of glycemic indices. In training and competition adequate amount of intake can be assured by micronutrient supplementation of functional foods. Potentially augmenting the physical performance, the food components considering caffeine, specific amino acids and carnitine is important.

14.4.1 OBESITY

Factors affecting weight management considers declining energy intake that can be accomplished by dipping the energy density of food, plummeting appetite, augmenting satiety or dropping fat absorption. Functional food components that includes chitosan, conjugated linoleic acid, caffeine, green tea, calcium, diglycerides and capsaicin, medium-chain triglycerides.

14.4.2 CARDIOVASCULAR DISEASE

14.4.2.1 IMPROVING LIPID PROFILE

Functional foods which are low in saturated and Trans fatty acids help to recommend optimal low-density lipoprotein cholesterol levels. Monounsaturated considering olive oil and polyunsaturated linoleic and alpha-linolenic acid can diminish plasma (LDL) level.

14.4.3 DIABETES MELLITUS

Increased risk of type II diabetes is linked with overweight and less physical activity. With the help of vegetables, whole grain foods, fruits, foods which are low in fat, starchy foods the problems can be lessened. Augmenting effects on insulin sensitivity and glucose metabolism can be taken by using soluble fibers like psyllium and inulin.

14.4.4 MUSCULOSKELETAL DISEASE

Osteoarthritis a debilitating chronic condition affecting older people is associated with adverse effect of using non-steroidal anti-inflammatory drugs and hence there is a necessity for safe and alternative dietary approaches. Substances such as chondroitin and glucosamine are used as nutraceuticals. There is also evidence of collagen hydrolysate in functional food components.

14.4.5 BONE HEALTH

Bone health is depended on the variety of nutrients, including vitamin D, calcium, vitamin K, magnesium, zinc, manganese, and copper. Vitamin D as function in bone growth and bone mineral density. High intake of vegetable and fruit intake is also linked with bone protection.

14.4.6 CANCER PREVENTION

A long-term and dynamic process, involving many complex factors known as metastasis had some epidemiological data have provided convincing evidence that dietary factors can influence carcinogenesis. Laboratory research can prevent cancer (Cencic and Chingwaru, 2010; Balsano and Alisi, 2009). Future acceptance of food bioactive components as chemopreventive agents will be proven by research (Pan et al., 2008).

14.5 FUNCTIONAL FOODS OF NATURAL ORIGIN

14.5.1 BROCCOLI TOMATO SALAD

Broccoli is a rich source of vitamin C, soluble fiber, diindolylmethane and selenium. The diindolylmethane found in broccoli is a potent innate immune system modulator with anti-bacterial, anti-viral, and anti-cancer effects. As public health recommendations intake of tomato and broccoli can be effective for slowing tumor growth than alone either and supports augment the intake of a variety of plant components (Verbeke et al., 2005). It can be served once a day in lunch (Canene-Adams et al., 2007).

14.5.2 TOMATO DRINK

Formulated tomato drink affects the level of carotenoid plasma and lymphocytes especially phytoene lycopene, phytofluene and beta-carotene. It also provides defense against DNA damage in lymphocytes. The Serving size of the tomato drink can be two glasses per day (Porrini et al., 2005).

14.5.3 OATMEAL

Beneficial effect of oatmeal in coronary heart disease (CHD) will reveal the link between the taking of wholegrain foods and CHD mortality, morbidity in participants earlier diagnosed with CHD or with existing risk factors for CHD was evaluated by Kelly et al. (2007). Oat bran meal high in beta-glucan had a low glycemic response and acted as an agent that decreased the postprandial glycemic response of an oral glucose load. Oatmeal can be consumed during breakfast with fat free milk for dietetics (Tapola et al., 2005).

14.5.4 APPLE PEEL AND JUICE

Constituents in apple juice include dihydrochalcones, hydroxycinnamic acids, flavonols (quercetin glycosides), oligomeric procyanidins and, catechins as well as anthocyanins in red apples and triterpenoids in apple peel. Previous scientific data suggests valuable things on health against cardiovascular disease (Keservani et al., 2016a), asthma and pulmonary (Keservani et al., 2015) dysfunction, diabetes, obesity (Jarouliya and Keservani 2019a), and cancer. The utility of apple component, apple juice, apple peel, apple extracts and its components was calculated by Gerhauser and accomplished that extracts and components as oligomeric procyanidins influence multiple mechanism of actions relevant for cancer prevention. Apple peel every day and apple juice thrice a week with fruit salad may be consumed (Gerhauser et al., 2008).

14.5.5 BERRIES

Berries are a natural source of anthocyanin, antioxidants (Singh et al., 2017a) and have a broad spectrum of therapeutic functions. This includes advancing age-induced oxidative stress, cardiovascular disorders, inflammatory responses, etc. The effect of an unique mixture of selected berry extracts known as 'optiberry' (wild bilberry, wild blueberry, cranberry, raspberry seeds strawberry, and elderberry) was reported by Zafra-Stone et al. (2007) and optiberry was found to have high antioxidant potency as demonstrated by its high oxygen radical absorption capacity value, against *Helicobacter pylori* a harmful pathogen responsible for various gastrointestinal diseases. Berries can be consumed every day with other fruits. Therefore, strawberries can be eaten regularly with other fruits or dessert as they reduce oxidative damage to LDL while maintaining blood lipids lowering and improving diet palatability (Jenkins et al., 2008).

14.5.6 ORANGE JUICE

Orange juice can be consumed twice a day (Guarnieri et al., 2007).

14.5.7 GRAPES AND GRAPE JUICE

Beneficial effects of grapefruit juice have been reported in the prevention of certain inflammatory-mediated diseases, including heart disease. The active components from grape extracts are polyphenols such as phenolic acids, resveratrol, flavonoids, and anthocyanins. All hold potent antioxidant properties and have been shown to reduce LDL-cholesterol oxidation and platelet aggregation. Consumption of grapes once a day and grape juice thrice a week is recommended (Leifert and Abeywardena, 2008; Singh et al., 2017b; Jarouliya and Keservani, 2019b).

14.5.8 PAPAYA JUICE

Benzyl ITC (BITC), a major ITC compound isolated from papaya (*Carica papaya* Linn.) fruit exhibits the relationship between cell cycle regulation and apoptosis induction suggested by Nakamura. Serving size may be one glass per day (Nakamura, 2009).

14.5.9 BLACKBERRY JUICE

A compound found in blackberries, cyanidin-3-glucoside (C3G), has chemopreventive and chemotherapeutic activity. It was investigated that a purified compound of anthocyanin inhibits tumor promoter-induced carcinogenesis and tumor metastasis *in vivo* (Ding et al., 2006).

14.5.10 CITRUS FRUIT SMOOTHIE

Soy milk and soy protein along with milkshakes can be used for people having lactose intolerance (Dunn, 2000).

14.5.11 COCOA POWDER

Cocoa powder is plentiful with polyphenols, such as catechins and procyanidins and it has inhibited oxidized LDL and atherogenesis in a variety of test models. The changes in Plasma LDL and HDL cholesterol and oxidized LDL concentrations in normal and hypercholesterolemic subjects after ingestion of different amounts of cocoa powder were studied by Baba et al. (2005). Polyphenolic substances contribute to lowering LDL cholesterol, raising HDL cholesterol and suppressing oxidized LDL (Zafra-Stone et al., 2007).

14.5.12 ALMOND

It is reported that almond intake is associated with an improvement in cancer and CVD. Almond peels are rich in polyphenols (ASP) which may contribute to these therapeutic benefits. It has been reported that these ASP acts as antioxidants and promote quinone reductase (QR) activity, but these actions are depended upon the method of extraction their dose, and interaction with antioxidant vitamins (Chen and Blumberg et al., 2008).

14.5.13 PEANUT AND PEANUT BUTTER

Consumption of peanuts, peanut butter and other nuts is associated with a lower risk of cholecystectomy (Chung-Jyi et al., 2004). Consistent evidence from major epidemiological and clinical studies shows that consuming peanuts regularly can help reduce coronary heart disease (CHD) risk (Kris-Etherton et al., 2001). Peanut consumers have also been shown to have significantly higher intakes of dietary fiber and vitamin E, both valuable nutrients that can help reduce the risk of CHD and lower cholesterol levels (Griel et al., 2004).

14.5.14 GREEN TEA

Investigation was held on free radical scavenging property blend (O^-) generating system *in vitro* by Nakagawa and Yokozawa. Mint added to the tea, makes it a strong diuretic beverage. Green tea may be consumed without milk everyday (Nakagawa and Yokozawa et al., 2002).

14.5.15 RED WINE

Hakimuddin et al. reported that the treatment of MCF-7 breast cancer cells with polyphenols present in red grape wine fraction results in disruption of calcium homeostasis and cell cycle arrest causes cytotoxicity. Red wine may be consumed occasionally or can be used as a flavoring agent for food preparation (He et al., 2008; Hakimuddin et al., 2006).

14.5.16 GARLIC

Garlic has an remarkably high concentration of sulfur compounds compared to other food plants. It also contains 26–30% (fresh weight) or up to 77% (dry weight) of a non-digestible carbohydrate-inulin. The results of the study carried out by Kanner et al. (1998) suggested that the inulin and garlic supplementation could stimulate the intestinal microflora growth, particularly in foods that contains high amount of fat, carbohydrate, and dietary fiber. Small piece of crushed garlic can be consumed along with cooked curries.

14.5.17 FISH OIL

Fish oils are accountable for maintaining normal lipid levels in blood and support the integrity and flexibility of the joints. LipoCell is an emulsification procedure for making nutritional oils that are soluble in water to enhance their absorption (Chung-Jyi et al., 2004). Two teaspoons are recommended as serving size of fish oil.

14.5.18 DARK CHOCOLATE

The influence of flavonoid-rich dark chocolate on aortic stiffness, endothelial function, oxidant status and wave reflections were investigated by Vlachopoulos et al. (2005). Dark chocolate can be consumed occasionally.

14.5.19 FERMENA

It is a revolutionary fermented Prebiotic Superfood (over 50 vegetables, fruits, nuts and grains fermented for about 6 months). It is a fermented blend of lots of beneficial bacteria and enzymes which help to promote maximum healthy absorption in the human system. Fermena should be taken alone or with food one to three times a day.

14.5.20 COW'S MILK AND YOGURT

For modulating mineralization, growth, and bone density, cow's milk is the main source. Protein, calcium and phosphorus augment the absorption and deposition of minerals in the protein matrix of the bone (Ghasemi and Kohnehrouz, 2016). On comparison to mother milk, cow's milk has alterations in the content of ameliorating factors of bone growth. Naturally fermented dairy foodstuffs contain probiotic flora (Ruchin, 2017). They prevent pathogen linkage, stimulate the multiplication of T and B lymphocytes, synthesis of immunoglobulin, and cytokine formation (Ould Yerou et al., 2017). Yoghurts are the best source of calcium, the essential nutrients that prevent osteoporosis (Ghasemi and Kohnehrouz, 2016; Zhang et al., 2018; Rezaei et al., 2016) (Table 14.1).

TABLE 14.1 List of Some Functional Ingredients, Benefits, and Sources

Functional Ingredient	Source/Food	Beneficial Effects
Xylofucans and alginic acid	*Sargassum vulgare*	Antiviral action
Flavonols, myricetin and quercetin anthocyanins, catechins, cyanidins.	Grape juices and peels, and wine (prepared from red grapes); berries and eggplant.	Antioxidant activity, superoxide scavengers, inhibits LDL oxidation.
Calcium	Milk and its derivatives	Reduces the colon cancer, decreases irritative and proliferative effects on intestine.
Capsaicin	Green and red peppers	Superoxide anions scavenger
Carnosoic acid and carnosol	Rosemary extracts	Inhibits oxidation of oils
Carotenoids and lycopene	Carrots, tomatoes, bell peppers, fruits, green leafy vegetables, and palm oils.	Antioxidant, anti-inflammatory, anti-carcinogenic actions, inhibits LDL oxidation, reduce the risk of cardiovascular diseases.
Conjugated linoleic acids	Turkey, lamb, beef	Reduces atherosclerosis breast cancer, and type II diabetes
Curcuminoids: Di feruloyl methane, *p*-hydroxycinnamoyl methane and bis *p*-hydroxycinnamoyl methane.	*Curcuma longa*	Decreases lipid peroxidation, avoid ischemic-hypoxic.
Genistein daidzein, and isoflavones.	Soybean and their derivatives	Inhibits lipid peroxidation, reduces cardiovascular and cancer risks.
Fibers (soluble and insoluble) phytochemicals.	Fortified juices, oat cereals, *Chondrus crispus* (algae).	decreases total and LDL and VLDL cholesterol.
Flavonoids (flavanones, flavones, flavonols and flavan-3-ols).	Celery, parsley, citrus, onions, tea, green beans, tomatoes, cocoa, apples, berries, certain beans, and chocolate.	Antiproliferative, anti-oxidant, anti-hypertensive, anti-thrombotic, anti-carcinogenic, cell cycle arrest, inhibition of phase-1 enzymes, improvement of vascular tone, inhibition of LDL-oxidation.
Gallic acid, quercetin, sulfides and thiols, and organosulfur compounds.	Garlic, broccoli, and onion	Reduces blood pressure
Lignans	Flax seeds, fruits, and vegetables	Estrogenic

TABLE 14.1 *(Continued)*

Functional Ingredient	Source/Food	Beneficial Effects
Monounsaturated fatty acids and Phenolic acids.	Coffee (Kesharwani et al., 2016), black tea, green tea, fruits, cereal bran, virgin olive oil.	Anti-inflammatory
Prebiotic	Garlic, banana, onions, milk, honey, artichokes.	Stimulate growth of helpful bacteria in the colon, thus improves host health.
Probiotics	Fermented dairy products	Support GIT health, boost immunity.
Ascorbic acid, resveratrol, various grape polyphenols and ginsenosides.	Red wines, grapes, ginseng, citrus fruits.	Reduces blood pressure, also prevents advanced malignance.
Sesaminol	Sesame seeds and sesame oil, rice, and rice oil.	Inhibits LDL oxidation and decreases cancer risk.
Stilbenes	Grapes, peanuts	Cardioprotective, antioxidant, lifespan extension.
Tocopherols, coenzyme Q10, tocotrienols, ω-3 fatty acids, phytosterols.	Nuts wheat, almonds, oils (from fish, rice, soy, and olive), fat and margarine.	Reduces blood cholesterol level, inhibits oxidation of lipids.

Source: Shahidi (2009); Holm (2003); Ferrari (2007); Gry et al. (2007), Plaza et al. (2008); and Patil et al. (2009).

14.6 PRODUCTION OF FUNCTIONAL FOODS

Approaches available for preparing functional food product:

- Identified as producing a detrimental effect when ingested (e.g., an allergenic protein).
- Elevating the content of a constituent present in nutrient to extent at which it will urged anticipated consequence [e.g., fortification with a micronutrient to achieve a daily intake that is higher than the recommended daily consumption but compatible with the dietary guidelines to reduce disease risk (Block, 1993)], or enhancing the content of a non-nutritive constituent to an extent known to produce a favorable consequence.
- Constituents showing beneficial effects but not found in food as inevitably a macronutrient or a micronutrient, can be introduced.
- Macronutrient (e.g., fats) having excessive use but having deleterious effects can be replaced by molecules having beneficial effects

[e.g., chicory inulin such as Rafticream (ORAFTI, Tienen, Belgium) (Franck-Frippiat, 1992)].
- Food components having functional effect or low risk potential should be of high bioavailability or stability.

The indication of such promising consequence must be relied on technology.

14.7 TRADITIONAL TECHNOLOGY ADAPTED FOR FOOD PROCESSING

14.7.1 *FORMULATION AND BLENDING*

Functional foods can be developed by using formulation and blending technology in food processing industries due to simplicity and economically sound. Vitamins deficiency, iodine and iron can be controlled by functional foods developed using this technology. In the early 1920s salt iodization was introduced (Burgi et al., 1990; Marine and Kimball, 1920). It was in common practice of fortification of cereal with a number of vitamins in the early 1940s (Kyritsi et al., 2011). Margarine was enriched with vitamin A (FAO and WHO, 2006; WHO, 2006), high cardiovascular risk consumed this after incorporation with phytosterols (Laforest et al., 2007). Folic acid reinforcement of wheat (Samaniego-Vaesken et al., 2010) became widespread later.

Vitamins, probiotics, bioactive peptides, antioxidants, etc., are considered in dietary functional nutrients and the list is growing gradually (Wildman, 2006). From the classical supplemented milks (Kim et al., 2010; Alzate et al., 2010) and yogurts (Alliet et al., 2007; Karaaslan et al., 2011; Zare et al., 2011), probiotics (Puccio et al., 2007), vitamins (Chávez-Servín et al., 2008), and long chain polyunsaturated fatty acids (Chávez-Servín et al., 2009) for essential nutrients to infants for ideal growth and development, until juices mainly augmented with flavonoids (González-Molina et al., 2009), vitamins (Rivas et al., 2007) and resveratrol (González-Barrio et al., 2009), snacks (Da Costa et al., 2009), pastas rich in legumes (Petitot et al., 2010) and meats enriched with large number of bioactive compounds (Jiménez-Colmenero et al., 2010; Zhang et al., 2010) are being flourished by formulation.

14.7.2 *TECHNIQUES FOR CULTIVATION AND ANIMAL BREEDING*

Nutrients obtained from agriculture and livestock are considered as primary sources needed by human. By including a diversified diet of vegetables, fruits, meat and fish, micronutrient deficiency can be addressed (FAO and WHO, 2001).

Biotechnology provides worthy substitution where significant and required nutrients in food products are not available from agronomic and breeding approaches (Zhao and Shewry, 2011).

Golden Rice has been considered as most significant upgraded crop product (Ye et al., 2000), rice endosperm has 1.6 μg/g total carotenoids. Golden Rice second generation (Golden Rice 2) (Paine et al., 2005), 37 μg/g of total carotenoids. Studies predicted that the b-carotene converted into retinol by incorporating with vitamin A, found in Golden Rice 2 (Tang et al., 2009) and in 2012 it was anticipated to reach market (Potrykus, 2010).

The reinforcement of vitamin A has been accomplished also in other food crops like maize (Zhu et al., 2008) and potato (Tanumihardjo et al., 2008). Folates have been nourished with tomato (De La Garza, Gregory and Hanson, 2007) and rice grain (Storozhenko et al., 2007), corn seeds and soybean seeds (Karunanandaa et al., 2005) in vitamin E. Fatty acids in oil seeds which were converted and stored in the form of triacylglycerol have been replaced by synthesizing LC-PUFAs in plants (reviewed by Venegas-Caleron et al., 2010).

Animal breeding also acquires the strong consideration to obtain upgraded food products. Matsushita et al. (2007) conceded out research on fatty acids profile and physic-chemical profile of milk samples of Saanen goats fed with vegetable oils like soybean, canola and sunflower. In addition (Laible, 2009), to enhance the victual quality of milk and it´s serving into dairy products many methodological suggestions have been made. Woods and Fearon (2009) reviewed the sources of fatty acids. Juniper et al. (2009) resolute the amount of total selenium and the ratio of consisting of selenomethionine and selenocysteine.

14.8 TECHNOLOGIES FOR PREPARING FUNCTIONAL FOODS HAVING LESS DETRIMENTAL EFFECTS

14.8.1 MICROENCAPSULATION

Microencapsulation is the technique in which the entrapment of small solid particles will be coated by using polymers (Thies, 1987). Polymeric matrix creates embedding effect around capsule, which fabricated a micro-environment inside for possible interaction between internal and external sides (Borgogna et al., 2010). This technique is also beneficial in providing shielding cover for a number of components like enzymes and hormones having biological interest from bacterial, yeast and animal origin cells (Thies, 2005). In the futuristic fields of biopharmaceutics and biomedicine, microencapsulation technique has been extensively studied and utilized, for executing research (Smidsrød and Skjak-Braek, 1990). For the production of valued nutraceuticals, microencapsulation technique offers advantages.

There have been numerous encapsulation processes deemed generally applicable for bioactive dietary components. The reason behind this is that each bioactive food constituents have its own distinctive molecular structure (Augustin and Hemar, 2009). Encapsulation procedures should also have particulars characteristic that remains unaffected by the surrounding environment along with the special qualities showed compatibility with the bioactive (Augustin and Hemar, 2009).

To maintain the stability and functionality of bioactive compounds, encapsulation techniques must involve all factors like particle size required, techniques used, etc. In encapsulation technique, a serious concern should be given on worthwhile delivery of bioactive food components. It should not be affected only unpredictable events but also protected from the adverse circumstances. Solvent for the bioactive components, are a substantial impediment to the effective delivery of bioactive food components (de Vos et al., 2010). Many food elements containing bioactive food component that has been introduced. As a result, the encapsulation approach must ensure that the bioactive component is protected during the whole production, storage, and transportation process (Gibbs et al., 1999). Another need is that the encapsulation method facilitates package loading (McClements et al., 2009a, b). The sort of molecule needed as a bioactive component and the exact product that serves as a vehicle will determine how 'efficient' this package load is. Big structures will necessitate a higher package efficiency than molecule structures, such as vitamins. When selecting an encapsulating, it is critical to select one that can be easily absorbed into the food without affecting the texture or taste. Last but not least, the encapsulation system may need to be designed so that the bioactive component is released at a precise place in the gastrointestinal tract (de Vos et al., 2010).

In four key research directions, the investigations address a wide range of questions and challenges connected to microencapsulation:

- materials for microencapsulation;
- microencapsulation wall (matrix) materials;
- microencapsulation procedures;
- encapsulated systems' properties and functioning.

Several biopolymers have already been spray dried as encapsulating materials (Gharsallaoui et al., 2007). Proteins, with their strong functional qualities and high nutritional value are used for the instigation of delivery systems (Chen et al., 2006). Because of its emulsification activity, proteins from soy beans were used in the microencapsulation, and were successful in the retention and protection against oxidation (Kim and Morr, 1996). By spray drying protein isolates from green pea and cowpea, Rodrigues Pereira et al. (2009) were able to extract ascorbic acid microparticles. Sansone et al. (2011) used a mixture of cellulose acetate phthalate (CAP) as a coating gastro resistant polymer and swelling or surfactant agents as enhancers of dissolving rate to generate naringenin and quercetin particles using a spray drying process. The occurrence of a conjugation of both the above materials in the formulation resulted in microparticles that were resistant to low pH of the stomach fluid and released full flavonoids into the intestinal environment. The powders' technological qualities, such as form and size, were improved by microencapsulation, which also provided long-term storage stability and preserved antioxidant capabilities.

Lipids, proteins, and carbohydrates are the most often used bioactive food components that are already encapsulated in commercial applications (Augustin and Hemar, 2009). Fatty acids, phospholipids, carotenoids, and oil-soluble vitamins are all examples of lipids (Hämäläinen et al., 2007; McClements et al., 2009a, b). They are poly-unsaturated fatty acids with a limited water solubility and are particularly susceptible to oxidation. They are currently frequently used that create an efficient oxygen barrier (de Vos et al., 2010). As a result, several various encapsulation technologies for lipids have been evolved in order to be able to use them in a wide range of food products (McClements et al., 2009a, b).

Peptides generated from food can operate as a growth factor, anti-hypertensive agent, antioxidant, or immunological regulator (Hartmann and Meisel, 2007; McClements et al., 2009a, b). In order to have a good health effect, some of these proteins must reach the site of uptake in the small bowel in an intact shape (de Vos et al., 2010). Most peptides must be hydrolyzed in the stomach and small intestine (McClements et al., 2009a, b). The type of protein, the targeted health effect, and the product that acts as a container for the bioactive protein all influence whether or not protein encapsulation should be considered (de Vos et al., 2010).

Bioactive carbohydrates contained in dietary fibers are among the carbohydrates that can benefit from microencapsulation (Redgwell and Fischer, 2005). The soluble non-digestible polysaccharides are the fibers or components that would benefit the most from encapsulation. These fibers have been included to help with cholesterol lowering, glycemic control, constipation prevention, prebiotic benefits, and even cancer prevention (McClements et al., 2009a, b; Redgwell and Fischer, 2005). The key problem in this area is to increase the amount of fiber in diet in order to obtain the aforementioned health benefits, rather than to focus fibers to specific areas of the gut (de Vos et al., 2010). The main encapsulation endeavor in this area is to increase the food load of fibers by compressing enough fibers into capsules without compromising product quality, such as texture, mouth feel, or flavor (de Vos et al., 2010).

Encapsulation techniques have also been frequently used to improve probiotic bacteria viability in commercial products.

Several authors investigated the presence of probiotic strains in simulated gastrointestinal circumstances (Mokarram et al., 2009), as well as liquid-based items such as dairy products (Kailasapathy, 2006). By encapsulating *Lactobacillus delbrueckii* ssp. bulgaricus within artificial sesame oil emulsions, Hou et al. established a technology to protect *Lactobacillus delbrueckii* ssp. bulgaricus in 2003. Ross et al. (2008) refined a microencapsulating method by combining non-fat milk cell suspension with sodium alginate solution to maximize strain survival in 2004. Mokarram et al. (2009) explored the effects of multi-stage coating on the characteristics of alginated beads as well as the survival of microencapsulated Lactobacillus bacteria. Weinbreck et al. explored the utilization of microencapsulation to conserve probiotic *Lactobacillus rhamnosus* GG (LGG) viability under exposure to harmfully high levels of water activity such as newborn formula powder in 2010.

Even if microencapsulation can boost probiotics' survival rate, this does not mean that we will see an increase in functional survival (de Vos et al., 2010). In recent years, it has become obvious that the presence of certain bioactive compounds or effector molecules in the cell membrane of probiotic bacteria determines their effects (van Baarlen et al., 2009). These effector molecules are (glyco) proteins that must be conserved in order for functional effects to be achieved (Konstantinov et al., 2008). Even more crucial than the survival of probiotic quantities is the survival of these effector molecules in the product and during passage through the gastrointestinal tract. The currently discovered effector molecules (Konstantinov et al., 2008) are vulnerable to acidic conditions of the small bowel. It will be a big challenge to preserve and protect these effector molecules (Ledeboer et al., 2006).

14.8.2 EDIBLE FILMS AND COATINGS

Foods to enhance the shelf life of the product and that can be consumed with or without additional removal are edible foods (Pavlath and Orts, 2009). Many items have edible films and coatings put to them to manage moisture transfer, gas exchange, and oxidation processes. Solvent removal is essential to achieve solid film formation and control of their properties (Hernández-Izquierdo & Krochta, 2008). Edible films can be prepared by two methods: a "wet process" in a film-forming solution (solution casting) before the solvent is evaporated, and a "dry process" in which some proteins and polysaccharides exhibit thermoplastic behavior like molding and extrusion (Liu et al., 2006).

Consumable films and coatings have a high potential for sustaining colorants, flavors, nutrients, spices, and antimicrobial compounds, which can extend product shelf life, reduce the risk associated with one or more body functions.

Mei and Zhao (2003) investigated the feasibility of using milk protein-based consumable films to deliver high calcium and vitamin E. In contrast, Park and Zhao (2004) found that increasing the content of mineral (5–20% w/v zinc lactate) or vitamin E in the film matrix increased the water barrier function of chitosan-based films. The addition of large amounts of calcium or vitamin E did, however, impair the tensile strength of the films. Gómez-Estaca et al. (2007) investigated the microbiological and oxidative stability of cold-smoked sardine using a functional edible film loaded with oregano or rosemary extract, a gelatine-chitosan film-coating, and/or high-pressure processing. Films coated with oregano or rosemary extract slowed lipid oxidation but did not inhibit microbiological growth. Gómez-Guillén et al. (2007) created edible films using tuna-fish gelatine and murta ecotype leaf extracts (Ugni molinae Turcz). Tuna-fish gelatine edible films were translucent and had good mechanical qualities as well as water vapor and UV radiation barrier properties.

Several researchers have attempted to add minerals, vitamins, and fatty acids into edible film and coating compositions in order to improve the nutritional value of select fruits and vegetables with low levels of these micronutrients. According to Tapia et al. (2008), adding ascorbic acid (1% w/v) to alginate and gel-based edible coatings helped to preserve the natural ascorbic acid content in fresh-cut papaya, preserving its nutritious quality throughout storage. According to Han (2002), chitosan-based coatings can hold high concentrations of calcium or vitamin E, effectively enhancing their content in fresh and frozen strawberries and red raspberries. Hernández-Muoz et al. (2006) found that strawberries coated with chitosan reserved more calcium gluconate (3079 g/kg dry matter) than strawberries immersed in calcium solutions (2340 g/kg).

The use of probiotics to create functional edible films and coatings has experienced little research. Tapia et al. (2007) effectively coated fresh-cut apple and papaya with alginate or gel film-forming solutions containing live 106 cfu/g bifidobacteria, resulting in the first edible films for probiotic coatings.

14.8.3 IMPREGNATION WITH VACCUM

Vacuum impregnation has long been seen to be a valuable tool for introducing desirable solutes into the porous structure of meals, as well as a convenient approach to change their original composition and generate new goods. This strategy can be used to incorporate physiologically active substances into fruit and vegetable products without affecting their integrity. It is distinguished from other processing methods by its so-called 'direct-formulation' (Mavroudis et al., 1998a, b; Torreggiani, 1993).

There are two techniques of using vacuum impregnation to create functional meals. On the one hand, various researches have used the vacuum impregnation approach to alter the original composition of porous foods to make them more attractive. Fito et al. (2001) employed vacuum impregnation to fortify fruits and vegetables with minerals. To obtain a 20–25% dietary reference intake (DRI) fortification in 200 g of samples, mathematical models were constructed to determine the concentration of different minerals in impregnation solutions. Experimental validation indicated that VI could be a viable approach for enriching fruits and vegetables, as predicted by the modeling. Gras et al. (2003) used VI with sucrose solutions to fortify calcium in eggplants, carrots, and oyster mushrooms, and discovered that raw material variability causes considerable changes in the final impregnation level. Using VI processing of high fructose corn syrup solution containing calcium and/or zinc in fresh-cut apples, strawberry slices, and full marionberry. Xie and Zhao (2003a, b) investigated calcium and zinc fortification of fruits in fresh-cut apples, strawberry slices, and entire marionberries. Cortés et al. (2007) created vitamin E-enriched apple products (100% DRI/200 g fresh apple) and tested the shelf life of the goods after drying at 40°C in terms of color, texture, and vitamin E stability under various storage circumstances.

Anino et al. (2006) investigated the ability of apple matrix to absorb calcium using two alternative impregnation methods (in vacuum or at atmospheric pressure) and analyzed the impact of these treatments on material compression behavior. Some researchers looked at the impact of mineral fortification in fruits on post-osmotic dehydration (Barrera et al., 2009; Moraga et al., 2009).

Betoret et al. (2003) used the VI approach to create probiotic-enriched dried fruits, making use of commercial apple juice carrying *Saccharomyces cerevisiae* or whole milk or apple juice having 107 or 108 cfu/ml of *Lactobacillus casei* (spp. rhamnosus). Dried apple samples were found to contain about 106 cfu/g *Lactobacillus casei* (spp. rhamnosus), a level comparable to that found in commercial dairy products.

Some researchers have recently focused on the protection that this technology can give for active substances. Watanabe et al. (2011) investigated the influence of sucrose solution impregnation on anthocyanin stability in strawberry jam. Strawberry jam was prepared with or derived from strawberries impregnated with sucrose ranging from 0.29 to 1.46 mol/L, and the anthocyanin stability in the jam was tested. The results revealed that impregnating the jam with sucrose prior to making it stabilized the anthocyanin in the jam more efficiently than just adding sucrose during the process.

14.9 FOOD SCIENCE WITH PURPOSE

A functional food's favorable impacts might range from maintaining a state of well-being and health to lowering the chance of pathologic outcomes.

The finding of a specified association between one or more constituents of a functional food and a function (i.e., genomic, cellular, biochemical, or physiologic) in the living host that is probably advantageous for health is the first stage in research and development of a functional nutrients. This is basic research, and it should lead to one or more hypotheses on how the identified interactions work, as well as the development and validation of relevant biomarkers.

With this as a backdrop, a functional effect may be identified, which must then be proven in appropriate models. This phase of functional food development finishes with a new theory about the functional effect's relevance to human health. The theory must be examined in properly designed nutritional studies with well selected volunteers, and the demonstration of consequence must be supported by a safety assessment, which is a fundamental requirement for the creation of functional foods. In any case, a functional food's health benefits will be restricted if it is not consumed as part of a balanced diet.

"Functional foods must remain foods; they are not pills or capsules, but components of a diet or part of a dietary pattern recognized as being advantageous for well-being and health," Pascal (1996) said in his presentation at the First East West Perspectives Conference on Functional Foods. Designing and developing functional nutrients is a scientific task that should be approached in a sequential manner. The method starts with basic scientific understanding about functions that are susceptible to food component modification, are important for maintaining well-being and health, and can be connected to a change in disease risk when altered. The next step is to apply this knowledge to the development of markers that can be linked to the important functions. The next step will be a new generation of hypothesis-driven human intervention studies that will use these established, relevant markers to set effective and safe intakes. Last but not least, innovative tools for human investigations should be developed that are less invasive and scalable.

14.10 CONCLUSIONS

Functional foods have become a well-researched science in recent years. Functional foods, when combined with a healthy lifestyle, can help with health and well-being. Functional foods have been found to benefit the development and functioning of the gut, neurological system, and bone, as well as immune modulation, in children. The following conditions may apply to their inclusion in the diet: (1) To include functional foods in one's diet that provide additional advantages based on one's physiological condition (pregnancy, lactation, physical activity, etc.). (2) To consume functional nutrients if intolerance (e.g., fish), insufficient intakes (e.g., vegetables), or shallow intakes prevent you from eating particular meals (e.g., dairy products and fish). In certain circumstances, adding omega-3 fatty acid-rich foods, vitamins or calcium-fortified milk and dairy products, or probiotics to the diet may assist in meeting particular nutritional needs. In chronic illnesses (osteoporosis, CVD, obesity, diabetes, metabolic syndrome, and so on), the need for particular nutrients may increase, and some bioactive compounds contained in functional meals may have health advantages. The final point to examine is the preventative or health-promoting effect of functional foods, which, when included in a balanced diet, can assist in achieving optimal nutrition. However, this would need a significant amount of additional study using well-controlled methodology.

Studies are needed to determine the interaction of functional foods with the standard diet, as well as the best dosing intervals, doses, and effectiveness in various groups. One of the most significant lessons we've learnt is that food is a source of energy for the body and that it must offer us with a variety of nutrients to assist the body function more efficiently. Unfortunately, having a healthy diet that isn't balanced isn't enough. This is the clearest illustration of the importance of quality above quantity. In a nutshell, functional foods are foods or chemicals that include biologically active constituents that have the capability to upgrade physical and mental health while also helping to avoid chronic illnesses in the long term. These foods are high in certain minerals, vitamins, fatty acids, dietary fiber, antioxidants, prebiotics, and probiotics, which help to balance the diet and give the body with all of the nutrients it requires and are no longer available in industrial agricultural goods.

KEYWORDS

- **early development and growth**
- **functional foods**
- **physical performance**
- **physiological condition**

REFERENCES

Alliet, P., Scholtens, P., Raes, M., Hensen, K., Jongen, H., Rummens, J. L., Boehm, G., & Vandenplas, Y., (2007). Effect of prebiotic galacto-oligosaccharide, long-chain fructo-oligosaccharide infant formula on serum cholesterol and triacylglycerol levels. *Nutrition (Burbank, Los Angeles County, Calif.), 23*(10), 719–723.

Alzate, A., Pérez-Conde, M. C., Gutiérrez, A. M., & Cámara, C., (2010). Selenium enriched fermented milk: A suitable dairy product to improve selenium intake in humans. *Int. Dairy J., 20*(11), 761–769.

Anino, S. V., Salvatori, D. M., & Alzamora, S. M., (2006). Changes in calcium level and mechanical properties of apple tissue due to impregnation with calcium salts. *Food Res. Int., 39*(2), 154–164.

Augustin, M. A., & Hemar, Y., (2009). Nano- and micro-structured assemblies for encapsulation of food ingredients. *Chem. Soc. Rev., 38*(4), 902–912.

Baba, S., Natsume, M., Yasuda, A., Seigo, B., Midori, N., Akiko, Y., Yuko, N., Takaaki, T., Naomi, O., Minoru, K., Kazuo, K., et al. (2007). Plasma LDL and HDL cholesterol and oxidized LDL concentrations are altered in normo- and hypercholesterolemic humans after intake of different levels of cocoa powder. *J. Nutr., 137*(6), 1436–1441.

Balsano, C., & Alisi, A., (2009). Antioxidant effects of natural bioactive compounds. *Curr. Pharm. Des., 15*(26), 3063–3073.

Barrera, C., Betoret, N., Corell, P., & Fito, P., (2009). Effect of osmotic dehydration on the stabilization of calcium-fortified apple slices (Var. Granny Smith): Influence of operating variables on process kinetics and compositional changes. *J. Food Eng., 92*(4), 416–424.

Betoret, N., Puente, L., Diaz, M. J., Pagán, M. J., García, M. J., Gras, M. L., Marto, J., & Fito, P., (2003). Development of probiotic enriched dried fruits by vacuum impregnation. *J. Food Eng., 56*(2), 273–277.

Block, G., (1993). Micronutrients and cancer: Time for action? *J. Natl. Cancer Inst., 85*(11), 846–848.

Borgogna, M., Bellich, B., Zorzin, L., Lapasin, R., & Cesàro, A., (2010). Food microencapsulation of bioactive compounds: Rheological and thermal characterization of non-conventional gelling system. *Food Chem., 122*(2), 416–423.

Burgi, H., Supersaxo, Z., & Selz, B., (1990). Iodine deficiency diseases in Switzerland 100 years after Theodor Kocher's survey: A historical review with some new goitre prevalence data. *Acta Endocrinologica., 123*, 577–590.

Canene-Adams, K., Lind, S. B. L., Wang, S., Jeffery, E. H., Clinton, S. K., & Erdman, J. W. Jr., (2007). Combinations of tomato and broccoli enhance anti-tumor activity in dunning r3327-h prostrate adenocarcinomas. *Cancer Res., 67*(2), 836–43.

Cencic, A., & Chingwaru, W., (2010). Antimicrobial agents deriving from indigenous plants. *Recent Pat. Food, Nutr. Agric., 2*, 83–92.

Chávez-Servín, J. L., Castellote, A. I., & López-Sabater, M. C., (2008). Vitamins A and E content in infant milk-based powdered formulae after opening the packet. *Food Chem., 106*(1), 299–309.

Chávez-Servín, J. L., Castellote, A. I., Martín, M., Chifré, R., & López-Sabater, M. C., (2009). Stability during storage of LC-PUFA-supplemented infant formula containing single cell oil or egg yolk. *Food Chem., 113*(2), 484–492.

Chen, C. Y., & Blumberg, J. B., (2008). *In vitro* activity of almond skin polyphenols for scavenging free radicals and inducing quinine reductase, *J Agric Food Chem., 56*(12), 4427–4434.

Chen, L., Remondetto, G. E., & Subirade, M., (2006). Food protein-based materials as nutraceutical delivery systems. *Trends Food Sci. Technol., 17*(5), 272–283.

Chung-Jyi, T., Michael, F. L., Frank, B. H., Walter, C. W., & Edward, L. G., (2004). Frequent nut consumption and decreased risk of cholecystectomy in women. *Am. J. Clin Nutr., 80*(1), 76–81.

Clarke, S. D., & Jump, D. B., (1994). Dietary polyunsaturated fatty acid regulation of gene transcription. *Annu. Rev. Nutr., 14*, 83–98.

Cortés, M., Osorio, A., & García, E., (2007). Air dried apple fortified with vitamin e using matrix engineering. *Revista de la Facultad de Química Farmacéutica., 14*(2), 17–26.

Da Costa, P. F. P., Ferraz, M. B. M., Ros-Polski, V., Quast, E., Collares, Q. F. P., & Steel, C. J., (2010). Functional extruded snacks with lycopene and soy protein. *Ciência e Tecnología de Alimentos., 30*(1), 143–151.

De La Garza, R. I. D., Gregory, J. F., & Hanson, A. D., (2007). Folate biofortification of tomato fruit. *Proceedings of the National Academy of Sciences of the United States of America, 104*, 4218–4222.

De Vos, P., Faas, M. M., Spasojevic, M., & Sikkema, J., (2010). Encapsulation for preservation of functionality and targeted delivery of bioactive food components. *Int. Dairy J., 20*, 292–302.

Ding, M., Feng, R., Wang, S. Y., Bowman, L., Lu, Y., Qian, S. Y., Castranova, V., Jiang, B. H., & Shi, X., (2006). Cyanidino-3 glucoside, a natural product derived from blackberry, exhibits chemopreventive and chemotherapeutic activity. *J. Biol. Chem., 281*(25), 359–368.

Dunn, A. V., (2000). Incorporating soy protein into a low-fat-low cholesterol diet, *Cleve Clin J Med., 67*(10), 767–772.

FAO, WHO, (2001). *Human Vitamin and Mineral Requirements*. Report of a Joint FAO/WHO Expert Consultation, Bangkok, Thailand. Food and Nutrition Division, FAO Rome.

FAO, WHO, (2006). *Guidelines on Food Fortification with Micronutrients*. Geneva: WHO Press..

Ferrari, C. K. B., (2007). Functional foods and physical activities in health promotion of aging people. *Maturitas, 58*(4), 327–339.

Fito, P., Chiralt, A., Betoret, N., Gras, M., Chafer, M., Martinez-Monzó, J., &rés, A., & Vidal, D., (2001). Vacuum impregnation and osmotic dehydration in matrix engineering application in functional fresh food development. *J. Food Eng., 49*, 175–183.

Franck-Frippiat, A., (1992). Rafticreaming: The new process allowing to turn fat into dietary fiber. *FIE Conference Proceedings.* Maarsen, Netherlands: FIE, 1903–1907.

Georgieva, N., & Kosev, V., (2018). Adaptability and stability of white lupin cultivars. *Banat's J. Biotechnol., 9*(18), 65–76.

Gerhauser, C., (2008). Cancer chemopreventive potential of apples, apple juice and apple components. *Planta Med., 74*(13), 1608–24.

Gharsallaoui, A., Roudaut, G., Chambin, O., Voilley, A., & Saurel, R., (2007). Applications of spray-drying in microencapsulation of food ingredients: An overview. *Food Res. Int., 40*(9), 1107–1121.

Ghasemi, E., & Kohnehrouz, B. B., (2016). Cloning the cotton rrn23–rrn5 region for developing an universal interfamily plastidial vector. *Banat's J. Biotechnol., 7*(14), 81–88.

Gibbs, B. F., Kermasha, S., Alli, I., & Mulligan, C. N., (1999). Encapsulation in the food industry: A review. *Int. J. Food Sci. Nutr., 50*(3), 213–224.

Gómez-Estaca, J., Montero, P., Giménez, B., & Gómez-Guillén, M. C., (2007). Effect of functional edible films and high-pressure processing on microbial and oxidative spoilage in cold-smoked sardine (*Sardina pilchardus*). *Food Chem., 105*(2), 511–520.

Gómez-Guillén, M. C., Ihl, M., Bifani, V., Silva, A., & Montero, P., (2007). Edible films made from tuna-fish gelatin with antioxidant extracts of two different murta ecotypes leaves (*Ugni molinae* Turcz). *Food Hydrocolloids, 21*, 1133–1143.

González-Barrio, R., Vidal-Guevara, M. L., Tomás-Barberán, F. A., & Espín, J. C., (2009). Preparation of a resveratrol-enriched grape juice based on ultraviolet C-treated berries. *Innovative Food Sci. Emerging Technol., 10*, 374–382.

González-Molina, E., Moreno, D. A., & García-Viguera, C., (2009). A new drink rich in healthy bioactives combining lemon and pomegranate juices. *Food Chem., 115*, 1364–1372.

Gras, M. L., Vidal, D., Betoret, N., Chiralt, A., & Fito, P., (2003). Calcium fortification of vegetables by vacuum impregnation. *J. Food Eng., 56*(2, 3), 279–284.

Griel, A. E., Eissenstat, B., Juturu, V., Hsieh, G., & Kris-Etherton P. M., (2004). Improved diet quality with peanut consumption. *J Am Coll Nutr., 23*, 1–9.

Gry, J., Black, L., Eriksen, F. D., Pilegaard, K., Plumb, J., & Rhodes, M., (2007). EuroFIR-BASIS a combined composition and biological activity database for bioactive compounds in plant-based foods. *Trends Food Sci. Tech., 18*, 434–444.

Guarnieri, S., Riso, P., & Porrini, M., (2007). Orange juice vs peroxide-induced DNA damage in mononuclear blood cells. *Br. J. Nutr., 97*(4), 639–643.

Hakimuddin, F., Paliyath, G., & Meckling, K., (2006). Treatment of MCF-7 breast cancer cells with a red grape wine polyphenol fraction results in disruption of calcium homeostasis and cell cycle arrest causing selective cytotoxicity. *J Agric Food Chem., 54*, 7912–7923.

Hämäläinen, M., Nieminen, R., Vuorela, P., Heinonen, M., & Moilanen, E., (2007). Anti-inflammatory effects of flavonoids: Genistein, kaempferol, quercetin and daidzein inhibit STAT-1 and NF-kappaB activations, whereas flavone, isorhamnetin, naringenin, and pelargonidin inhibit only NF-kappaB activation along with their inhibitory effect on iNOS expression and NO production in activated macrophages. *Mediators Inflammation, 2007*, 45673.

Han, J., (2002). Protein-based edible films and coatings carrying antimicrobial agents. In: Gennadios, A., (ed.), *Protein-Based Films and Coatings* (pp. 485–498). Florida: CRC Press.

Hartmann, R., & Meisel, H., (2007). Food-derived peptides with biological activity: From research to food applications. *Curr. Opin. Biotechnol., 18*, 163–169.

He, S., Sun, C., & Pan, Y., (2008). Red wine polyphenols for cancer prevention. *Int. J. Mol. Sci., 9*(5), 842–853.

Hernandez-Izquierdo, V. M., & Krochta, J. M., (2008). Thermoplastic processing of proteins for film formation: A review. *J. Food Sci., 73*(2), R30–R39.

Hernández-Muñoz, P., Almenar, E., Ocio, M. J., & Gavara, R., (2006). Effect of calcium dips and chitosan coatings on postharvest life of strawberries (*Fragaria_ananassa*). *Postharvest Biol. Technol., 39*, 247–253.

Holm, F., (2003). *New Functional Food Ingredients Cardiovascular Health* (pp. 8–31). Food Group Denmark, Skødstrup, Denmark.

Hou, R. C. W., Lin, M. Y., Wang, M. M. C., & Tzen, J. T. C., (2003). Increase of viability of entrapped cells of *Lactobacillus delbrueckii* ssp. bulgaricus in artificial sesame oil emulsions. *J. Dairy Sci., 86*, 424–428.

Jarouliya, U., & Keservani, R. K., (2019a). Pathways leading to child obesity. In: Debasis, B., (ed.), *Global Perspectives on Childhood Obesity: Current Status, Consequences and Prevention* (2nd Edn., pp. 137–146). Academic Press, Elsevier, chapter 12. ISBN: 9780128128404.

Jarouliya, U., & Keservani, R. K., (2019b). "Protein function as cell surface and nuclear receptor in human diseases. In: Sanjay, B., & Debarshi, K. M., (eds.), *Medicinal Chemistry with Pharmaceutical Product Development* (pp. 1–32). Academic Press, CRC Press, Taylor & Francis Group. Chapter 1. ISBN: 9781771887106.

Jenkins, D. J., Nguven, T. H., Kendall, C. W., Faulkner, D. A., Bashyam, B., Kim, I. J., et al. (2008). The effect of strawberries in a cholesterol lowering dietary portfolio. *Metabolism, 57*(12), 1636–44.

Jiménez-Colmenero, F., Sánchez-Muniz, F. J., & Olmedilla-Alonso, B., (2010). Design and development of meat-based functional foods with walnut: Technological, nutritional and health impact. *Food Chem., 123*, 959–967.

Juniper, D. T., Phipps, R. H., Ramos-Morales, E., & Bertin, G., (2009). Effects of dietary supplementation with selenium enriched yeast or sodium selenite on selenium tissue distribution and meat quality in lambs. *Anim. Feed Sci. Technol., 149*, 228–239.

Kailasapathy, K., (2006). Survival of free and encapsulated probiotic bacteria and their effect on the sensory properties of yoghurt. *LWT – Food SciTech., 39*, 1221–1227.

Kannar, D., Mohandoss, P., Wattanapenpaiboon, N., & Wahlqvist, M. L., (1998). The prebiotic effect of garlic and inulin. *Proc. Nutr. Soc. Australia, 22*, 285.

Karaaslan, M., Ozden, M., Vardin, H., & Turkoglu, H., (2011). Phenolic fortification of yogurt using grape and callus extracts. *LWT – Food Sci Tech., 44*, 1065–1072.

Karunanandaa, B., Qi, Q. G., Hao, M., Baszis, S. R., Jensen, P. K., Wong, Y. H. H., et al. (2005). Metabolically engineered oilseed crops with enhanced seed tocopherol. *Metab. Eng., 7*, 384–400.

Kelly, S. A., Summerbell, C. D., Brynes, A., Whittaker, V., & Frost, G., (2007). Wholegrain cereals for coronary heart disease. *Cochrane Database Syst. Rev., 18*(2), CD005051.

Keservani, R. K., & Sharma, A. K., (2014). Flavonoids: Emerging trends and potential health benefits. *Journal of Chinese Pharmaceutical Sciences, 23*(12), 815.

Keservani, R. K., Kesharwani, R. K., & Sharma, A. K., (2015). Pulmonary and respiratory health: Antioxidants and nutraceuticals. In: Debasis, B., Harry, G. P., & Anand, S., (eds.), *Nutraceutical and Functional Foods in Human Life and Disease Prevention* (pp. 279–295). Boca Raton, Fl. CRC Press, Taylor and Francis, Chapter 17. ISBN: 9781482237214.

Keservani, R. K., Kesharwani, R. K., Sharma, A. K., Vyas, N., & Chadoker, A., (2010b). Nutritional supplements: An overview. *International Journal of Current Pharmaceutical Review and Research, 1*(1), 59–75.

Keservani, R. K., Kesharwani, R. K., Vyas, N., Jain, S., Raghuvanshi, R., & Sharma, A. K., (2010a). Nutraceutical and functional food as future food: A review. *Der Pharmacia Lettre, 2*(1), 106–116.

Keservani, R. K., Sharma, A. K., & Kesharwani, R. K., (2016a). Nutraceutical and functional foods for cardiovascular health. In: *Food Process Engineering* (pp. 257–278, 291–312). Apple Academic Press, CRC Press, (pp.). ISBN: 9781771884020.

Keservani, R. K., Sharma, A. K., & Kesharwani, R. K., (2016b). Medicinal effect of nutraceutical fruits for the cognition and brain health. *Scientifica*, 10. Article ID 3109254. doi: 10.1155/2016/3109254.

Keservani, R. K., Sharma, A. K., & Kesharwani, R. K., (2017). An overview and therapeutic applications of nutraceutical and functional foods. *Recent Advances in Drug Delivery Technology*, 160–201. ISBN:9781522507543.

Keservani, R. K., Sharma, A. K., & Kesharwani, R. K., (2020). "*Nutraceuticals and Dietary Supplements: Applications in Health Improvement and Disease Management*. CRC Press. ISBN: 9781771888738.

Kesharwani, R. K., Singh, P., & Keservani, R. K., (2016). Green coffee bean extract and chlorogenic acids: Chemistry and novel antioxidant benefits. In: Bagchi, D., Moriyama, H., & Swaroop, A., (eds.), *Green Coffee Bean Extract in Human Health* (pp. 1–17). Boca Raton, Taylor and Francis, USA. Chapter 1. ISBN: 9781498716376.

Kim, J., Ko, Y., Park, Y. K., Kim, N. I., Ha, W. K., & Cho, Y., (2010). Dietary effect of lactoferrin-enriched fermented milk on skin surface lipid and clinical improvement of acne vulgaris. *Nutrition, 26*, 902–909.

Kim, Y. D., & Morr, C. V., (1996). Microencapsulation properties of gum Arabic and several food proteins: Spray dried orange oil emulsion particles. *J. Agric. Food Chem., 44*(5), 1314–1320.

Kobayashi, E., Sato, Y., Umegaki, K., & Chiba, T., (2017). Analysis of safety alerts associated with dietary supplements from Japan and overseas. *Food Hygiene and Safety Science, 59*(2), 93–99.

Konstantinov, S. R., Smidt, H., De Vos, W. M., Bruijns, S. C., Singh, S. K., Valence, F., et al. (2008). S layer protein A of *Lactobacillus acidophilus* NCFM regulates immature dendritic cell and T cell functions. *Proceedings of the National Academy of Sciences of the United States of America, 105*, 19474–19479.

Krasaekoopt, W., Bhandari, B., & Deeth, H., (2004). The influence of coating materials on some properties of alginate beads and survivability of microencapsulated probiotic bacteria. *Int. Dairy J., 14*, 737–743.

Kris-Etherton, P. M., Zhao, G., Binkoski, A. E., Coval, S. M., & Etherton, T. D., (2001). The effects of nuts on coronary heart disease risk. *Nutr. Rev., 59*(4), 103–111.

Kumar, A., & Senapati, B. K., (2015). Genetic analysis of character association for polygenic traits in some recombinant inbred lines (ril's) of rice (*Oryza sativa* L.) *Banat's J. Biotechnol., 6*(11), 90–99.

Kyritsi, A., Tzia, C., & Karathanos, V. T., (2011). Vitamin fortified rice grain using spraying and soaking methods. *Food Sci. Tech., 44*, 312–320.

Laforest, L., Moulin, P., Schwalm, M. S., Le Jeunne, P., Chretin, S., Kitio, B., et al. (2007). Use of margarine enriched in phytosterols by patients at high cardiovascular risk and treated by hypolipidemic drugs. *Nutr., Metab. Cardiovasc. Dis., 17*, 657–665.

Laible, G., (2009). Enhancing livestock through genetic engineering – Recent advances and future prospects. Comparative immunology. *Microbiol. Infect. Dis., 32*, 123–137.

Ledeboer, A. M., Nauta, A., Sikkema, J., Laulund, E., Niederberger, P., & Sijbesma, W., (2006). Survival of probiotics. *European Dairy Magazine, 4*, 31, 32.

Leifert, W. R., & Abeywardena, M. Y., (2008). Cardioprotective actions of grape polyphenols. *Nutr Res., 28*(11), 729–737.

Liu, L., Kerry, J. F., & Kerry, J. P., (2006). Effect of food ingredients and selected lipids on the physical properties of extruded edible films/casings. *Int. J. Food Sci. Technol., 41*, 295–302.

Marine, D., & Kimball, O. P., (1920). Prevention of simple goitre in man. *Arch. Intern. Med., 25*, 661–672.

Matsushita, M., Tazinafo, N. M., Padre, R. G., Oliveira, C. C., Souza, N. E., & Visentainer, J. V., (2007). Fatty acid profile of milk from Saanen goats fed a diet enriched with three vegetable oils. *Small Rumi., 72*, 127–132.

Mavroudis, N. E., Gekas, V., & Sjoholm, I., (1998a). Osmotic dehydration of apples—Effects of agitation and raw material characteristics. *J. Food Eng., 35*, 191–209.

Mavroudis, N. E., Gekas, V., & Sjoholm, I., (1998b). Osmotic dehydration of apples—Shrinkage phenomena and the significance of initial structure on mass transfer rates. *J Food Eng., 38*, 101–123.

McClements, D. J., Decker, E. A., & Park, Y., (2009a). Controlling lipid bioavailability through physicochemical and structural approaches. *Crit. Rev. Food Sci. Nutr., 49*, 48–67.

McClements, D. J., Decker, E. A., Park, Y., & Weiss, J., (2009b). Structural design principles for delivery of bioactive components in nutraceuticals and functional foods. *Crit. Rev. Food Sci. Nutr., 49*, 577–606.

Mei, Y., & Zhao, Y., (2003). Barrier and mechanical properties of milk protein-based edible films incorporated with nutraceuticals. *J Agric Food Chem., 51*, 1914–1918.

Mokarram, R. R., Mortazavi, S. A., Habibi, N. M. B., & Shahidi, F., (2009). The influence of multi stage alginate coating on survivability of potential probiotic bacteria in simulated gastric and intestinal juice. *Food Res. Int., 42*, 1040–1045.

Moraga, M. J., Moraga, G., Fito, P. J., & Martínez-Navarrete, N., (2009). Effect of vacuum impregnation with calcium lactate on the osmotic dehydration kinetics and quality of osmodehydrated grapefruit. *J. Food Eng., 90*, 372–379.

Nakagawa, T., & Yokozawa, T., (2002). Direct scavenging of nitric oxide and superoxide by green tea. *Food Chem Toxicol., 40*(12), 1745–1750.

Nakamura, Y., (2009). Chemoprevention by isothiocyanates: Molecular basis of apoptosis induction. *Forum Nutr., 61*, 170–181.

National Research Council, (1989). *Past Experience with Genetic Modification of Plants and Their Introduction into the Environment* (pp. 16–36). Washington DC: National Academy Press.

Ould, Y. K., Meddah, B., Touil, A. T., & Sarsar, F., (2017). Laurus nobilis from Algeria and immune response. *Banats. J. Biotechnol., 8*(15), 119–122.

Paine, J. A., Shipton, C. A., Chaggar, S., Howells, R. M., Kennedy, M. J., Vernon, G., et al. (2005). Improving the nutritional value of golden rice through increased pro-vitamin A content. *Nature Biotechnol., 23*, 482–487.

Pan, M. H., Ghai, G., & Ho, C. T., (2008). Food bioactives, apoptosis, and cancer. *Mol. Nutr. Food Res., 52*, 43–52.

Park, S., & Zhao, Y., (2004). Incorporation of a high concentration of mineral or vitamin into chitosan-based films. *J. Agric. Food Chem., 52*, 1933–1939.

Pascal, G., (1996). Functional foods in the European Union. *Nutr. Rev., 54*(suppl), S29–32.

Patil, B. S., Jayaprakasha, G. K., Murthy, K. N. C., & Vikram, A., (2009). Bioactive compounds: Historical perspectives, opportunities and challenges. *J. Agr. Food Chem., 57*, 8142–8160.

Pavlath, A. E., & Orts, W., (2009). Edible films and coatings: Why, what and how? In: Embuscado, M. E., & Huber, K. C., (eds.) *Edible Films and Coatings for Food Applications* (pp. 1–23). New York: Springer. (Chapter 1).

Petitot, M., Boyer, L., Minier, C., & Micard, V., (2010). Fortification of pasta with split pea and faba bean flours: Pasta processing and quality evaluation. *Food Res Int., 43*, 634–641.

Plaza, M., Cifuentes, A., & Ibàñez, E., (2008). In the search of new functional food ingredients from algae. *Trends Food Sci. Tech., 19*, 31–39.

Porrini, M., Riso, P., Brusamolino, A., Berti, C., Guarnieri, S., & Visioli, E., (2005). Daily intake of a formulated tomato drink affects carotenoid plasma and lymphocyte concentration and improves cellular antioxidant protection, *Br J Nutr., 93*(1), 93–99.

Potrykus, I., (2010). Regulation must be revolutionized. *Nature, 466*, 561.

Redgwell, R. J., & Fischer, M., (2005). Dietary fiber as a versatile food component: An industrial perspective. *Mol. Nutr. Food Res., 49*, 521–535.

Rezaei, A., Akhshabi, S., & Sadeghi, F., (2016). Evaluation of exon 17 of insulin receptor (INSR) gene and its relationship with diabetes type 2 in an Iranian population. *Banat's J. Biotechnol., 7*(13), 61–69.

Rivas, A., Rodrigo, D., Company, B., Sampedro, F., & Rodrigo, M., (2007). Effects of pulsed electric fields on water-soluble vitamins and ACE inhibitory peptides added to a mixed orange juice and milk beverage. *Food Chem., 104*, 1550–1559.

Roberfroid, M. B., (1996). Functional effects of food components and the gastrointestinal system: Chicory fructooligosaccharides. *Nutr Rev., 54*(suppl), S38–42.

Rodrigues, P. H. V., Saraiva, K. P., Carvalho, L. M. J., &rade, L. R., Pedrosa, C., & Pierucci, A. P. T. R., (2009). Legumes seeds protein isolates in the production of ascorbic acid microparticles. *Food Res. Int., 42*, 115–121.

Ross, G. R., Gusils, C., & Gonzalez, S. N., (2008). Microencapsulation of probiotic strains for swine feeding. *Biol Pharm Bull., 31*(11), 2121–2125.

Ruchin, A. B., (2017). The effects of illumination on the early development of tailed and tailless amphibians. *Banat's J. Biotechnol., 8*(15), 113–118.

Samaniego-Vaesken, M. L., Alonso-Aperte, E., & Varela-Moreiras, G., (2010). Analysis and evaluation of voluntary folic acid fortification of breakfast cereals in the Spanish market. *J. Food Compos. Anal., 23*, 419–423.

Sansone, F., Picerno, P., Mencherini, T., Villecco, F., D'Ursi, A. M., Aquino, R. P., et al. (2011). Flavonoid microparticles by spray-drying: Influence of enhancers of the dissolution rate on properties and stability. *J. Food Eng., 103*, 188–196.

Shahidi, F., (2009). Nutraceuticals and functional foods: Whole versus processed foods. *Trends Food Sci. Tech., 20*(9), 376–387.

Singh, P., Kesharwani, R. K., & Keservani, R. K., (2017a). Antioxidants and vitamins: Roles in cellular function and metabolism. In: *Sustained Energy for Enhanced Human Functions and Activity* (pp. 385–407). Academic Press. ISBN:9780128054130.

Singh, P., Kesharwani, R. K., & Keservani, R. K., (2017b). Protein, carbohydrates, and fats: Energy metabolism. In: *Sustained Energy for Enhanced Human Functions and Activity* (pp. 103–115). Academic Press. ISBN:9780128054130.

Smidsrød, O., & Skjak-Braek, G., (1990). Alginate as immobilization matrix for cells. *Trends Biotechnol., 8*, 71–78.

Storozhenko, S., De Brouwer, V., Volckaert, M., Navarrete, O., Blancquaert, D., Zhang, G. F., et al. (2007). Folate fortification of rice by metabolic engineering. *Nature Biotechnol., 25*, 1277–1279.

Tang, G. W., Qin, J., Dolnikowski, G. G., Russell, R. M., & Grusak, M. A., (2009). Golden rice is an effective source of vitamin A. *Am. J. Clin. Nutr., 89*, 1776–1783.

Tanumihardjo, S. A., Bouis, H., Hotz, C., Meenakshi, J. V., & McClafferty, B., (2008). Biofortification of staple crops: An emerging strategy to combat hidden hunger. *Compr. Rev. Food Sci. Food Saf., 7*, 329–334.

Tapia, M. S., Rojas-Grau, M. A., Carmona, A., Rodriguez, F. J., Soliva-Fortuny, R., & Martin-Belloso, O., (2008). Use of alginate and gellan-based coatings for improving barrier, texture and nutritional properties of fresh-cut papaya. *Food Hydrocolloids, 22*, 1493–1503.

Tapia, M. S., Rojas-Graü, M. A., Rodríguez, F. J., Ramírez, J., Carmona, A., & MartinBelloso, O., (2007). Alginate – and gellan based edible films for probiotic coatings on fresh-cut fruits. *J. Food Sci., 72*, 190–196.

Tapola, N., Karvonen, H., Niskanen, L., Mikola, M., & Sarkkinen, E., (2005). Glycemic responses of oat bran products in type 2 diabetic patients. *Nutr. Metab. Cardiovasc Dis., 15*(14), 255–261.

Thies, C., (1987). Microencapsulation. In: Mark, H. F., Bikales, N. M., Overberger, C. G., Menges, G., & Kroschwitz, J. I., (eds.). *Encyclopedia of Polymer Science and Engineering* (pp. 724–745). New York: John Wiley and Sons.

Thies, C., (2005). A survey of microencapsulation processes. In: Benita, S., (ed.), Microencapsulation (pp. 1–20). New York: Marcel Dekker Inc.

Van, B. P., Troost, F. J., Van, H. S., Van, D. M. C., De Vos, W. M., De Groot, P. J., et al. (2009). Differential NF-kB pathways induction by *Lactobacillus plantarum* in the duodenum of healthy humans correlating with immune tolerance. *Proceedings of the National Academy of Sciences of the United States of America, 106*, 2371–2376.

Venegas-Caleron, M., Sayanova, O., & Napier, J. A., (2010). An alternative to fish oils: Metabolic engineering of oil-seed crops to produce omega-3 long chain polyunsaturated fatty acids. *Prog. Lipid Res., 49*, 108–119.

Verbeke, W., (2005). Consumer acceptance of functional foods: Socio-demographic, cognitive and attitudinal determinants. *Food Quality Prefer., 16*(1), 45–57.

Vlachopoulos, C., Aznaouridis, K., Alexopoulos, N., Economou, E., Andreadou, I., & Stefenadis, C., (2005). Effect of dark chocolate on arterial function in healthy individuals. *Am. J. Hypertens., 18*(6), 785–91.

Watanabe, Y., Yoshimoto, K., Okada, Y., & Nomura, M., (2011). Effect of impregnation using sucrose solution on stability of anthocyanin in strawberry jam. *Food Sci. Tech., 44*, 891–895.

Weinbreck, F., Bodnár, I., & Marco, M. L., (2010). Can encapsulation lengthen the shelf-life of probiotic bacteria in dry products? *Int. J. Food Microbiol., 136*, 364–367.

Wildman, R. E. C., (2006). Classifying nutraceuticals. In: *Handbook of Nutraceuticals and Functional Foods* (2nd edn., pp. 13–31), Wildman Ed., CRC Publisher.

Woods, V. B., & Fearon, A. M., (2009). Dietary sources of unsaturated fatty acids for animals and their transfer into meat, milk and eggs: A review. *Livest Sci., 126*, 1–20.

Xie, J., & Zhao, Y., (2003a). Nutritional enrichment of fresh apple (Royal Gala) by vacuum impregnation. *Int. J. Food Sci. Nutr., 54*, 387–398.

Xie, J., & Zhao, Y., (2003b). Improvement of physicochemical and nutritional qualities of frozen marionberry by vacuum impregnation pre-treatment with cryoprotectants and minerals. *J. Hortic. Sci. Biotechnol., 78*, 248–253.

Ye, X. D., Al-Babili, S., Kloti, A., Zhang, J., Lucca, P., Beyer, P., et al. (2000). Engineering the provitamin A (beta-carotene) biosynthetic pathway into (carotenoid-free) rice endosperm. *Science, 287*, 303–305.

Zafra-Stone, S., Yasmin, T., Bagchi, M., Chatterjee, A., Vinson, J. A., & Bagchi, D., (2007). Berry anthocyanins as novel antioxidants in human health and disease prevention, *Mol., Nutr. Food Res., 51*(6), 675–83.

Zare, F., Boye, J. I., Orsat, V., Champagne, C., & Simpson, B. K., (2011). Microbial, physical and sensory properties of yogurt supplemented with lentil flour. *Food Res. Int., 44*(8), 2482–2488.

Zhang, Q., Li, L., Lan, Q., Li, M., Wu, D., & Chen, H., (2018). Protein glycosylation: A promising way to modify the functional properties and extend the application in food system. *Critical Reviews in Food Science and Nutrition, 22*, 1–28.

Zhang, W., Xiao, S., Samaraweera, H., Lee, E. J., & Ahn, D. U., (2010). Improving functional value of meat products. *Meat Sci., 86*, 15–31.

Zhao, F. J., & Shewry, P. R., (2011). Recent developments in modifying crops and agronomic practice to improve human health. *Food Policy, 36*, S94–S10.

Zhu, C. F., Naqvi, S., Breitenbach, J., Sandmann, G., Christou, P., & Capell, T., (2008). Combinatorial genetic transformation generates a library of metabolic phenotypes for the carotenoid pathway in maize. *Proceedings of the National Academy of Sciences of the United States of America, 105*, 18232–18237.

Ziegler, R. G., (1991). Vegetables, fruits, carotenoids and the risk of cancer. *Am. J. Clin. Nutr., 53*(suppl), 251S–259S.

CHAPTER 15

OVERVIEW OF FUNCTIONAL FOODS AND THEIR IMPACT ON OSTEOPOROSIS

TANMAY SARMA,[1] TRISHNA DAS,[1] DIPANKAR SAHA,[1] SMRITI REKHA CHANDA DAS,[1] SHANTI BHUSHAN MISHRA,[2] AMIT KUMAR SINGH,[2] and INDRANIL CHANDA[3]

[1]*Girijananda Chowdhury Institute of Pharmaceutical Science, Guwahati, Assam, India*

[2]*United Institute of Pharmacy, Prayagraj, Uttar Pradesh, India*

[3]*Assam Science and Technology University, Jalukbari, Guwahati, Assam, India*

ABSTRACT

Osteoporosis is an enervating disease that affects many older people. Fragility fractures are the basic signs of osteoporosis. There are many nutrients required for optimal bone health throughout the lifecycle. Bone is a living tissue and requires all the nutrients that other tissues need. In addition, bones have exceptional requirements for calcium, phosphorus and protein, as these make up its main structural components. Other bone nutrients such as vitamin D, Mg and vitamin K are sometimes added to promote optimal bone health. This chapter provides an overview of current opinions on various nutritional aspects of functional foods and their implications in the management of osteoporosis.

Applications of Functional Foods in Disease Prevention. Raj K. Keservani & Eknath D. Ahire, PhD (Eds.)

15.1 INTRODUCTION

15.1.1 WHAT IS FUNCTIONAL FOOD?

Traditionally, food has two functions: it provides humans with the nutrients they require for their metabolism (nutritional function), and it contributes to their well-being through its flavor (sensory orhedonistic function). A new, possible purpose for food has evolved in recent years: that of fulfilling a certain "physiological" function. While food that serves this purpose – referred to as "functional" food – is taken as part of a regular diet, it has health benefits that go beyond standard nutritional benefits. As a result, it is similar to – but distinct from – ideas like food supplements and nutraceuticals. The demand for health-promoting foods and drinks has risen dramatically in recent decades, corresponding with increased healthcare costs, higher life expectancy, and a desire for a higher standard of living (Tur and Bibiloni, 2016). The first documented evidence of functional meals may be found in China, dating back to 1000 BC. Foods and plants have long been thought to have medicinal or therapeutic characteristics throughout Asia, although such views have been dismissed as anecdotal and dependent on popular traditions (Verbeke, 2005). Because all foods give flavor, scent, and nutritional value, they are all functional to some level. Foods, on the other hand, are now being studied in depth for additional physiological effects that may help to minimize the risk of chronic disease or otherwise improve health (Hasler, 2002). There is no commonly agreed definition of functional foods. However, it may be described as foods with a long history of use, whereas nutraceuticals are foods originating from many edible sources but ingested in medicinal forms such as tablets, capsules, or pills (Tur and Bibiloni, 2016; Keservani et al., 2010a, b, 2017, 2020).

'Designer food,' 'nutritional foods,' 'nutraceutical foods,' 'therapeutic foods,' 'healthy foods,' and 'fortified food' are synonyms for foods fortified or boosted with nutrient content already present in them or other complementary nutrient, or they can be understood as "foods that are tailor-made to meet any special criteria in terms of functionality, nutrition, convenience, and therapeutic aspects (Shandilya, 2017). Foods with proven health effects have become more popular as a result of factors such as increased public awareness of human health, the expansion of the food business, and medicine and fields of natural science exploring the relationship between nutrition and health (Vukasović, 2017).

Functional foods are substances that give health advantages beyond their real nutritional value. Some varieties contain vitamins or other added substances aimed to promote health. The concept of functional food was first developed in Japan in the 1980s, when the Ministry of Health and Family Welfare, faced with rising healthcare costs, established a regulatory system to approve certain foods with documented health benefits in the hopes of improving the nation's aging population's health (Lafarga et al., 2020). The word subsequently spread from Japan to the European Union and eventually to the United States of America, where it was extremely profitable (Martirosyan and Singh, 2015).

The Japanese government financed three research programs: the analysis and development of food functions, the study of physiological regulators of food functions, and the analysis of functional foods and molecular design. Those were nationwide initiatives aimed at lowering the rising costs of health care. As a result, the Food for Specified Health Use (FOSHU) program was developed in 1991, with the expectation that these foods would have specified health impacts due to the inclusion of appropriate elements or the elimination of allergens (Tur and Bibiloni, 2016). Functional foods are currently recognized as a distinct category in Japan, and the Japanese functional food market is one of the most advanced in the world. The Japanese government is investing a significant amount of money in the promotion of functional foods.

Casimir Funk proposed the 'Vitamin E' idea in a landmark study published in 1912. He argued that sickness was caused by the absence of certain minute components in meals rather than by the presence of germs. He established a theory and notion that has had a direct impact on the creation of functional foods (Henry, 2010).

In general, the following related ideas are frequently conflated with functional foods:

1. **Nutraceuticals:** Products derived from food that are offered as pills, powders, or other dosage forms not typically associated with food that have been found to have therapeutic or protective physiological characteristics against chronic illnesses.
2. **Pharma Foods:** An universally accepted term to designate products on the food-medicine borderline, which first emerged in Japan; they are also called 'border foods' as they claim drug-like preventive and therapeutic qualities (Tur and Bibiloni, 2016).

Food that are considered as functional food are:

i. **Fruits:** Berries, kiwi, pears, peaches, apples, oranges, bananas.
ii. **Vegetables:** Broccoli, cauliflower, kale, spinach, zucchini.

iii. **Nuts:** Almonds, cashews, pistachios, macadamia nuts, Brazil nuts.
iv. **Seeds:** Chia seeds, flaxseeds, hempseeds, pumpkin seeds.
v. **Legumes:** Black beans, chickpeas, navy beans, lentils.
vi. **Wholegrains:** Oats, barley, buckwheat, brown rice, couscous.
vii. **Seafood:** Salmon, sardines, anchovies, mackerel, cod.
viii. **Fermented foods:** Tempeh, kombucha, kimchi, kefir, sauerkraut.
ix. **Herbs and spices:** Turmeric, cinnamon, ginger, cayenne pepper.
x. **Beverages:** Coffee, green tea, black tea (Santos-Buelga et al., 2019) (Figure 15.1).

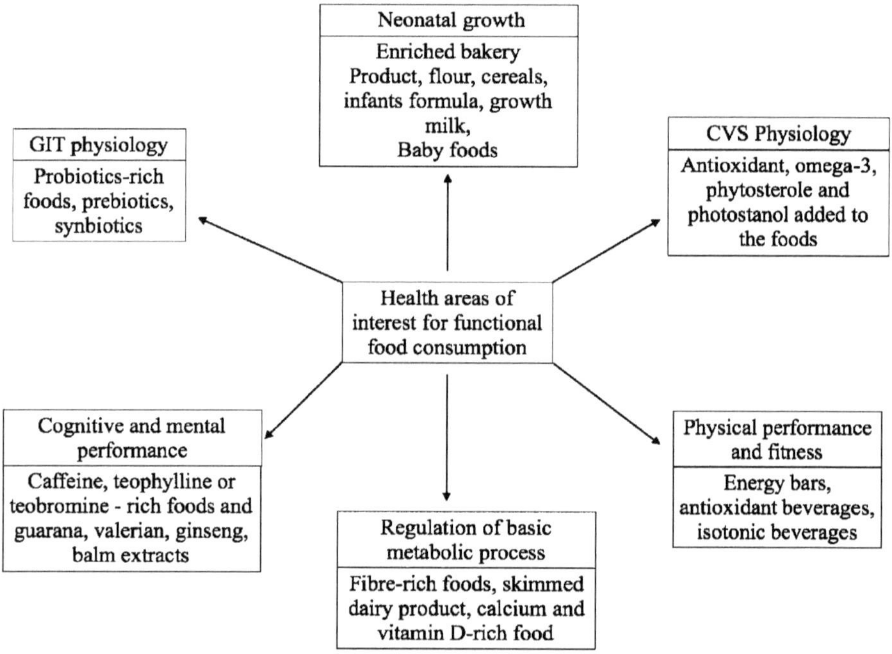

FIGURE 15.1 The importance of health in functional food intake.
Source: Adapted from Tur and Bibiloni, 2016; Kim et al., 2021

15.2 FUNCTIONAL FOOD REGULATIONS

The Food and Drug Administration (FDA) regulates foods in the United States under the Federal Food, Drug, and Cosmetic Act of 1938. The term "functional foods" has been defined. However, the Nutrition Labeling and Education Act of 1990 includes both conventional foods and foods for special dietary use. Four categories of claims can be used by food manufacturers on food labels to communicate health information to consumers.

These include the following:

- Nutrient content claims;
- Structure/function claims;
- Health claims;
- Qualified health claims.

All four types of claims on functional foods are allowed by the FDA if they fulfill the established requirements for each claim. FDA (Food and Drug Administration) of the United States Traditional meals and dietary supplements can both make claims.

Such claims are permitted based on considerable scientific agreement or on an official declaration by a scientific body of the US Government or the National Academy of Sciences. FDA (Food and Drug Administration) of the United States Traditional meals and dietary supplements can both make claims (Sikand et al., 2015).

15.3 FUNCTIONAL FOODS BENEFITS

- Providing antioxidants (like carotenoids, flavonoids, lycopene, anthocyanin and polyphenols) that fight free radical damage.
- Reducing inflammation.
- Aiding in disease prevention, such as reducing the risk for cardiovascular diseases, neurological conditions, depression, or cancer.
- Supporting gut health and therefore enhancing the immune system.
- Providing live microbial cultures, also called probiotic bacteria.
- Providing "prebiotics" that help feed probiotics (Alkhatib et al., 2017).

15.4 FUNCTIONAL FOOD CONSUMPTION WORLDWIDE

The largest functional food markets are in Japan, Korea, and the United States, followed by Europe. Spain, Finland, Cyprus, the Netherlands, and Sweden are among the European Union's top consumers of functional foods (Tur and Bibiloni, 2016; Kim et al., 2021) (Figures 15.2–15.4).

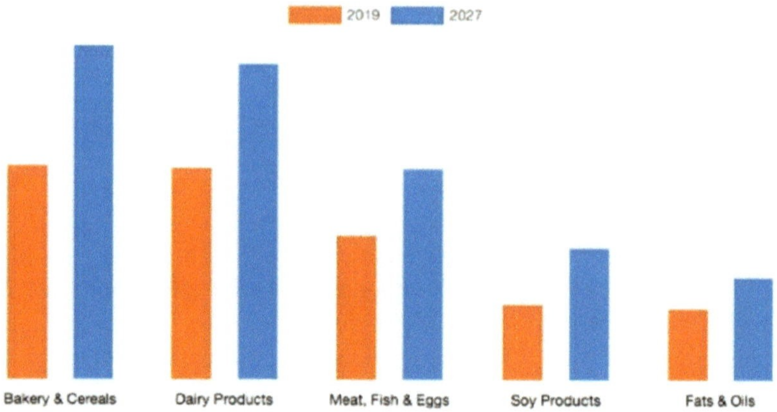

FIGURE 15.2 Functional food market by ingredients.

Source: Adapted from Tur and Bibiloni, 2016; Kim et al., 2021

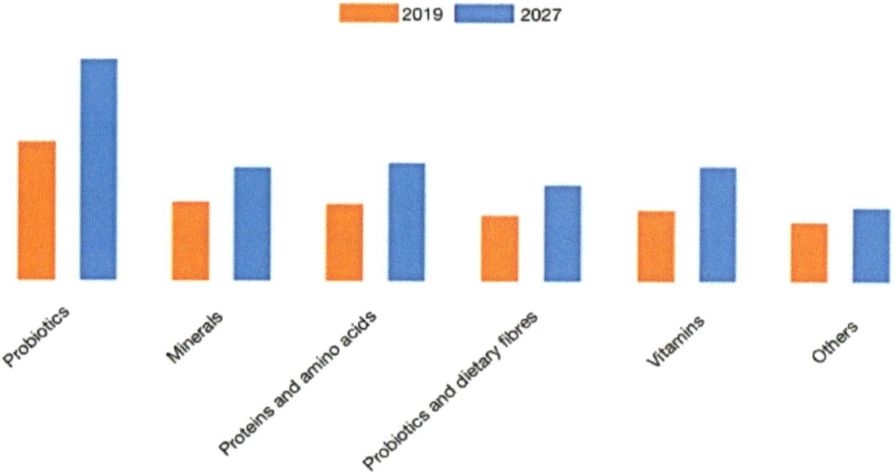

FIGURE 15.3 Functional food market by product.

Source: Adapted from Tur and Bibiloni, 2016; Kim et al., 2021

15.5 WHAT IS OSTEOPOROSIS?

Osteoporosis is a prevalent disorder that affects 30% of women and 12% of men at some point in their lives, according to the World Health Organization (WHO). It is defined as a bone mineral density (BMD) T-score less than –2.5 as evaluated by dual-emission x-ray absorptiometry (DXA) (Armas and Recker, 2012). As mentioned in National Institutes of Health consensus

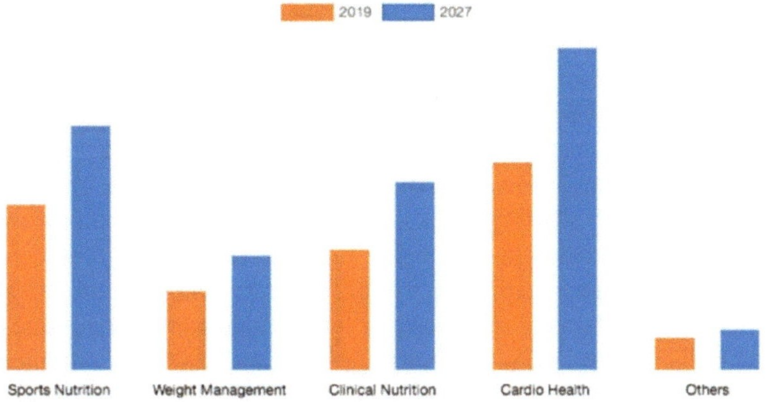

FIGURE 15.4 Functional food market by applications.

Source: Adapted from Tur and Bibiloni, 2016; Kim et al., 2021

conferences conducted in 1984 and 2001, shifting concepts about the pathophysiology of osteoporosis are reflected in changing definitions of osteoporosis. "Primary osteoporosis is an age-related condition characterized by reduced bone density and increased susceptibility to fractures in the absence of other identifiable sources of bone loss," according to a 1984 definition. "Osteoporosis is described as a skeletal condition characterized by decreased bone strength predisposing a person to an elevated risk of fracture," it was modified to in 2001. The combination of bone density and bone quality is predominantly reflected in bone strength (Morii and Genant, 1998; Bowring and Francis, 2011). Osteoporosis is a metabolic bone disease caused by osteoclastic bone resorption that is not balanced by osteoblastic bone synthesis at the cellular level. As a result, bones become weak and brittle, putting them at risk of fracture (Samwald et al., 2020). Year after year, the old population, defined as people aged 65 and up, grows. In 2019, there were 703 million old people globally, with Eastern and South-Eastern Asia (260 million) having the biggest population, followed by Europe and Northern America (over 200 million). By 2050, the world's older population will have tripled to 1.5 billion. The elderly accounted for 6% and 9% of the total population in 1990 and 2019, respectively, and this ratio is predicted to rise to 16% in 2050. By 2050, one out of every six people will be 65 years old or older.

Because the elderly population is rising, certain efforts to improve their quality of life, such as addressing the public health concern of osteoporosis, must be done. Women are more susceptible to osteoporosis than males, since around 50 years of age, women's bone density begins to decline, but men's bone density does not. One in every three women over the age of 50 and one in every five men will suffer from osteoporosis at some point in their lives. [Harper, 2019]. Data on the prevalence of osteoporosis among Indian women come from small-group studies done across the country. Estimates imply that 46 million women have the disease (Khadilkar and Mandlik, 2015). Osteoporosis affected around one-tenth of women in their 60s, one-fifth of women in their 70s, two-fifths of women in their 80s, and two-thirds of women in their 90s (Kanis et al., 2008). According to a review based on recent research, the elderly person requires functional foods containing calcium, magnesium, vitamin D, phytoestrogens, omega-3 polyun saturated fatty acids, polyphenols, carotenoids, lutein, zeaxanthin, synbiotics, plant sterols, dietary fiber, probiotic and stanols, as well as food containing trace elements like copper, zinc, boron, etc., to reduce the risk of osteoporosis. Antioxidants have also been proven to reduce oxidative stress, potentially lowering the risk of osteoporosis (Domazetovic, 2017).

15.6 PATHOPHYSIOLOGY OF OSTEOPOROSIS

Osteoporosis is a classic multifactorial illness, with a complex interaction of genetic, intrinsic, external, and lifestyle variables all contributing to a person's risk of developing the disease. Traditional patho physiologic models usually stressed endocrine processes as significant drivers of postmenopausal osteoporosis, such as estrogen shortage and secondary hyperparathyroidism in the elderly due to estrogen deficit, lower food consumption, and widespread vitamin D inadequacy (Clarke and Khosla, 2010). However, in recent years, it has been established that the pathophysiological pathways that contribute to the beginning of osteoporosis are significantly more complex. The following paragraphs will go over some of the mechanisms (Figure 15.5).

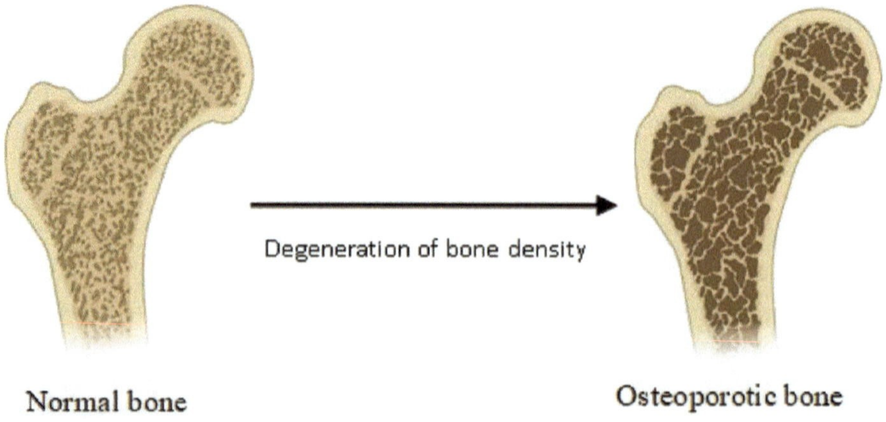

FIGURE 15.5 Process of osteoporosis.

15.7 OSTEOIMMUNOLOGY

Although Arron and Choi originated the word "osteo immunology" in the English literature in 2000, one writer had already used the German term "osteo immunologie" in 1997 (Arron and Choi, 2000). The immune system's role in osteoporosis-related bone loss is less well known. The loss of estrogen associated with the cessation of ovarian activity has been known since almost eight decades to be a crucial event in inducing rapid bone loss in early menopause (ALBRIGHT, 1941) and, as proposed by the unitary model for the pathophysiology of primary or involutional osteoporosis, also in slow bone loss in late post menopause and in elderly men it is referred to as age-related bone loss (Riggs, 1998). The effects of estrogen deficiency are partially mediated by estrogen receptor-mediated direct regulation of osteoblast, osteoclast, and osteocyte physiology. In particular, estrogen deficiency increases the number of osteoclasts while decreasing the number of osteoblasts, resulting in an imbalanced activity of the basic multicellular unit favoring bone resorption (Clarke and Khosla, 2010). However, it is now well known that the effects of estrogen deprivation, particularly on bone resorption, are primarily indirect and occur through the release of bone-active cytokines. Inflammatory cytokines are among the osteoclastogenic cytokines, suggesting that immune system-bone tissue interactions play a role in the pathogenesis of osteoporosis. The discovery of higher levels of TNF, IL-1, IL-6, or IL-17 in the first 10 years following menopause and in osteoporotic postmenopausal women compared to non-osteoporotic post-menopausal women supports a role for proinflammatory cytokines in osteoporosis (Pacifici, 2018; Zhao et al., 2016). Furthermore, the phenomenon of "inflammaging" in aged persons, which is characterized by a low-grade inflammatory condition and increasing levels of proinflammatory markers (FRANCESCHI et al.) has in several studies been linked to bone loss and fracture risk (Cauley et al., 2007; Ding et al., 2008; Nakamura et al., 2010).

T-cells are thought to be a primary source of proinflammatory cytokines in osteoporosis (Pietschmann et al., 2015; Rauner et al., 2006). When comparing postmenopausal women with osteoporotic fracture stoage-matched controls, researchers discovered that the proportion of CD8+ cells that express TNF is increased (Pietschmann et al., 2001). D'Amelio et al. postulated a higher production of TNF in T-cells and monocytes (D'Amelio et al., 2008) and Zhao et al. (2016) described an upregulated expression of IL-17 in CD4+ cells of osteoporotic postmenopausal women. However, recently a significant association has been shown between estrogen depletion, T-cell-dependent inflammation, and osteoporosis. In ovariectomized mice, Cline-Smith et al. described for the first time a biochemical mechanism by which estrogen loss promotes a low-grade inflammation by T-cells during the acute phase of bone loss. They present evidence that bone marrow dendritic cells BMDCs) play a key role in the generation of proinflammatory cytokine-producing T-cells. In their investigation, ovarian ovariectomy (OVX) increased the number of BMDCs and, as a result, the amount of IL-7 and IL-15 generated by these cells. Antigen-independent production of IL-17A and TNF is induced by IL-7 and IL-15 in a sub population of memory T cells (TMEM). Furthermore, T-cell-specific ablation of IL15RA reduced IL-17A and TNF production as well as an increase in bone resorption or bone loss following OVX, indicating that the hypothesized pathway is important in OVX-related bone loss. Interestingly, the authors hypothesize that acritical role for TMEM activation in postmenopausal bone loss might explain differences in osteoporosis susceptibility within a community. Individual differences in TMEM reflect varying lifetime exposures to commensal and pathogenic bacteria, and may be linked to greater or lower levels of TNF and IL-17A, as well as bone loss caused by menopause (Cline et al., 2020).

Regulatory T (Treg) cells are another kind of T-cell that has been shown to play a role at the immune-skeletal interface. Their main function is to suppress multiple types of immune cells and prevent excessive immune reactions, inflammation, and tissue damage. They are identified by the expression of the transcription factor FOXP3. Their anti-osteoclastogenic significance in bone biology is undeniable (Bozec and Zaiss, 2017). As a result, transferring Treg cells into T-cell deficient animals resulted in increased bone density and are duction in the number of osteoclasts. Furthermore, Foxp 3 transgenic mice were protected against OVX-induced bone loss, indicating that Treg cells play a role in bone loss in the context of estrogen deprivation (Zaiss et al., 2010). Consistent with this, estrogen has been shown to stimulate proliferation and differentiation of Treg cells (Tai et al., 2007). The discovery that B-cells generate RANKL and OPG, and so function as regulators of the RANK/RANKL/OPG axis, supports a role for B-cells in the pathogenesis of osteoporosis (Walsh and Choi, 2014). Infact, B-cells produce more RANKL by in postmenopausal women (Fatourechi et al., 2003) and B-cellablation of RANKL in mice partially protects from trabecular bone loss after ovariectomy (Onal et al., 2012). The findings of a worldwide gene expression investigation by Pineda et al. Supports the significance of B-cells in bone metabolism and osteoporosis. When they compared gene expression in OVX mice and control mice, they discovered numerous B-cell biology-related pathways among the top canonical pathways disrupted Pineda et al., 2014). More recent research evaluated global gene expression in B-cells extracted from OVX and control mice's bone marrow (Panach et al., 2017). In a second stage, they looked at the relationship between polymorphisms in selected differentially expressed genes and bone mineral density (BMD) and the risk of osteoporosis in postmenopausal women, and discovered a significant link between single nucleotide polymorphisms (SNPs) in CD80 and BMD and the risk of osteoporosis. This chemical may have an indirect relationship to BMD via its costimulatory action in T-cell activation or a direct link via the documented inhibitory impact on osteoclast formation (Bozec et al., 2014). To summarize, there is compelling evidence that B-cells play a role in osteoporosis development. However, the specific mechanism connecting estrogen deprivation to B-cells and postmenopausal women's bone loss is unknown.

15.8 CELLULAR SENESCENCE AND OSTEOPOROSIS

Farr et al. made a significant addition to our knowledge of the function of senescence in the development of osteoporosis only a few years ago. They discovered that B and T cells, myeloid cells, osteoprogenitors, osteoblasts, and osteocytes are among the categories of cells in the bone microenvironment that become senescent with age, and that senescent myeloid cells and osteocytes produce more important SASP factors than younger cells (Farr et al., 2016). They also found that senescent cells accumulated in bone biopsy samples from older postmenopausal women compared to younger premenopausal women (Franceschi et al., 2006). Farr et al. established a causal relationship between cellular senescence and age-related bone loss in a more recent investigation. They found that removing senescent cells or inhibiting their SASP reduced age-related bone loss in mice by reducing trabecular and cortical bone resorption and enhancing or preserving bone formation on endocortical and trabecular surfaces (Farr et al., 2017). Furthermore, the observed bone sparing effect was mediated in part by the removal of senescent osteocytes, which is consistent with the discovery of an elevated SASP specifically in osteocytes (Farr et al., 2016; Piemontese et al., 2017). The removal of senescent osteocytes had a role in the observed bone sparing effect (Farr et al., 2016). To summarize, significant progress has been achieved in analyzing the importance of senescence and its processes in age-related osteoporosis. Targeting cellular senescence with senolytics and senostatics has emerged asapotentiallyeffectivetherapeuticmethodforthetreatmentofage-relatedosteoporosis, based on these novel discoveries (Table 15.1).

TABLE 15.1 T-Scores and WHO Diagnostic Criteria for Osteoporosis

Interpretation	**T-Score***
Normal	−1.0 and higher
Osteopenia	−1.0 to −2.5
Osteoporosis	−2.5 and lower
Severe osteoporosis	−2.5 and lower with one or more fragility fractures
Interpretation	T-score*
Normal	−1.0 and higher
Osteopenia	−1.0 to −2.5
Osteoporosis	−2.5 and lower
Severe osteoporosis	−2.5 and lower with one or more fragility fractures

**Reference values vary by geographical location.*

Source: Tu, et al., 2018.

15.9 ROLE OF HEALTHY DIET AND OSTEOPOROSIS

Calcium absorption is reduced and calcium excretion is increased with a normal European diet that is heavy in animal protein and low in salt, alcohol, and caffeine. The diet in Asian nations is high in vegetable protein, but it is also high in calcium and vitamin D (Kostecka, 2014). According to research, a sufficient amount of multi-role vitamin D, along with a calcium-rich diet, can help prevent osteoporosis. Osteoporosis is a major concern for Chinese-American women. Increasing their calcium and vitamin D intake by eating more calcium-rich foods, many of which are also fortified with vitamin D, is a healthy method to do so (Lv and Brown, 2011). Dieticians should be directed to those at high risk of osteoporosis, and general practitioners should counsel their patients on the kind, dosage, and safety of dietary supplements. Bone mineralization and good bone density are aided by a well-balanced diet that includes a healthy ratio of natural calcium to phosphorus and the appropriate amounts of vitamin D, even following menopause. In a Polish study of kids between the ages of 10 and 12, it was discovered that in 46% of the participants, insufficient calcium consumption led to an inaccurate mineral ratio in the diet (0.62 compared to their recommended 1). This finding shows that these youngsters will have a higher chance of getting osteoporosis as adults (Orkusz, 2022).

TABLE 15.2 Medications Approved by the U.S. Food and Drug Administration for Osteoporosis

Indication	Typical Dosage	Medication	Fracture Type	Route	Monthly Cost*
Prevention	0.625 mg daily	Estrogen†, with or without progesterone	Hip, vertebral, nonvertebral	Oral	With progesterone: $40 Without progesterone: $47
Prevention and treatment	70 mg weekly	Alendronate (Fosamax)	Hip, vertebral, nonvertebral	Oral	Tablet: $87, $77 Solution: $96
–	150 mg monthly	Ibandronate (Boniva)	Vertebral	Oral	100
–	35 mg weekly	Risedronate (Actonel)	Hip, vertebral, nonvertebral	Oral	92
–	60 mg daily	Raloxifene (Evista)	Vertebral	Oral	108
Treatment	3 mg every three months for four doses	Ibandronate	Increases bone mineral density, but fracture endpoint note evaluated	Intravenous	162‡
–	5 mg annually for three doses	Zoledronic acid (Reclast)	Hip, vertebral, nonvertebral	Intravenous	104
–	200IU daily	Calcitonin (Miacalcin)	Vertebral	Nasal	$126
–	20 mcg daily up to two years	Teriparatide (Forteo)	Vertebral, nonvertebral	Subcutaneous	$675

* *Estimated cost to the pharmacist based on average wholesale prices (rounded to the nearest dollar)* per *Red Book: Pharmacy's Fundamental Reference* (Red Book, 2008). Cost to the patient will be higher, depending on the prescription filling fee. Costs are for brand name drugs only, unless otherwise indicated.

† *Hormone therapy is only appropriate in women with significant vasomotor symptoms for whom benefit outweighs risk.*

‡ *Costper 3-mg dose (given once every three months). Scientists have been demonstrated that various functional food factors stimulate osteoblastic bone formation and suppress osteoclastic bone resorption, thereby increasing bone mass. Some of the important functional food factors are discussed in the following paragraphs.*

15.10 FUNCTIONAL FOOD FACTORS

15.10.1 CALCIUM

The results of studies assessing the effectiveness of calcium on fracture risk are inconsistent, but one subgroup from a recent meta-analysis found that older women who adhered to calcium supplementation for at least 80% of the time had lower fracture rates (Prince et al., 2006). All women with osteoporosis should consume at least 1,200 mg of calcium every day (Cosman et al., 2014; Klibansk, 2001). Most postmenopausal women don't get enough calcium from their diets; therefore, supplementation is necessary. A single dosage of calcium supplement should include 500 mg or less of elemental calcium for effective absorption, necessitating numerous daily doses. Calcium carbonate is the cheapest, however it must be taken with food since it requires acid for absorption. Calcium citrate is more costly than calcium carbonate and does not need to be taken with food. Constipation and gastrointestinal distress are possible side effects of any calcium supplementation. When administered with calcium, the absorption of several drugs, including levothyroxine, fluoroquinolones, tetracycline, phenytoin (Dilantin), angiotensin-converting enzyme inhibitors, iron, and bisphosphonates, can be considerably reduced. Calcium supplements should be taken several hours before or after these drugs (Mauck and Clarke, 2006). Peak BMD is a significant pinnacle from which future losses can be predicted. Low calcium intake throughout skeletal growth is expected to reduce peak BMD and increase the risk of fracture later in life (Barrett-Connor, 2000). During the first year of life, an infant's skeletal mass doubles, and roughly 37% of total skeletal mass is accumulated during adolescence.

Food we should consume at least once every week to enrich our diet with calcium. Several studies utilizing milk as a calcium source have yielded similar findings. A doubling of milk consumption (from 150 to 300 mL/d) significantly enhanced total body BMD and BMC in a sample of white girls (mean age, 12.2 years) (Huncharek et al., 2008). Chan et al. looked at pubertal females who were 11 years old on average (Tanner Stage 11). The calcium intake of one group was boosted with dairy products (1200 mg per day), whereas the other group ate their regular diet. When compared to the control group, the dairy-supplemented group ingested more calcium and exhibited substantial improvements in total body BMD ($P<0.001$) and lumbar spine BMD ($P< 0.001$) (Chan et al., 1995).

15.11 ROLE OF CALCIUM DURING PRE-MENOPAUSE

Calcium supplementation has been shown in at least one research to help premenopausal women maintain their BMD. Over the course of three years, Baran et al. supplemented 30- to 40-year-old premenopausal women with dairy foods. Calcium intakes were tracked using three-day meal histories and 24-hour urine calcium collections. The amount of calcium consumed in dairy products had no effect on blood calcium, urine calcium-to-creatinine ratios, or PTH levels. Vertebral bone density did not change in the calcium-supplemented group, but it was considerably lower in the un supplemented group at 30 and 36 months. Calcium supplementation in estrogen-repleted premenopausal women reduced vertebral bone loss of roughly 1% per year (Baran et al., 1990).

15.11.1 VITAMIN D

Vitamin D has several metabolic functions, including maintaining blood calcium levels in a normal range in the plasma, increasing calcium absorption through the intestine, and increasing stem cell mobilization to become osteoblasts (Ross et al., 2011; Holick, 1996). A drop in blood ionized calcium stimulates the parathyroid glands to boost PTH production and secretion, increasing proximal renal tubular calcium reabsorption and enhancing the generation of 1,25-(OH) vitamin D. PTH also interacts with 1,25-(OH) vitamin D to enhance the maturation of osteoblasts, allowing calcium and phosphorus to be mobilized from the bone, via the osteoprotoger in ligand. Vitamin D hormone synthesis is carefully controlled and is influenced by the body's calcium and vitamin D requirements (Johnston et al., 1992).

Vitamin D comes in a variety of ways. The plant sterol ergocalciferol (vitamin D2) is converted to cholecalciferol (vitamin D3) in the human body. In most tissues of the body, 7-dehydrocholesterol is an immediate precursor for cholesterol production. When 7-dehydro cholesterol is exposed to ultraviolet Brays, a membrane-enhanced enzyme converts it to vitamin D. Pre-vitamin D eventually undergoes a conformational shift, allowing vitamin D to depart the skin cell and enter the extracellular space, where it is attached to vitamin D-binding protein before entering the circulation (Ringe, 2020). Vitamin D is 25-hydroxylated in the liver, resulting in the predominant circulating form, 25OH vitamin D. Vitamin D generation through UV activation in the skin provides for over 90% of the vitamin D supply. Vitamin

D is asteroid-thyroid hormone super family member that works through a nuclear vitamin D receptor (Bergman et al., 2009). The parathyroid gland, pancreas, hematopoetic cells, slun keratinocytes, and reproductive organs have all been found to have vitamin D receptors. The relevance of vitamin D receptors in bone is a point of contention (Eisman, 2009).

15.11.2 FOOD SOURCES OF VITAMIN D

Vitamin D can be measured in either international units or micrograms in foods and supplements. In bioassays with rats and chicks, one international unit of vitamin D is defined as the activity of 0.025 pg of cholecalciferol. A microgram of vitamin D is equal to 40 international units (IU). One IU of vitamin D equals 0.005 pg of 25-Oh vitamin D, which is five times more active than vitamin D. Vitamin D is mostly obtained through sunlight. Liver oils and fortified foods are natural sources of vitamin D. Milk in the United States is fortified with 400 IU of cholecalciferol or ergocalciferol every quart. However, survey shave revealed that the levels of vitamin contained in milk may be highly varied (Peacock, 2009).

15.12 ROLE OF VITAMIN D IN OSTEOPOROSIS

In postmenopausal (type I) and senile (type 11) osteoporosis, there are good reasons to take vitamin D and calcium supplements. One-third of menopausal osteoporosis patients are thought to be in a negative calcium balance, with 75% having substantial calcium malabsorption and 60% having low 1,25-(OH) vitamin D concentrations (Hill and Aspray, 2017). Interleukin-1 and tumor necrosis factor levels are raised during early postmenopausal osteoporosis (type I) due to estrogen deprivation. As a result of the bone resorption, serum calcium levels rise slightly, suppressing PTH release. Because PTH and phosphate govern vitamin D synthesis in the kidneys, a lower PTH level means a lower 1,25-(0H) vitamin D level, making calcitriol treatment a viable option for early postmenopausal women. A separate mechanism is proposed for senile osteoporosis (type 11). Secondary hyperparathyroidism develops as calcium absorption decreases with age. The therapy of choice for secondary hyper parathyroidism is hence the correction of calcium malabsorption. Vitamin D levels of 1,25-(OH) have been measured in older people. Many “normal old” people have normal blood 25-OH vitamin D levels, 1,25-(OH) vitamin D levels in the normal range, and elevated serum PTH levels, as well as markers of increased bone resorption and osteoporosis, according to Akesson et al. (1997). Age-related vitamin D resistance and a decrease in vitamin synthesis in the kidney are also mentioned. Calcitriol, according to the authors, is the best therapy for secondary hyperparathyroidism (Akesson et al., 1997).

15.12.1 MAGNESIUM

There is a clear link between osteoporosis and magnesium (Mg). In humans, magnesium shortage impairs parathyroid hormone (PTH) production and function, and in animal models, it causes osteopenia and increased bone fragility. Hypomagnesemia is common in patients with malabsorption disorders, owing to intestinal magnesium malabsorption (Mg). Hypomagnesemia has been found in up to 40% of individuals with malabsorption disorders (Booth et al., 1963; Rude et al., 1998). However, because only 1% of body Mg is in the extracellular fluid compartment, the serum Mg concentration does not correctly reflect Mg status. In other patient populations at high risk for Mg depletion, such as diabetics and alcoholics, the incidence of intracellular Mg^{2+} depletion has been demonstrated to be substantially higher than hypomagnesemia (Nadler et al., 1992; Ryzen et al., 1987). Mg^{2+} shortage can have a variety of effects on bone and mineral metabolism. Hypomagnesemic individuals have been shown to have inhibited parathyroid hormone (PTH) secretion as well as skeletal and renal resistance to PTH activity (Rude et al., 1985; ESTEP et al., 1969) Reduced blood 1,25-dihydroxy vitamin D concentrations and vitamin D end-organ resistance have also been documented (Rude et al., 1985). Mg deficiency causes poor bone development, decreased osteoblast activity, and decreased bone production in animal models, according to histomorphometric research. Increased bone fragility has been discovered through bio mechanical testing (Kenney et al., 1994).

When Mg deprived patients are given Mg supplementation, serum PTH concentrations rise dramatically, according to a study. This shows that these people had functional hypoparathyroidism, which, when combined with the direct effect of Mg shortage on bone metabolism, might have resulted in poor bone remodeling. The study's most intriguing finding is that Mg repletion causes an increase in BMD in the lumbar spine and proximal femur. The considerable association between this increase in BMD and the rise in intracellular free Mg 2+ clearly implies that Mg repletion contributed to the increase in bone density, either directly or indirectly (through an increase in

PTH). Mg treatment has been found to improve bone density in postmenopausal women in the past (Grimnes et al., 2011).

15.12.2 *PHOSPHATE*

Phosphate is abundant in the diet due to meat, wheat, and dairy products, and it is increasingly being added to foods as food additives in recent decades (Riis et al., 1994). However, the effect of dietary phosphate in bone health is not well understood. While phosphate is a necessary mineral component of hydroxyapatite, the primary structural constituent of bone, the acid-ash hypothesis claims that dietary phosphate, which is a sign of acid metabolism, is harmful to bone (Riggs et al., 1982; Raisz, 1990). Phosphate, along with calcium, is the most prevalent mineral in bone tissue. Phosphate in high concentrations in cell culture medium inhibits bone resorption while promoting bone mineralization and collagen production (Raisz, 1990). Reduced osteoclast production and mature osteoclast activity are both involved in this process (Yates et al., 2009). However, phosphate intake with the daily diet is generally abundant. The study found that when phosphate dosages rose, urine calcium dropped, regardless of whether the phosphate supplement was neutral/alkaline or acidic. Urine calcium dropped by 0.008 mmol/day for every mmol/day increase in phosphate dosage ($p<0.037$) when the phosphate supplement was neutral/alkaline. If the phosphate was an acidic salt, the extent of change in urine calcium in reaction to it was much larger ($p<0.001$). When the phosphate was acidic, urine calcium dropped by 0.07 mmol/day for every mmol/day of phosphate supplement ($p<0.001$) (Fenton et al., 2009). A sustained rise in parathyroid hormone (PTH) and a decrease in cholecalciferol (vitamin D3) synthesis in the kidney come from dietary patterns rich in P and low in calcium, alterations that are not conducive to achieving peak bone mass. Bone density has been found to be negatively associated to high dietary phosphate consumption (Metz et al., 1993).

15.13 ROLE OF FUNCTIONAL FOOD AS TRACE ELEMENT IN OSTEOPOROSIS

Apart from the nutrients mentioned above, there are indications that various trace elements (zinc, copper, manganese, and boron) may affect bone metabolism. Strause discovered that calcium alone and calcium combined with zinc, copper, and manganese slowed the rate of bone loss in 59 postmenopausal women over the course of a two-year trial. Surprisingly, the calcium-trace element group was the only one that outperformed the placebo group. This finding implies that specific trace elements may help to promote bone metabolism (Strause et al., 1994).

15.13.1 COPPER

Some research has found a link between serum copper and bone density. Copper is found in the largest amounts in organ meats, nuts and seeds, chocolate, and shellfish (Ma and Betts, 2000). Only a few research have looked into the link between copper consumption and osteoporosis and bone mineral density (BMD) (Fan et al., 2021). Copper is a trace element that is crucial to human health (Melton et al., 1989). Copper is required for bone formation, development, and maintenance, as well as the health of the brain, cardiovascular system, and many other bodily organs (Ensrud et al., 2009). It is a vital trace mineral that can only be obtained from food. Copper is essential for collagen and elastin cross-linking. Collagen is a fundamental component of bone tissue's extracellular matrix, and it works along with elastin to produce bone matrix (Grimston et al., 1993; Conde et al., 2015). Copper also increases osteoblast proliferation and function, as well as the differentiation of bone marrow mesenchymal stem cells into osteogenic and adipogenic cells (BMSCs) (Bailey et al., 2019). Prematurely born newborns with indications of disuse osteoporosis have been shown to have copper deficiency. After copper supplementation, this was no longer an issue (Gybina et al., 2009; Schmidt et al., 1991). As a result, copper deficiency has been linked to duction in bone production. Furthermore, because copper, as a trace metal element, cannot be produced in the human body, it is critical to investigate the link between copper consumption and osteoporosis (Fan et al., 2021). However, one investigation found that copper shortage exacerbated experimental bone loss in rats, with supplementing having no effect. As a result, the true involvement of copper in the pathophysiology of osteoporosis is unknown (Hardcastle et al., 2010).

15.13.2 ZINC

Zinc is known as an essential nutritional factor in the growth of human and animals. The richest food sources of zinc include oysters and meat (e.g., beef, veal, pork, and lamb) (Ma and Betts, 2000). Bone growth retardation is a frequent symptom of dietary zinc insufficiency in a variety of diseases. Zinc concentration in the bones has been observed to decline with age, skeletal unloading, and postmenopausal circumstances, implying a role in bone disease.

Zinc has been shown to enhance osteoblastic bone formation and mineralization; the metal boosts cellular protein synthesis by directly activating aminoacyl-tRNA synthetase, a rate-limiting enzyme in the translational phase of protein synthesis (Zheng et al., 2014). Zinc has been demonstrated to increase gene expression of the Runt-related transcription factor 2 (Runx2) transcription factor, which is linked to osteoblast development. Zinc has also been demonstrated to reduce osteoclastic bone resorption by preventing the development of osteoclast-like cells in bone marrow cells and inducing apoptotic cell death in mature osteoclasts (Ma and Yamaguchi, 2001). The receptor activator of nuclear factor (NF)-jB ligand (RANKL)-induced osteoclastogenesis is suppressed by zinc. The zinc transporter is expressed in both osteoblastic and osteoclastic cells. Transcription is aided by zinc protein. Increased bone mass is caused by dietary zinc consumption. The zinc compound b-alanyl-L-histidine to zinc (AHZ) is made up of zinc chelated to b-alanyl-L-histidine. The stimulating impact of AHZ on bone growth is stronger than zinc sulfate's. Zinc acexamate has also been found to have a strong bone-building impact. Aging, skeletal unloading, aluminum bone toxicity, calcium- and vitamin D insufficiency, adjuvant arthritis, estrogen shortage, diabetes, and fracture healing all benefit from the oral use of AHZ or zinc acexamate. Zinc compounds might be developed as a novel supplemental component for osteoporosis prevention and treatment (Yamaguchi, 2009).

15.13.3 BORON

Boron may have a role in bone production and development by regulating osteoblast and/or osteoclast activity, as well as serum steroid hormone levels and calcium metabolism (Betz et al., 2015; Hunt, 2012; Mogoşanu et al., 2015). Boron deficiency causes aberrant limb development, delayed growth plate maturation, and reduced bone strength, bone volume percentage, and trabecular thickness in animals (Nielsen and Stoecker, 2009). Coffee, milk, apples, dry and cooked beans, and potatoes are the major sources of boron in people's diets in the United States, owing to the high consumption of these items (RAINEY et al., 1999). Fruits and fruit juices account for 38% of boron consumption in toddlers, whereas milk and cheese account for 19% (HUNT et al., 2001). Milk and cheese products provide for 18%–20% of boron consumption in teenagers, whereas drinks, particularly instant coffee, are the most common dietary source of boron in adults (Betz et al., 2015). In postmenopausal women, a low-boron diet (0.25 mg boron/2,000 kcal) may increase urinary calcium and magnesium excretion while lowering estrogen blood concentrations (Nielsen et al., 1988). In men and women, low boron intakes (0.23 mg boron/2,000 kcal) appear to lower plasma calcium and serum 25-hydroxy vitamin D levels while increasing serum calcitonin and osteocalcin levels (Samman et al., 1998). These changes could affect bone mineral density. Animals that receive boron supplementation are compared to animals that ingest normal or low levels of boron, and the supplementation increases various measures of bone strength (Chapin, 1997). Boron intakes (mean of 0.9 mg/day) were not significantly linked with bone mineral density in the lumbar spine or femoral regions in an observational study of 134 Korean women (average age 41 years) (Kim et al., 2008). 3 mg/day boron supplementation for 10 months significantly reduced serum phosphorus levels and increased serum magnesium levels in sedentary females in a placebo-controlled clinical trial of 17 female athletes (mean age 19.8 years) and 11 sedentary females (mean age 20.3 years); such changes are often associated with increased bone mineral density (Meacham et al., 1994) (Table 15.3).

TABLE 15.3 Boron Content of Selected Foods

Food	Milligrams (mg)
Prune juice, 1 cup	1.43
Avocado, raw, cubed, ½cup	1.07
Raisins, 1.5 ounces	0.95
Peaches, 1 medium	0.80
Grape juice, 1 cup	0.76
Apples, 1 medium	0.66
Pears, 1 medium	0.50
Peanuts, roasted, salted, 1 ounce	0.48
Beans, refried, ½ cup	0.48
Peanut butter, 2 tablespoons	0.46

Source: Adapted from Meacham & Hunt (1998).

15.14 ROLE OF DIETARY FACTOR AS A FUNCTIONAL FOOD IN OSTEOPOROSIS

15.14.1 PHYTOESTROGEN

Phytoestrogens, vitamin K, and non-digestible carbohydrates have all attracted a lot of interest recently, both scientifically and commercially. As a result, these dietary aspects will be discussed in more depth. Estrogen action is necessary for women to avoid bone loss. The combination of estrogen and calcium has been found to provide full protection against bone loss (Ettinger, 1987). Estrogens are key bone-forming hormones. It is hypothesized that their lack might induce osteoporosis, whereas a phytoestrogen-rich diet can prevent it (Wuttke et al., 2002; Cassidy, 2003; Branca and Lorenzetti, 2005). Protein synthesis, osteoproteger in/receptor activation of nuclear factor-kappa Blig and ratio, and mineralization by osteoblast-like cells are all promoted by phytoestrogens in vitro. Phytoestrogens can suppress osteoclast differentiation and activation, production of tartrate resistant acid phosphatase, and pyridinoline compound secretion. As a result, phytoestrogens promote bone production and raise alkaline phosphatase, osteocalcin, osteopontin, and 1(I) collagen levels, as well as bone mineral density. According to the findings of mechanistic investigations, phytoestrogens reduce bone resorption and increase bone growth (Chiang and Pan, 2013). Through inducing osteoclastogenic inhibitor osteoproteger in and blocking NF-kappa B signaling, the soy phytoestrogen genistein has been shown to be a particularly potent enhancer of osteoblastic differentiation and maturation and an inhibitor of osteoclast formation and bone resorption, even though these effects are likely not mediated via estrogen receptors (Ming et al., 2012). Nonetheless, available clinical data is in conclusive and does not support soy's or red clover's bone-protective effects (Lagari and Levis, 2010; Powles et al., 2008).

15.14.2 VITAMIN K

The effects of vitamin K antagonists on human fetuses and experimental animals provided the first evidence for vitamin K's critical role in bone growth. Bone deformity, stunted development, and osteomalacia have all been reported as side effects (Lietman et al., 1975; Howe and Webster, 1994). Poor levels of vitamin K in the blood and serum have been linked to an increased risk of fractures and the development of low bone mass (HART et al., 1985). Poor dietary vitamin K consumption was linked to both low bone density and the prevalence of atherosclerotic calcifications in epidemiological investigations. It is found that 113 atherosclerotic postmenopausal women had a 7% lower bone mineral density and significantly higher calcium deposition in the arteries than controls in other epidemiological research. Bone mass, vascular calcium deposition, and vitamin K status indicators all showed a significant relationship (Jie et al., 1996). A 10-year follow-up of the Nurses' Health Study, which included 270 hip fractures, discovered an inverse connection between lettuce intake (the principal phylloquinone providing food) and the risk of hip fractures. Vitamin K comes in two different forms for oral use. It is administered in the form of synthetic vitamin K1 in Western Europe and America (phylloquinone, also naturally occurring in green vegetables like broccoli and spinach). Synthetic vitamin K2 (menaquinone, which is also found naturally in dairy products and generated by endogenous intestinal microbes) is utilized in Asia/Japan. The majority of investigations on the effect of vitamin K on bone mass have been conducted in Asia and have used vitamin K2 (Feskanich et al., 1999).

15.14.3 OMEGA-3 FATTY ACID

The three primary families of polyunsaturated fatty acids (PUFA) are oleic acid ($n-9$), linoleic acid ($n-6$), and α-linolenic acid (ALA) ($n-3$). Linoleic acid ($n-6$) and α-linolenic acid ($n-3$) are necessary fatty acids since they cannot be produced by the body (Schlemmer, 1998). Flaxseed, walnuts, green leafy vegetables, and rapeseed are some of the plants that contain ALA (Stark et al., 2008). While EPA and DHA are active components of fish oil and may be found in marine fish (Wang et al., 2014). Sardines, herring, and Pollock roe are the main sources of EPA, whereas salmon roe, flying fish, *Cirrhinus mrigata*, and *Catla catla* are the main sources of DHA (Bandarra et al., 2018). The imbalance between bone production by osteoblast cells and bone resorption by osteoclast cells causes bone mineral loss. Omega-3 fatty acids have been demonstrated to improve bone health by lowering osteoporosis risk by inhibiting osteoclast activity (Sun et al., 2003). M-CSF, MITF, and RANK expression have been discovered to be inhibited by fish oil treatment, which inhibits bone resorption (Nakanishi et al., 2013). Supplementing maternal and perinatal omega-3 fatty acids has been demonstrated to be beneficial.

In Korean postmenopausal women, a high erythrocyte level of DHA and EPA was observed to reduce the incidence of osteoporosis (Moon, 2012). A research done in China discovered that eating a lot of fish was linked to a lower incidence of osteoporosis (Chen et al., 2009). Furthermore, omega-3 fatty acids were found to improve bone mineral density in both lumbar and femoral bones in older females (Kruger et al., 1998). By boosting procollagen and osteocalcin, which are mineralization indicators, fish oil supplemented osteoporotic patients demonstrated improved calcium absorption and stimulation of osteoblastic activity (van Papendorp et al., 1995). In MC3T3-E1 osteoblast cells, omega-3 fatty acids also boost ALP activity. Furthermore, treatment of linolenic acid and EPA with estrogen resulted in greater calcium content in the femur bone and reduced urine excretion of deoxypyridinoline and hydroxypyridinoline in female Sprague Dawley rats when compared to the sham-operated group (Schlemmer et al., 1999).

15.5 CONCLUSION

The development of functional foods provides an unique opportunity to contribute to the improvement of the quality of the food offered to consumers who want to benefit their health and well-being. Only a rigorous scientific method with extremely significant findings can ensure the success of this new nutrition profession. It is clearly understandable that role of functional food plays a crucial role in bone health and to maintain the BMD. After completion of an extensive literature search, it was found that functional food is not just a food for nutritional benefits. It is much more our belief. In this chapter, it was clearly mentioned that functional food plays a vital role in osteoporosis. Functional food determines bone faith not only in the adulthood, but also in the neonatal stages. It was found that functional food may be obtained from various sources. As the definition states, functional food is a real ingredient of human development. Although the concept is new in India, but the tremendous benefits of functional food really attracts a ton of consumers in its new market, which is also supported by the statistics. So, it can be concluded that nutritional knowledge of food is most like to have direct co-relation with health benefits. As people are educating themselves about functional food, more and more new utilities of functional food are coming in focus in recent years.

KEYWORDS

- **bone nutrients**
- **fragility fractures**
- **functional foods**
- **osteoporosis**

REFERENCES

Akesson, K., Lau, K., & Baylink, D., (1997). Rationale for active vitamin D analog therapy in senile osteoporosis. *Calcif. Tissue Int., 60*(1), 100–105.

Albright, F., Smith, P., & Richardson, A., (1941). Postmenopausal osteoporosis. *JAMA, 116*(22), 2465.

Alkhatib, A., Tsang, C., Tiss, A., Bahorun, T., Arefanian, H., Barake, R., Khadir, A., & Tuomilehto, J., (2017). Functional foods and lifestyle approaches for diabetes prevention and management. *Nutrients, 9*(12), 1310.

Armas, L., & Recker, R., (2012). Pathophysiology of osteoporosis. *Endocrinol. Metab. Clin. North Am., 41*(3), 475–486.

Arron, J. R., & Choi, Y., (2000). Bone versus immune system. *Nature, 408*(6812), 535–536.

Bailey, R., Zou, P., Wallace, T., McCabe, G., Craig, B., Jun, S., Cauley, J., & Weaver, C., (2019). Calcium supplement use is associated with less bone mineral density loss, but does not lessen the risk of bone fracture across the menopause transition: Data from the study of women's health across the nation. *JBMR Plus, 4*(1), e10246.

Bandarra, N., Marçalo, A., Cordeiro, A., & Pousão-Ferreira, P., (2018). Sardine (*Sardina pilchardus*) lipid composition: Does it change after one year in captivity? *Food Chem., 244*, 408–413.

Baran, D., Sorensen, A., Grimes, J., Lew, R., Karellas, A., Johnson, B., & Roche, J., (1990). Dietary modification with dairy products for preventing vertebral bone loss in premenopausal women: A three-year prospective study. *J. Clin. Endocrinol. Metab., 70*(1), 264–270.

Barrett-Connor, E. *WHO Scientific Group on the Prevention and Management of Osteoporosis (2000: Geneva, Switzerland).* Prevention and management of osteoporosis: Report of a WHO scientific group. https://apps.who.int/iris/handle/10665/42841 (accessed on 22 February 2023).

Bergman, C., Gray-Scott, D., Chen, J., & Meacham, S., (2009). What is next for the dietary reference intakes for bone metabolism related nutrients beyond calcium: Phosphorus, magnesium, vitamin D, and fluoride? *Crit. Rev. Food Sci. Nutr., 49*, 136–144.

Betz, J., Blackman, M., Coates, P., Cragg, G., Levine, M., Moss, J., & White, J., (2015). *Encyclopedia of Dietary Supplements* (2nd edn.).

Booth, C., Babouris, N., Hanna, S., & MacIntyre, I., (1963). Incidence of hypomagnesaemia in intestinal malabsorption. *BMJ, 2*(5350), 141–144.

Bowring, C., & Francis, R., (2011). National osteoporosis society's position statement on hormone replacement therapy in the prevention and treatment of osteoporosis. *Menopause Int., 17*, 63–65.

Bozec, A., & Zaiss, M., (2017). T regulatory cells in bone remodeling. *Curr. Osteoporos. Rep., 15*(3), 121–125.

Bozec, A., Zaiss, M., Kagwiria, R., Voll, R., Rauh, M., Chen, Z., Mueller-Schmucker, S., et al., (2014). T cell costimulation molecules CD80/86 inhibit osteoclast differentiation by inducing the IDO/tryptophan pathway. *Sci. Transl. Med., 6*(235).

Branca, F., & Lorenzetti, S., (2005). Health effects of phytoestrogens. *Forum of Nutrition*, 100–111.

Cassidy, A., (2003). Potential risks and benefits of phytoestrogen-rich diets. *Int. J. Vitam. Nutr. Res., 73*(2), 120–126.

Cauley, J., Danielson, M., Boudreau, R., Forrest, K., Zmuda, J., Pahor, M., Tylavsky, F., et al., (2007). Inflammatory markers and incident fracture risk in older men and women: The health aging and body composition study. *J. Bone Miner. Res., 22*(7), 1088–1095.

Chan, G., Hoffman, K., & McMurry, M., (1995). Effects of dairy products on bone and body composition in pubertal girls. *J. Pediatr., 126*(4), 551–556.

Chapin, R., (1997). The effects of dietary boron on bone strength in rats. *Fundam. Appl. Toxicol., 35*(2), 205–215.

Chen, Y., Ho, S., & Lam, S., (2009). Higher sea fish intake is associated with greater bone mass and lower osteoporosis risk in postmenopausal Chinese women. *Osteoporos Int., 21*(6), 939–946.

Chiang, S., & Pan, T., (2013). Beneficial effects of phytoestrogens and their metabolites produced by intestinal microflora on bone health. *Appl. Microbiol. Biotechnol., 97*(4), 1489–1500.

Clarke, B., & Khosla, S., (2010). Physiology of bone loss. *Radiol. Clin. N., 48*(3), 483–495.

Cline-Smith, A., Axelbaum, A., Shashkova, E., Chakraborty, M., Sanford, J., Panesar, P., Peterson, M., et al., (2020). Ovariectomy activates chronic low-grade inflammation mediated by memory T cells, which promotes osteoporosis in mice. *J. Bone Miner. Res., 35*(6), 1174–1187.

Conde, J., Scotece, M., Abella, V., López, V., Pino, J., Gómez-Reino, J., & Gualillo, O., (2015). Basic aspects of adipokines in bone metabolism. *Clin. Rev. Bone Miner. Metab., 13*(1), 11–19.

Cosman, F., De Beur, S., LeBoff, M., Lewiecki, E., Tanner, B., Randall, S., & Lindsay, R., (2014). Clinician's guide to prevention and treatment of osteoporosis. *Osteoporos Int., 25*(10), 2359–2381.

D'Amelio, P., Grimaldi, A., Di Bella, S., Brianza, S., Cristofaro, M., Tamone, C., Giribaldi, G., et al., *(*2008*)*. Estrogen deficiency increases osteoclastogenesis up-regulating T cells activity: A key mechanism in osteoporosis. *Bone, 43*(1), 92–100.

Ding, C., Parameswaran, V., Udayan, R., Burgess, J., & Jones, G., (2008). Circulating levels of inflammatory markers predict change in bone mineral density and resorption in older adults: A longitudinal study. *J. Clin. Endocrinol. Metab., 93*(5), 1952–1958.

Domazetovic, V., Marcucci, G., Iantomasi, T., Brandi, M. L., & Vincenzini, M. T., (2017). Oxidative stress in bone remodeling: Role of antioxidants. *Clinical Cases in Mineral and Bone Metabolism, 14*(2), 209.

Eisman, J., (2009). Vitamin D receptor gene alleles and osteoporosis: An affirmative view. *J. Bone Miner. Res., 10*(9), 1289–1293.

Ensrud, K., Palermo, L., Black, D., Cauley, J., Jergas, M., Orwoll, E., Nevitt, M., et al., (2009). Hip and calcaneal bone loss increase with advancing age: Longitudinal results from the study of osteoporotic fractures. *J. Bone Miner. Res., 10*(11), 1778–1787.

Estep, H., Shaw, W., Watlington, C., Hobe, R., Holland, W., & Tucker, S., (1969). Hypocalcemia due to hypomagnesemia and reversible parathyroid hormone unresponsiveness. *J. Clin. Endocrinol. Metab., 29*(6), 842–848.

Ettinger, B., (1987). Postmenopausal bone loss is prevented by treatment with low-dosage estrogen with calcium. *Ann. Intern. Med., 106*(1), 40.

Fan, Y., Ni, S., & Zhang, H., (2021). Associations of copper intake with bone mineral density and osteoporosis in adults: Data from the national health and nutrition examination survey. *Biol. Trace Elem. Res.*, 10.1007/s12011-021-02845-5.

Farr, J., Fraser, D., Wang, H., Jaehn, K., Ogrodnik, M., Weivoda, M., Drake, M., et al., (2016). Identification of senescent cells in the bone microenvironment. *J. Bone Miner. Res., 31*(11), 1920–1929.

Farr, J., Xu, M., Weivoda, M., Monroe, D., Fraser, D., Onken, J., Negley, B., et al., (2017). Targeting cellular senescence prevents age-related bone loss in mice. *Nat. Med., 23*(9), 1072–1079.

Fatourechi, G. E., Khosla, S., Sanyal, A., Boyle, W. J., Lacey, D. L., & Riggs, B. L., (2003). Role of RANK ligand in mediating increased bone resorption in early postmenopausal women. *J. Clin. Invest., 111*(8), 1221–30.

Fenton, T., Lyon, A., Eliasziw, M., Tough, S., & Hanley, D., (2009). Phosphate decreases urine calcium and increases calcium balance: A meta-analysis of the osteoporosis acid-ash diet hypothesis. *Nutrition Journal, 8*(1).

Feskanich, D., Weber, P., Willett, W., Rockett, H., Booth, S., & Colditz, G., (1999). Vitamin k intake and hip fractures in women: A prospective study. *Am. J. Clin. Nutr., 69* (1), 74–79.

Franceschi, C., Bonafè, M., Valensin, S., Olivieri, F., De Luca, M., Ottaviani, E., & De Benedictis, G., (2006). Inflamm-aging: An evolutionary perspective on immunosenescence. *Ann. N. Y. Acad. Sci., 908*(1), 244–254.

Grimnes, G., Joakimsen, R., Figenschau, Y., Torjesen, P., Almås, B., & Jorde, R., (2011). The effect of high-dose vitamin d on bone mineral density and bone turnover markers in postmenopausal women with low bone mass—A randomized controlled one-year trial. *Osteoporos Int., 23*(1), 201–211.

Grimston, S., Willows, N., & Hanley, D., (1993). Mechanical loading regime and its relationship to bone mineral density in children. *Med. Sci. Sports Exerc., 25*(11), 1203–1210.

Gybina, A., Tkac, I., & Prohaska, J., (2009). Copper deficiency alters the neurochemical profile of developing rat brain. *Nutr. Neurosci., 12*(3), 114–122.

Hardcastle, A., (2010). Nutritional aspects of osteoporosis. *Bone, 47*, S16.

Harper, S., (2019). The convergence of population ageing with climate change. *J. Popul. Ageing, 12*, 401–403.

Hart, J., Shearer, M., Klenerman, L., Catterall, A., Reeve, J., Sambrook, P., Dodos, R., et al., (1985). Electrochemical detection of depressed circulating levels of vitamin K1 in osteoporosis. *J. Clin. Endocrinol. Metab., 60*(6), 1268–1269.

Hasler, C., (2002). Functional foods: Benefits, concerns and challenges—A position paper from the American council on science and health. *J. Nutr., 132*(12), 3772–3781.

Henry, C., (2010). Functional foods. *Eur. J. Clin. Nutr., 64*(7), 657–659.

Hill, T., & Aspray, T., (2017). The role of vitamin D in maintaining bone health in older people. *Ther. Adv. Musculoskelet. Dis., 9*(4), 89–95.

Holick, M., (1996). Vitamin D and bone health. *J. Nutr., 126*(suppl 4), 1159–1164.

Howe, A., & Webster, W., (1994). Vitamin K – its essential role in craniofacial development. *Aust. Dent. J., 39*(2), 88–92.

Huncharek, M., Muscat, J., & Kupelnick, B., (2008). impact of dairy products and dietary calcium on bone-mineral content in children: Results of a meta-analysis. *Bone, 43*(2), 312–321.

Hunt, C., & Meacham, S., (2001). Aluminum, boron, calcium, copper, iron, magnesium, manganese, molybdenum, phosphorus, potassium, sodium, and zinc. *J. Am. Diet. Assoc., 101*(9), 1058–1060.

Hunt, C., (2012). Dietary boron: Progress in establishing essential roles in human physiology. *J. Trace Elem. Med. Biol., 26*(2, 3), 157–160.

Jie, K., Bots, M., Vermeer, C., Witteman, J., & Grobbee, D., (1996). Vitamin K status and bone mass in women with and without aortic atherosclerosis: A population-based study. *Calcif. Tissue Int., 59*(5), 352–356.

Johnston, C., Miller, J., Slemenda, C., Reister, T., Hui, S., Christian, J., & Peacock, M., (1992). Calcium supplementation and increases in bone mineral density in children. *N. Engl. J. Med., 327*(2), 82–87.

Kanis, J., Burlet, N., Cooper, C., Delmas, P., Reginster, J., Borgstrom, F., & Rizzoli, R., (2008). European guidance for the diagnosis and management of osteoporosis in postmenopausal women. *Osteoporos Int., 19*(4), 399–428.

Kenney, M., McCoy, H., & Williams, L., (1994). Effects of magnesium deficiency on strength, Mass, and composition of rat femur. *Calcif. Tissue Int., 54*(1), 44–49.

Keservani, R. K., Kesharwani, R. K., Sharma, A. K., Vyas, N., & Chadoker, A., (2010b). Nutritional supplements: An overview. *International Journal of Current Pharmaceutical Review and Research, 1* (1), 59–75.

Keservani, R. K., Kesharwani, R. K., Vyas, N., Jain, S., Raghuvanshi, R., & Sharma, A. K., (2010a). Nutraceutical and functional food as future food: A review. *Der Pharmacia Lettre, 2*(1), 106–116.

Keservani, R. K., Sharma, A. K., & Kesharwani, R. K., (2017). An overview and therapeutic applications of nutraceutical and functional foods. *Recent Advances in Drug Delivery Technology*, 160–201. ISBN:9781522507543.

Keservani, R. K., Sharma, A. K., & Kesharwani, R. K., (2020). *Nutraceuticals and Dietary Supplements: Applications in Health Improvement and Disease Management*. CRC Press. ISBN: 9781771888738.

Khadilkar, A., & Mandlik, R., (2015). Epidemiology and treatment of osteoporosis in women: An Indian perspective. *Int. J. Women's Health,* 841.

Kim, M., Bae, Y., Lee, Y., & Choi, M., (2008). Estimation of boron intake and its relation with bone mineral density in free-living Korean female subjects. *Biol. Trace Elem. Res., 125*(3), 213–222.

Kim, S., Lim, S., & Kim, S., (2021). Real-time analysis and predictability of the health functional food market using big data. *Food Sci. Biotechnol., 30*(13), 1667–1674.

Klibanski, A., (2001). NIH consensus development panel on osteoporosis prevention, diagnosis, and therapy. osteoporosis prevention, diagnosis, and therapy. *JAMA, 285*(6), 785–795.

Kostecka, M., (2014). The role of a healthy diet in the prevention of osteoporosis in the perimenopausal period. *Pak. J. Med. Sci., 30*(4), 763–768.

Kruger, M., Coetzer, H., De Winter, R., Gericke, G., & Van, P. D., (1998). Calcium, gamma-linolenic acid and eicosapentaenoic acid supplementation in senile osteoporosis. *Aging. Clin. Exp. Res., 10*(5), 385–394.

Lafarga, T., Fernández-Sevilla, J., González-López, C., & Acién-Fernández, F., (2020). Spirulina for the food and functional food industries. *Int. Food Res. J., 137*, 109356.

Lagari, V., & Levis, S., (2010). Phytoestrogens and bone health. *Curr Opin Endocrinol Diabetes Obes., 17*(6), 546–553.

Lietman, P., Pettifor, J., & Benson, R., (1975). Congenital malformations associated with the administration of oral anticoagulants during pregnancy. *J. Pediatr., 86*(3), 459–462.

Lv, N., & Brown, J., (2011). Impact of a nutrition education program to increase intake of calcium-rich foods by Chinese-American women. *J. Am. Diet. Assoc., 111*(1), 143–149.

Ma, J., & Betts, N., (2000). Zinc and copper intakes and their major food sources for older adults in the 1994–96 continuing survey of food intakes by individuals (CSFII). *J. Nutr., 130*(11), 2838–2843.

Ma, Z., & Yamaguchi, M., (2001). Role of endogenous zinc in the enhancement of bone protein synthesis associated with bone growth of newborn rats. *J. Bone Miner. Metab., 19*(1), 38–44.

Martirosyan, D., & Singh, J., (2015). A new definition of functional food by FFC: What makes a new definition unique? *Funct. Foods Health Dis., 5*(6), 209.

Mauck, K., & Clarke, B., (2006). Diagnosis, screening, prevention, and treatment of osteoporosis. *Mayo Clin. Proc., 81*(5), 662–672.

Meacham, S., & Hunt, C., (1998). Dietary boron intakes of selected populations in the United States. *Biol. Trace Elem. Res., 66*(1–13), 65–78.

Meacham, S., Taper, L., & Volpe, S., (1994). Effects of boron supplementation on bone mineral density and dietary, blood, and urinary calcium, phosphorus, magnesium, and boron in female athletes. *Environ. Health Perspect., 102*(suppl 7), 79–82.

Melton, L., Kan, S., Frye, M., Wahner, H., O'Fallon, W., & Riggs, B., (1989). Epidemiology of vertebral fractures in women. *Am. J. Epidemiol., 129*(5), 1000–1011.

Metz, J., Anderson, J., & Gallagher, P., (1993). Intakes of calcium, phosphorus, and protein, and physical activity level are related to radial bone mass in young adult women. *Am. J. Clin. Nutr., 58*(4), 537–542.

Ming, L., Chen, K., & Xian, C., (2012). Functions and action mechanisms of flavonoids genistein and icariin in regulating bone remodeling. *J. Cell. Physiol., 228*(3), 513–521.

Mogoşanu, G., Biţă, A., Bejenaru, L., Bejenaru, C., Croitoru, O., Rău, G., Rogoveanu, O., et al., (2015). Calcium fructoborate for bone and cardiovascular health. *Biol. Trace Elem. Res., 172*(2), 277–281.

Moon, H., Kim, T., Byun, D., & Park, Y., (2012). Positive correlation between erythrocyte levels of N3 polyunsaturated fatty acids and bone mass in postmenopausal Korean women with osteoporosis. *Ann. Nutr. Metab., 60*(2), 146–153.

Morii, H., & Genant, H., (1998). Statement on the diagnosis and management of osteoporosis from the consensus development conference at the second international conference on osteoporosis, Osaka 1997. *J. Bone Miner. Metab., 16*, 206–214.

Nadler, J., Malayan, S., Luong, H., Shaw, S., Natarajan, R., & Rude, R., (1992). Intracellular free magnesium deficiency plays a key role in increased platelet reactivity in type II diabetes mellitus. *Diabetes Care, 15*(7), 835–841.

Nakamura, K., Saito, T., Kobayashi, R., Oshiki, R., Oyama, M., Nishiwaki, T., Nashimoto, M., & Tsuchiya, Y., (2010). C-reactive protein predicts incident fracture in community-dwelling elderly Japanese women: The Muramatsu study. *Osteoporos Int., 22*(7), 2145–2150.

Nakanishi, A., Iitsuka, N., & Tsukamoto, I., (2013). Fish oil suppresses bone resorption by inhibiting osteoclastogenesis through decreased expression of M-CSF, PU.1, MITF And RANK in ovariectomized rats. *Mol. Med. Rep., 7*(6), 1896–1903.

Nielsen, F., & Stoecker, B., (2009). Boron and fish oil have different beneficial effects on strength and trabecular microarchitecture of bone. *J. Trace Elem. Med. Biol., 23*(3), 195–203.

Nielsen, F., Hunt, C., Mullen, L., & Hunt, J., (1988). Effect of dietary boron on mineral, estrogen, and testosterone metabolism in postmenopausal women. *Maturitas, 10*(3), 245.

Onal, M., Xiong, J., Chen, X., Thostenson, J., Almeida, M., Manolagas, S., & O'Brien, C., (2012). Receptor activator of nuclear factor Kb ligand (RANKL) protein expression by b lymphocytes contributes to ovariectomy-induced bone loss. *J. Biol. Chem., 287*(35), 29851–29860.

Orkusz, A., (2022). An assessment of the nutritional value of the preschool food rations for children from the Wroclaw District, Poland—The case of a big city. *Nutrients, 14*(3), 460.

Pacifici, R., (2018). Bone remodeling and the microbiome. *Cold Spring Harb. Perspect. Med., 8*(4), 31203.

Panach, L., Serna, E., Tarín, J., Cano, A., & García-Pérez, M., (2017). A translational approach from an animal model identifies CD80 as a candidate gene for the study of bone phenotypes in postmenopausal women. *Osteoporos Int., 28*(8), 2445–2455.

Peacock, M., (2009). Vitamin D receptor gene alleles and osteoporosis: A contrasting view. *J. Bone Miner. Res., 10*(9), 1294–1297.

Piemontese, M., Almeida, M., Robling, A., Kim, H., Xiong, J., Thostenson, J., Weinstein, R., et al., (2017). Old age causes de novo intracortical bone remodeling and porosity in mice. *JCI Insight, 2*(17).

Pietschmann, P., Grisar, J., Thien, R., Willheim, M., Kerschan-Schindl, K., Preisinger, E., & Peterlik, M., (2001). Immune phenotype and intracellular cytokine production of peripheral blood mononuclear cells from postmenopausal patients with osteoporotic fractures. *Exp. Gerontol., 36*(10), 1749–1759.

Pietschmann, P., Mechtcheriakovà, D., Meshcheryakova, A., Föger-Samwald, U., & Ellinger, I., (2015). Immunology of osteoporosis: A mini-review. *Gerontology, 62*(2), 128–137.

Pineda, B., Serna, E., Laguna-Fernández, A., Noguera, I., Panach, L., Hermenegildo, C., Tarín, J., et al., (2014). Gene expression profile induced by ovariectomy in bone marrow of mice: A functional approach to identify new candidate genes associated to osteoporosis risk in women. *Bone, 65*, 33–41.

Powles, T., Howell, A., Evans, D., Mccloskey, E., Ashley, S., Greenhalgh, R., Affen, J., et al., (2008). Red clover isoflavones are safe and well tolerated in women with a family history of breast cancer. *Menopause Int., 14*(1), 6–12.

Prince, R., Devine, A., Dhaliwal, S., & Dick, I., (2006). Effects of calcium supplementation on clinical fracture and bone structure. *Arch. Intern. Med., 166*(8), 869.

Rainey, C., Nyquist, L., Christensen, R., Strong, P., Culver, B., & Coughlin, J., (1999). Daily boron intake from the American diet. *J. Am. Diet. Assoc., 99*(3), 335–340.

Raisz, L., (1990). Recent advances in bone cell biology: Interactions of vitamin D with other local and systemic factors. *Bone and Mineral, 9*(3), 191–197.

Rauner, M., Sipos, W., & Pietschmann, P., (2006). Osteoimmunology. *Int. Arch. Allergy Immunol., 143*(1), 31–48.

Red Book: Pharmacy's Fundamental Reference, 2008. Thomson PDR, Montvale, NJ.

Riggs, B., Khosla, S., & Melton, L., (1998). An unitary model for involutional osteoporosis: Estrogen deficiency causes both type I and type II osteoporosis in postmenopausal women and contributes to bone loss in aging men. *J. Bone Miner. Res., 13*(5), 763–773.

Riggs, B., Wahner, H., Seeman, E., Offord, K., Dunn, W., Mazess, R., Johnson, K., & Melton, L., (1982). Changes in bone mineral density of the proximal femur and spine with aging. *J. Clin. Investig., 70*(4), 716–723.

Riis, B., (1994). Premenopausal bone loss: Fact or artifact? *Osteoporos Int., 4*(S1), S35–S37.

Ringe, J., (2020). Plain vitamin D or active vitamin D in the treatment of osteoporosis: Where do we stand today? *Arch. Osteoporos, 15*(1).

Ross, A. C., Taylor, C. L., & Yaktine, A. L., (2011). *Dietary Reference Intakes for Calcium and Vitamin D*. Institute of Medicine (US) Committee to Review Dietary Reference Intakes for Vitamin D and Calcium. National Academies Press (US).

Rude, R., Adams, J., Ryzen, E., Endres, D., Niimi, H., Horst, R., Haddad, J., & Singer, F., (1985). Low serum concentrations of 1,25-dihydroxy vitamin D in human magnesium deficiency. *J. Clin. Endocrinol. Metab., 61*(5), 933–940.

Rude, R., Kirchen, M., Gruber, H., Stasky, A., & Meyer, M., (1998). Magnesium deficiency induces bone loss in the rat. *Mineral and Electrolyte Metabolism, 24*(5), 314–320.

Rude, R., Oldham, S., & Singer, F., (1976). Functional hypoparathyroidism and parathyroid hormone end-organ resistance in human magnesium deficiency. *Clin. Endocrinol. News, 5*(3), 209–224.1.

Ryzen, E., Nelson, T. A., & Rude, R. K., (1987). Low blood mononuclear cell patients. *West. J. Med., 147*, 549–553.

Samman, S., Naghii, M., Wall, P., & Verus, A., (1998). The nutritional and metabolic effects of boron in humans and animals. *Biol. Trace Elem. Res., 66*(1–3), 227–235.

Samwald, U. F., Dovjak, P., Semrad, U. A., Schindl, K. K., & Pietschmann, P., (2020). Osteoporosis: Pathophysiology and therapeutic options. *EXCLI Journal, 19*, 1017–1037.

Santos-Buelga, C., González-Paramás, A., Oludemi, T., Ayuda-Durán, B., & González-Manzano, S., (2019). Plant phenolics as functional food ingredients. *Adv. Food Nutr. Res.*, 183–257.

Schlemmer, C., Coetzer, H., Claassen, N., Kruger, M., Rademeyer, C., Van, J. L., & Smuts, C., (1998). Ectopic calcification of rat aortas and kidneys is reduced with n-3 fatty acid supplementation. *Prostaglandins, Leukotrienes and Essential Fatty Acids, 59*(3), 221–227.

Schmidt, H., Herwig, J., & Greinacher, I., (1991). Skeletal changes in preterm infants with copper deficiency. *RöFo – Advances in the Field of X-Rays and Imaging Processes, 155*(07), 38–42.

Shandilya, U., (2017). Functional foods and their benefits: An overview. *J. Nutri. Health Food Eng., 7*(4), 353–356.

Sikand, G., Kris-Etherton, P., & Boulos, N., (2015). Impact of functional foods on prevention of cardiovascular disease and diabetes. *Curr. Cardiol. Rep., 17*(6), 39.

Stark, A., Crawford, M., & Reifen, R., (2008). Update on alpha-linolenic acid. *Nutr. Rev., 66*(6), 326–332.

Strause, L., Saltman, P., Smith, K., Bracker, M., & Andon, M., (1994). Spinal bone loss in postmenopausal women supplemented with calcium and trace minerals. *J. Nutr., 124*(7), 1060–1064.

Sun, D., Krishnan, A., Zaman, K., Lawrence, R., Bhattacharya, A., & Fernandes, G., (2003). Dietary N-3 fatty acids decrease osteoclastogenesis and loss of bone mass in ovariectomized mice. *J. Bone Miner. Res., 18*(7), 1206–1216.

Tai, P., Wang, J., Jin, H., Song, X., Yan, J., Kang, Y., Zhao, L., et al., (2007). Induction of regulatory T cells by physiological level estrogen. *J. Cell. Physiol., 214*(2), 456–464.

Tu KN, Lie JD, Wan CKV, Cameron M, Austel AG, Nguyen JK, Van K, Hyun D. Osteoporosis: A Review of Treatment Options. P T. 2018 Feb;43(2):92-104. PMID: 29386866; PMCID: PMC5768298.

Tur, J., & Bibiloni, M., (2016). Functional foods. *Encyclopedia of Food and Health*, 157–161.

Van, P. D., Coetzer, H., & Kruger, M., (1995). Biochemical profile of osteoporotic patients on essential fatty acid supplementation. *Nutr. Res. Rev., 15*(3), 325–334.

Verbeke, W., (2005). Consumer acceptance of functional foods: Socio-demographic, cognitive and attitudinal determinants. *Food Qual. Prefer., 16*(1), 45–57.

Vukasović, T., (2017). Functional foods in line with young consumers: Challenges in the marketplace in Slovenia. *Developing New Functional Food and Nutraceutical Products*, 391–405.

Walsh, M., & Choi, Y., (2014). Biology of the RANKL-RANK-OPG system in immunity, bone, and beyond. *Front. immunol., 5*, 511.

Wang, J., Eguchi, K., Matsumoto, S., Fujiu, K., Komuro, I., Nagai, R., & Manabe, I., (2014). The Ω-3 polyunsaturated fatty acid, eicosapentaenoic acid, attenuates abdominal aortic aneurysm development via suppression of tissue remodeling. *PLoS One, 9*(5), 96286.

WHO Scientific Group on The Assessment of Osteoporosis at Primary Health Care Level Summary Meeting Report Brussels, Belgium. https://apps.who.int/iris/handle/10665/42841 (accessed on 22 February 2023).

Wuttke, W., Jarry, H., Westphalen, S., Christoffel, V., & Seidlová-Wuttke, D., (2002). Phytoestrogens for hormone replacement therapy? *J. Steroid Biochem. Mol. Biol., 83*(1–5), 133–147.

Yamaguchi, M., (2009). Role of nutritional zinc in the prevention of osteoporosis. *Mol. Cell. Biochem., 338*(1, 2), 241–254.

Yates, A., Oreffo, R., Mayor, K., & Mundy, G., (2009). Inhibition of bone resorption by inorganic phosphate is mediated by both reduced osteoclast formation and decreased activity of mature osteoclasts. *J. Bone Miner. Res., 6*(5), 473–478.

Zaiss, M., Sarter, K., Hess, A., Engelke, K., Böhm, C., Nimmerjahn, F., Voll, R., Schett, G., & David, J., (2010). Increased bone density and resistance to ovariectomy-induced bone loss in foxp3-transgenic mice based on impaired osteoclast differentiation. *Arthritis Rheumatol., 62*(8), 2328–2338.

Zhao, R., Wang, X., & Feng, F., (2016). Upregulated cellular expression of IL-17 by CD4+ T-cells in osteoporotic postmenopausal women. *Ann. Nutr. Metab., 68*(2), 113–118.

Zheng, J., Mao, X., Ling, J., He, Q., & Quan, J., (2014). Low serum levels of zinc, copper, and iron as risk factors for osteoporosis: A meta-analysis. *Biol. Trace Elem. Res., 160*(1), 15–23.

INDEX

G

H

I

K

L

T

U